Nicholas Senn

Surgical Bacteriology

Nicholas Senn

Surgical Bacteriology

ISBN/EAN: 9783337106010

Printed in Europe, USA, Canada, Australia, Japan

Cover: Foto ©berggeist007 / pixelio.de

More available books at **www.hansebooks.com**

SURGICAL

BACTERIOLOGY.

BY

N. SENN, M.D., Ph.D.,

Professor of Surgery in Rush Medical College, Chicago, and in the Chicago Polyclinic; Attending Surgeon to the Milwaukee Hospital, Consulting Surgeon to the Milwaukee County Hospital and to the Milwaukee County Insane Asylum; Honorary Fellow College of Physicians of Philadelphia; Permanent Member of the German Congress of Surgeons; Corresponding Member Harveian Society, London; Honorary Member of La Academia de la Medicina de Mexico, of the D. Hayes Agnew Surgical Society of Philadelphia, of the Ohio State Medical Society, and of the Minnesota State Medical Society; Member of the American Surgical Association, of the American Medical Association, of the British Medical Association, of the Wisconsin State Medical Society, and of the Brainard Medical Society, etc.

SECOND EDITION, THOROUGHLY REVISED.

PHILADELPHIA:
LEA BROTHERS & CO.
1891.

Entered according to the Act of Congress, in the year 1891, by
LEA BROTHERS & CO.,
in the Office of the Librarian of Congress at Washington.

PREFACE TO SECOND EDITION.

THE speedy exhaustion of the first edition is a pleasing evidence of the interest the medical profession of this country is taking in the subject of surgical bacteriology.

In the preparation of the second edition the author has aimed to add new facts illustrative of the relations of pathogenic micro-organisms to the various surgical lesions, and eight new illustrations have been inserted in the text descriptive of microbes not illustrated in the first edition. The book has also been divided into chapters, which it is hoped will prove useful for a better classification of the material and for more ready reference.

N. SENN.

CHICAGO, MARCH, 1891.

PREFACE TO FIRST EDITION.

WITHIN a few years bacteriology has revolutionized surgical pathology. All wound complications and most of the acute and chronic inflammatory lesions which come under the treatment of the surgeon are caused by microörganisms; hence the necessity of a proper recognition of the importance of bacteriology as an integral part of the science and practice of modern surgery.

It has been the endeavor of the author to present the different subjects contained in this work in as concise a manner as possible, and at the same time to omit nothing which might be deemed necessary to impart a fair knowledge of the subject.

The illustrations are reproductions from *Lehrbuch der pathologischen Anatomie*, by Prof. Klebs, Jena, 1887.

N. SENN.

MILWAUKEE, APRIL, 1889.

LIST OF PLATES.

I. Facing page 40
II. Facing page 82
III. Facing page 112
IV. Facing page 130
V. Facing page 158
VI. Facing page 158
VII. Facing page 160
VIII. Facing page 180
IX. Facing page 200
X. Facing page 202
XI. Facing page 206
XII. Facing page 230
XIII. Facing page 242

CONTENTS.

	PAGE
INTRODUCTION	17

CHAPTER I.
HEREDITARY TRANSMISSION OF MICROBIC DISEASES   19

CHAPTER II.
DO PATHOGENIC MICROÖRGANISMS EXIST IN THE HEALTHY BODY?   26

CHAPTER III.
SOURCES OF INFECTION   29

CHAPTER IV.
LOCALIZATION OF MICROBES   37

CHAPTER V.
ELIMINATION OF PATHOGENIC MICROÖRGANISMS   56

CHAPTER VI.
ANTAGONISM AMONG MICROÖRGANISMS   68

CHAPTER VII.
INFLAMMATION   71

CHAPTER VIII.
SUPPURATION   79

CHAPTER IX.
GANGRENE   106

CHAPTER X.
SEPTICÆMIA . . . 110

CHAPTER XI.
PYÆMIA 124

CHAPTER XII.
ERYSIPELAS . . 129

CHAPTER XIII.
ERYSIPELOID . . 139

CHAPTER XIV.
NOMA . . 141

CHAPTER XV.
TETANUS 142

CHAPTER XVI.
TUBERCULOSIS . . 156

CHAPTER XVII.
CLINICAL FORMS OF SURGICAL TUBERCULOSIS . 170

CHAPTER XVIII.
ANTHRAX . 199

CHAPTER XIX.
GLANDERS (MALLEUS HUMIDUS) . . 207

CHAPTER XX.
ACTINOMYCOSIS HOMINIS . . 213

CHAPTER XXI.
GONORRHŒA . . . 217

CHAPTER XXII.
SYPHILIS . 240

CHAPTER XXIII.
ON THE ALLEGED MICROBIC ORIGIN OF TUMORS . 249

SURGICAL BACTERIOLOGY.

INTRODUCTION.

The recent advances in surgical pathology laid the foundation for the wonderful achievements of modern surgery. In proportion as our knowledge of the etiology and pathology of the different surgical lesions approaches perfection, will surgery assume the dignity of an exact science and a true art. The future progress in surgery will be characterized by original research in the elucidation of as yet obscure etiological and pathological questions. If the nature of a disease is well understood, rational suggestions as to proper treatment will follow as a natural consequence. During the last fifteen years there has been more real advance made in surgical pathology than during twenty centuries preceding them. Bacteriology opened a new era for surgical pathology. The knowledge which has accumulated from the bacteriological investigations of disease has opened new fields of usefulness for the surgeon. Many diseases heretofore uselessly treated with drugs by the physician are now successfully treated by surgical measures.

At this time, surgical pathology has almost become synonymous with surgical bacteriology. Text-books on surgical pathology of only a few years ago are consulted in vain for information on many subjects which now attract universal attention. Owing to the activity which is manifested everywhere in the investigation of the microbic cause of disease, the many discoveries which are being made in rapid succession, works on pathology soon become old, and are consigned to the shelves of the antiquarian almost before they have left the press. The author of one of the best text-books on general pathology found it necessary to prepare a lengthy appendix while the book was going through the press, in order to escape the criticism of the reviewer that it did not represent modern ideas. We live in an age of independent thought and investigation. No discovery is accepted unchallenged, and all new claims are subjected to the crucial test of criticism based on original research. Reputation no longer carries with it the weight of authority, unless the views advanced can be corroborated by the independent work of others.

In the preparation of this work it has been my principal object to gather from the current medical literature, in compact form, the result of the best work by the ablest men on the various subjects which will be discussed in this book. In all instances in which the opinions of authors and investigators came in conflict it has been my aim to represent their ideas without an effort to harmonize them. In the consideration of disputed points, I have not hesitated in stating my own convictions, which were based either upon the results obtained from independent original research, clinical observation, or a careful study of the literature on the subject at my command. Crookshank's *Introduction to Practical Bacteriology*, New York, 1887, and Klebs's *Lehrbuch der pathologischen Anatomie*, I. Theil, Jena, 1887, are the only text-books I consulted. From these I have freely quoted in the descriptions of the different microörganisms.

CHAPTER I.

HEREDITARY TRANSMISSION OF MICROBIC DISEASES.

THAT many of the infectious surgical diseases are hereditary has been known for a long time, and many theories have been advanced at different times in the past in explanation of their occurrence. The modern views of this subject may be narrowed down to two suppositions: 1. Transmission from parents to child of a predisposition to certain diseases. 2. Transmission from parents to fœtus of the essential cause of the disease. The supposed hereditary predisposition is interpreted as meaning anatomical or physiological defects in the tissues, which render the organism susceptible to the action of subsequent specific causes. The existence of minute anatomical defects of bloodvessels, lymphatic glands and vessels, connective-tissue spaces, etc., are looked upon as conditions which favor localization of floating microbes, which find their way into the body during post-natal life. An inherited defective vital resistance on the part of the tissues to the action of pathogenic bacteria is also considered by many in the light of an hereditary influence in the causation of disease. The conditions are recognized, but no satisfactory, demonstrative, or experimental proofs of their existence have been furnished, and yet the immunity of some animals to certain diseases cannot be explained in any other way than in attributing to the tissues anatomical or physiological properties which protect the organism against the action of certain microörganisms which, in other animals not so protected, produce a fatal disease. Clinical experience has also shown that a great difference is found among different persons in reference to susceptibility to the same form of infection. In many persons, for instance, inoculation with a pure culture of tubercle bacilli would be a perfectly harmless procedure; in some it would produce a local, latent tuberculosis; while in a few, rendered more susceptible to this form of infection by antecedent hereditary or acquired causes, the inoculation of the same number of bacilli would be followed by rapid and extensive destruction of tissue, and death by early and diffuse dissemination. The same can be said of nearly all, if not all infectious diseases. If their existence has not been demonstrated, we are, nevertheless, forced to recognize the influence of certain as yet unknown conditions inherent in the tissues, and often traceable to an hereditary cause, which favor or resist the action of microbic causes. During the last few years great progress has been made in showing that hereditary disease, in many instances, at least, is due to a more direct cause—the transmission from mother to fœtus of the essential cause of the disease. This method of infection is not only interesting from a scientific standpoint, but is of the greatest practical importance, alike to the surgeon and physician, in regard to prophylaxis, diagnosis, and treatment of many forms of disease. Although our knowledge of the intra-uterine origin of microbic diseases is yet imperfect, there can be no doubt that future study and research will clear up many existing dark points, and furnish a satisfactory demonstrative explanation of the direct and indirect hereditary influences in the causation of disease. I shall in this connection refer particularly to the clinical observations and experimental work which tend to prove the direct transmission of pathogenic microbes from the mother to the fœtus.

1. CLINICAL EVIDENCES.—Lebedeff ("Ueber die intra-uterine Uebertragbarkeit des Erysipels," *Zeitschrift für Geburtshülfe*, B. xii. No. 2) reports a case of premature birth which occurred eight days after the mother had recovered from erysipelas. The child died ten minutes after birth, and the author found Fehleisen's streptococcus in the lymphatic vessels, in the diseased skin, and in the umbilical cord, but not in the placenta. The author believes that the streptococci were transported from the lymphatic vessels of the lower extremities of the mother through the lymphatics of the uterus into the placental vessels, and from the maternal into the fœtal circulation.

Curt Jani (*Lancet*, September 4, 1886, p. 455) has examined the healthy sexual organs of nine male phthisical patients for tubercle bacilli. No bacilli were found, in any of these, in the semen from the vesiculæ seminales; but, on the other hand, in five out of eight cases a few were found in the testicle, and in four out of six in the prostate. The testicles appeared healthy in structure. He further examined two women who died of pulmonary phthisis, the ovaries in both presenting negative results. In one case of chronic pulmonary phthisis, with extensive intestinal tuberculosis, he examined the Fallopian tubes and found tubercle bacilli. He believes that the tuberculous virus can be transmitted from parents to offspring in one of two ways: 1. Through the semen of the male. 2. Through the migration of bacilli into the uterus from the abdominal cavity. Infection of the impregnated ovum by the placental circulation he thinks must be unusual, because the examination of the body of a woman, five months pregnant, who died from acute miliary tuberculosis, in whom infection took place through the growth of a caseous mass in a pulmonary vein, showed that there were no bacilli either in the placental attachment, in the lungs, or in any of the organs usually the seat of localization in the embryo. He, however, considers that it is by no means certain that in chronic miliary tuberculosis deposits may not form in the neighborhood of the placenta, and thus infect the fœtal organism.

That pathogenic microörganisms may exist in the blood of apparently healthy mothers without doing any harm is well illustrated by children who have been born suffering from acute suppurative osteomyelitis, while the mother, through whose blood only the microörganisms could have come, showed no evidences of disease. Rosenbach reports such a case in his article on acute osteomyelitis.

Sangalli (Virchow u. Hirsch's *Jahresbericht*, 1883, B. 1, p. 383) found the bacilli of anthrax in the blood of a fœtus from a woman who had died of carbuncle. In opposition to the views of Golgi and others, he affirms that the transmission of the disease from the mother to the fœtus in utero could only have taken place by the passage of the bacilli or spores from the maternal to the fœtal circulation through the placenta.

Ahlfeld and Marchand (*Centralblatt f. Bakteriologie u. Parasitenkunde*, No. 14, 1887) report the case of a woman who presented no symptoms of disease except a moderate pallor and tympanitic distention of the abdomen. After a normal labor she gave birth to her second child; eight hours after delivery the patient died in collapse, for which no explanation could be found. The autopsy revealed anthrax as the cause of death. The child died four days after birth from the same cause. The mother, as was later ascertained, contracted the disease in sorting horsehair, and the child was infected directly through the placental circulation.

Netter ("Transmission de la pneumonie et de l'infection pneumonique de la mère au fœtus," *Compt. rend.*, March 9, 1889) reports a carefully-observed case of the direct transmission of the diplococcus of pneumonia from mother to fœtus. The mother was a VI.-para, pregnant eight months when she was attacked with croupous pneumonia, which terminated on the seventh day in recovery. On the ninth day after the attack she was delivered of a living child. The child died on the fifth day after birth. The autopsy revealed lobar pneumonia, involving the right upper lobe, double fibrinous pleuritis, pericarditis, suppurative meningitis, and otitis media on both sides. Bacteriological examination of the different inflammatory products, as well as of the blood taken from the left ventricle, showed the presence of Fränkel's diplococcus of pneumonia.

2. EXPERIMENTAL PROOF.—Bollinger was the first to assert that the bacillus of anthrax could pass from mother to fœtus through the placenta.

Strauss and Chamberland ("Sur le passage de la bacteridie charbonneuse de la mère au fœtus," *Gazette hebdom. de Méd. et de Chir.*, 1883, No. 10) experimented on guinea-pigs to prove that intra-uterine transmission of anthrax from mother to offspring is possible. Gravid animals were inoculated and the fœtuses examined immediately after death. Blood taken from the cavities of the heart and the liver, examined under the microscope, never showed bacilli. Cultivation experiments were made with the fœtal blood in veal bouillon, and these demonstrated that in some instances the blood of all fœtuses from the same mother contained bacilli; sometimes all cultures remained sterile; while in some the blood of only one fœtus would yield a positive result. From these experiments the authors came to the conclusion that the tissues of the placenta offer no insurmountable obstacle to the passage of the bacillus of anthrax from the maternal into the fœtal circulation.

Max Wolff ("Ueber Vererbung von Infectious-krankheiten," Virchow's *Archiv*, B. 112, pp. 136–202) has studied the subject of intra-uterine transmission of infective diseases by a series of carefully-conducted experiments. He worked with cultures of anthrax bacilli and examined the blood and internal organs of the fœtuses under the microscope, and tested their sterility by cultivation and inoculation experiments. The pregnant animals which were inoculated with a pure culture of anthrax bacilli manifested symptoms of the disease in from thirty-six hours to three days. The blood and internal organs of 29 fœtuses contained no bacilli, as was shown by most careful microscopical examination, after single and double staining.

The placenta in each instance contained numerous bacilli, while the villi of the chorion showed no trace of them. From these 29 fœtuses 156 cultivation experiments were made with tissue from the kidney, liver, and lungs, and of this number a positive result was obtained in only 6. 14 guinea-pigs and 16 white mice were inoculated with fœtal tissue, with the result that in only 3 cases was the disease transmitted in this manner. The author is inclined to the belief that the successful cultivations and inoculations were due to contamination with the maternal tissues. He affirms that the placenta constitutes an impervious barrier to the passage of micro-organisms from the maternal into the fœtal circulation.

Koubassoff (*Centralblatt f. d. med. Wissenschaften*, Jan. 6, 1886) came to more positive results in his experiments with anthrax bacillus. In all of his experiments, the fœtuses of the infected animals contracted the disease in utero. He also found that time played an important part as far as the number of bacilli in the fœtus was concerned, as the longer the period which intervened between the inoculation and the death of the mother, the more numerous were the bacilli in the fœtal organs, showing that the migration of microbes from the maternal to the fœtal side of the placenta is continuous. Inoculation experiments with the attenuated virus proved that intra-uterine transmission took place more slowly. Inoculation of gravid animals with a very strong anthrax vaccine nearly always proved fatal to the fœtus. Prophylactic vaccination of the fœtus through the mother proved insufficient against subsequent inoculations with the attenuated virus. In another paper the same author (*Centralblatt f. Gynäkologie*, Dec. 5, 1885) has published some important observations on the same subject, made in Pasteur's laboratory, in which he showed that in pigs the microbes of malignant œdema and tuberculosis pass directly from mother to offspring through the placenta.

Kroner ("Ueber den gegenwärtigen Stand der Frage des Ueberganges pathogener Mikroörganismen von Mutter auf Kind," *Breslauer ärzt. Zeitschr.*, 1886, No. 11) published his experiments on rabbits with the bacteria of sepsis. The results obtained were not constant. In all, six rabbits were inoculated at different stages of gestation. In the blood of the fœtuses no bacteria could be found. Inoculation with the blood of the fœtuses yielded, among a number of negative, four positive results. After a careful consideration of the subject, based upon his own experience and the observations of others, he came to the conclusion that a number of microbic diseases are communicable from mother to child through the placental circulation.

One of the strongest evidences of direct transmission of pathogenic microbes from mother to fœtus through the placental circulation, has been furnished by Johne (*Lancet*, March 6, 1886). An eight months' fœtus was taken from a cow the subject of advanced tuberculosis. No tuberculous products were found in the placenta or the uterus; but in the lower lobe of the right lung of the fœtus a nodule, the size of a pea, was detected, containing four caseous centres. The bronchial glands were the seat of tubercular adenitis. The liver contained numerous miliary tubercles. All the lesions presented under the microscope the characteristic histological structure of tubercle.

Levy (*Journal des Connaissances Médicale*, Janvier, 1890, abstract in *American Journal Medical Sciences*, August, 1890) reports the case of a woman, aged thirty years, who had died from double fibrinous pneumonia, complicated by pleurisy and pericarditis. By aspiration sero-purulent fluid was removed from the thoracic cavities. This fluid gave cultures showing the diplococcus of Fränkel-Weichselbaum; inoculations from this fluid showed the presence and potency of the germ. This woman's child was born thirty-six hours before its mother's death, and died two days later from hemorrhagic catarrhal pneumonia and fibrinous lobar pneumonia. Autopsy demonstrated the infectiousness and that it had persisted at least thirty-six hours before the child was born. Cultures made with fluids removed from the left heart ventricle and from the right lung, demonstrated the presence of the diplococcus; the microörganisms were especially numerous in the blood. The conclusions are that the child was infected from the mother's pneumonia.

That syphilis is a microbic disease can no longer be doubted, and that it is one of the diseases which is most frequently transmitted from parents to offspring is well known. Some interesting observations have recently been made on the etiology and transmissibility of syphilis by Disse and Taguchi, of Japan (*Centralblatt f. Gynäkologie*, 1888, No. 11), and if future research should corroborate their claims, their researches will constitute an important contribution to our knowledge of the heredity of this disease. They discovered in syphilitic lesions, isolated, and cultivated from them, a diplococcus. With a pure culture of this microbe they inoculated gravid dogs, and found that the microbe permeated the placental tissues, and entered directly into the circulation of the embryo. The pups suffered from all the characteristic lesions of hereditary syphilis as observed in man. The lesions commonly found were pneumonia, disease of the liver, bones, and kidneys. In all of these organs the same diplococcus was found as was used in the inoculation experiments.

Romeo Mangeri, of Catania, believes that direct transmission of microbes from mother to fœtus through normal placental vessels is impossible. As the result of an extensive study of the literature of the subject and of original experiments, he has come to the conclusion that no formed elements naturally pass out of the mother's blood into the fœtal circulation. Cinnabar, Indian ink, carmine, and other finely-divided pigment materials were injected into the jugular veins of animals advanced in pregnancy, but in no case could any trace of the substance employed be found in the fœtus. According to his belief, passage of formed elements can only occur when the placenta becomes diseased by inflammation, or is partially detached so that the walls of the villi are destroyed. Only under this condition he maintains can pathogenic microörganisms be transmitted from the mother into the fœtal blood.

Most all authors agree that, when extravasations, or other pathological processes occur in the placental attachment, the direct entrance of microbes from the mother into the fœtal circulation is not only a possible, but a probable occurrence. Abnormality in the placental circulation must, therefore, be recognized as a condition which favors the transmission of microörganisms from mother to fœtus.

All of the foregoing observations and experiments furnish substantial proof that in some infectious diseases heredity is traceable to direct transmission of the specific microbes floating in the circulation of the mother to the fœtus, through the thin wall which separates the maternal from the fœtal blood. It is no more difficult to explain the migration of microbes through such a thin septum than their transportation from one tissue to another, and from organ to organ in other parts of the body, more especially as the anatomical conditions for mural implantation in the placental vessels are most favorable for such an occurrence.

CHAPTER II.

DO PATHOGENIC MICROÖRGANISMS EXIST IN THE HEALTHY BODY?

IT still remains a disputed question whether pathogenic microorganisms can exist in the body without giving rise to disease. It has been definitely ascertained by experimental research that many of the pathogenic germs are harmless as long as they remain in the circulating blood, and that their specific pathogenic action only becomes evident after localization takes place in some part of the body, in a soil prepared by injury, or disease, for their arrest and multiplication. It has also been definitely settled by clinical experience that pathogenic spores may remain in the healthy body, in a latent condition, for an indefinite period of time, until by some accidental pathological changes the tissues in which they may exist have been prepared for their growth. Numerous experiments will be cited elsewhere, in which injections of pure cultures directly into the circulation produced no ill effects in healthy animals, but when previous to the injection, or soon after, an injury was inflicted in some part of the body, localization occurred at the seat of trauma, and in the culture soil thus prepared the microbes produced their specific pathogenic effects. From these remarks it is reasonable to assume that pathogenic germs may exist in the healthy body without necessarily giving rise to disease, especially if, as is well known, they are being constantly eliminated through the excretory organs. Some of the arguments for and against this theory will now be introduced.

Fodor ("Bacterien im Blute lebender Thiere," *Archiv f. Hygiene*, B. iv. p. 129) introduced directly into the circulation of rabbits pathogenic bacteria, in order to study their effects on the tissues and manner of elimination. As a rule, he found that they had completely disappeared from the blood after twenty-four hours. No culture experiments were made less than four hours after inoculation. He believes that the microbes are removed by the red corpuscles, or that they are digested by the leucocytes. He affirms that, as a rule, pathogenic germs are not present in the healthy organism, as he found the blood of healthy rabbits, without exception, sterile; and only in exceptional cases was he able to demonstrate the presence of bacteria in animals killed, even where the examination was postponed until putrefaction had set in.

Zahor ("Ueber das Vorkommen von Spaltpilzen im normalen thierischen Körper," *Wiener med. Jahrbücher*, p. 343, 1886) examined the blood, testicle, heart, and spleen of a healthy rabbit, and found in fresh as well as in hardened sections, after staining with methyl-violet, cocci, and here and there rods. The same examinations with like results were made on the organs of a young cat. Under strict antiseptic precautions he removed a testicle from another healthy animal, and embedded it at once in paraffin, which, under the microscope, was also seen to contain bacteria.

Bizzozero (Virchow's *Jahresbericht*, 1887, B. 1, p. 283) could not detect bacteria of any kind in animals soon after birth, but in the lymph follicles of the vermiform process in healthy rabbits he found numerous bacteria which could be readily stained with Gram's solution, and which, in form and size, corresponded with the schizomycetes which are always contained in the intestinal contents of these animals. The microbes were seen mostly in the protoplasm of cells, a condition which would indicate that they are transferred from the intestinal canal into the closed lymph follicles through the medium of migrating cells.

Hauser (Vorkommen von Microörganismen im lebenden Gewebe gesunder Thiere, *Archiv f. experimentelle Pathologie und Pharmacologie*, B. xx. pp. 162–202) has made a number of carefully-conducted experiments to show that no microbes exist in healthy animals. The experiments consisted principally in procuring tissues prone to fermentation, as parts of internal organs, blood, etc., and protecting them against infection from without. He kept the specimens in rarefied air, in filtered air, hydrogen, oxygen, carbonic acid gas, and water, and in various artificial culture soils, at a temperature favorable to putrefaction, but in all instances in which the specimens remained uncontaminated no putrefactive changes were observed. By this method he believed he was able to demonstrate that tissues taken from healthy animals immediately after death contained no putrefactive bacteria, since it is well known that if the specimens were not perfectly sterile putrefaction would have taken place. The author did not only appear to demonstrate that living tissues contained no microörganisms, but he also ascertained that the preserved sterile organs in time underwent a sort of regressive metamorphosis similar to that which takes place in the body in the absence of microörganisms, and, what is of especial interest, that the products of such processes of resolution possess no poisonous properties whatever.

Fodor (*Deutsche med. Wochenschrift*, 1887) has shown that the power of the blood to destroy bacteria is not diminished by a moderate degree of anæmia, but is lessened when diluted with water, as when this is done the microbes are destroyed more slowly, and with greater difficulty. Anthrax bacilli injected in large quantities directly into the circulation disappeared from the blood in four hours; but within twenty to twenty-four hours they reappeared and the animal died. As the bacilli disappeared from the circulating blood localization in internal organs took place, and their reappearance in the circulation only implies an enormous increase in number, and invariably resulted in the death of the animal. Anthrax is a blood disease; but the blood is rather the protector of the organism against pathogenic microbes.

Watson Cheyne ("On Suppuration and Septic Diseases," *British Medical Journal*, March 3, 1888) found in his experiments on the presence or absence of microörganisms in the living tissues that, while germs were absent when the animal was in a good state of health, yet if the vitality of the animal was depressed, say, by administering large doses of phosphorus for some time, organisms could be found at times in the blood and tissues of the body. Again, it has been found that, while some microörganisms when introduced into the living body in small number, disappear after a short time, when a large quantity of the cultivation is introduced the tissues of the body are injured by the preëxisting ptomaïnes and the germs retain their vitality and often produce their specific pathogenic effects. Diminution of the force of the circulation also plays an important part in the production of microbic diseases. Thus, according to Cornil, a septic nephritis is readily produced by ligating the renal arteries for some time and then, after removal of the ligature, injecting pyogenic organisms into the blood. Heubner made the same experiment on the bladder and found that by interrupting the arterial supply for a certain time coagulation-necrosis occurred in the protoplasm of the epithelial cells which furnished the most favorable conditions for septic infection when septic germs were introduced into the circulation.

Cornil also found that if a simple nephritis is induced by the administration of cantharides, the pathological conditions thus produced furnished the most favorable soil for septic nephritis, when the animal was infected with septic germs.

CHAPTER III.

SOURCES OF INFECTION.

PETTENKOFER divided the pathogenic bacteria into two classes —the *endo-* and *ecto-genous*. Under the former head were classified all bacteria which were supposed to have an endogenous origin, while the ectogenous variety included all bacteria which enter the organism from without. Syphilis and tuberculosis were regarded as endogenous processes, but since it has been made possible to cultivate the tubercle bacilli outside of the body and to produce with them typical tubercular lesions, this disease cannot be attributed any longer to an endogenous origin, and it will not be long before future research will transfer syphilis from the endogenous to the ectogenous variety for the same reason. Even in the most marked cases of so-called auto-infection the microbes must have entered the organism at some previous time from without, and all such affections are in every sense of the word ectogenous processes. In surgery it is of special importance that the endogenous origin of infective diseases should be no longer recognized, and that their causes should always be sought for outside of the body.

Bacteriology has rendered the term *miasma* obsolete. All infective diseases are now traced to an organic contagium. All infective diseases in the strict sense of the word are contagious, as they can only arise by the entrance of pathogenic microbes from without, and this can only take place if they are brought in contact with an absorbing surface. Of all substances which serve as a carrier of microbes, the atmospheric air is the most important, because it is present everywhere and no one can exclude himself from it. In a dry state, pathogenic germs move with the currents of air and attach themselves again to the solid or fluid substances with which they come in contact. Although most of the microbes under ordinary circumstances do not reproduce themselves outside the body, their resistance to heat and cold, moisture and dryness, is so great that they retain their disease-producing qualities often for an indefinite period of time, and after their entrance into the body and meeting with a proper nutrient medium they exert their specific pathogenic effects.

Sternberg ("The Thermal Death-point of Pathogenic Organisms," *American Journal Med. Sciences*, vol. 94, p. 146) has found by a series of very interesting experiments that different pathogenic germs possess varying degrees of resistance to heat. The following were his results in studying the microörganisms which more particularly interest the surgeon. The time of exposure was ten minutes, unless otherwise indicated by figures in parentheses:

	Fahrenheit.
Bacillus anthracis (Chaveau)	129.2°
Bacillus anthracis spores	212.0
Bacillus tuberculosis (Schill and Fischer)	212.0 (4 m.)
Staphylococcus pyogenes aureus	136.4
Staphylococcus pyogenes albus	143.6
Staphylococcus pyogenes citreus	143.6
Streptococcus erysipelatosus	129.2
Gonococcus	140.0

Of course, the same difference of resistance to cold and other destructive influences must exist among them.

The spores possess still greater resisting powers, and consequently, in places where the destructive influences are sufficiently severe to destroy microbes, infection may still take place from this source.

J. Geppert ("*Zur Lehre von den Antisepticis*," *Berl. klin. Wochenschrift*, Sept. 9, 1889) while making the experiments for the purpose of verifying the statements of Woronzoff and others, that the spores of the bacillus anthracis are rendered harmless by immersion in a 1 to 10,000 solution of sublimate for fifteen minutes, and by a 1 to 2000 solution in one minute, found a source of fallacy in their method. This consists in impregnating a thread with the spores and immersing it in the antiseptic solution for the desired time, and after washing it in alcohol or water, using it to inoculate nutrient media or animals, the extreme difficulty of removing all traces of the chemical used becomes apparent, while its importance is shown by the statement of Koch, that the presence of sublimate in the strength of 1 to 100,000 will retard the growth of anthrax spores. Geppert's method consists in using a suspension of spores in distilled water, which is then filtered through glass wool until the liquid is perfectly clear. After subjecting the spores for a definite length of time to the action of the antiseptic he used some indifferent chemical to precipitate the antiseptic. In the case of sublimate he used ammonium sulphide. He made inoculations with spores exposed to the action of the sublimate before and after precipitation. When he used spores exposed to the action of a sublimate solution 1 to 1000, without precipitation, he obtained cultures in all cases after an exposure for three minutes, once after seven minutes, also after ten minutes. In using spores of solutions where precipitation had been practised he always obtained numerous and large colonies after exposure to the action of the antiseptic for fifteen minutes, less after thirty minutes, and very few after an hour's exposure, and only in exceptional cases were positive results obtained after action of the antiseptic for two or three hours, and with one exception never after seven to twenty-four hours' exposure. Combinations with other antiseptics had no retarding effect. From his experiments he was led to conclude that spores exposed for some time to a sublimate solution will not grow upon a soil containing the antiseptic in a very diluted form, while spores similarly treated will not grow upon a soil upon which spores not so treated will grow.

By improved methods of investigation it has recently been made possible not only to demonstrate the presence of microbes in the atmospheric air, but also to prove their identity and approximately to determine their number in a certain volume of air. From the ingenious experiment made by Miquel ("Des bactéries atmosphériques," *Comptes rend.*, t. 91, No. 1) it appears that the air picks up in its passage across the city of Paris in the course of half an hour three times the number of microbes it contains when blowing over the country. On the basis of a series of carefully-made observations to determine the number of microbes which a certain volume of air contains at different seasons of the year, he found that a cubic centimetre of air contains on an average two hundred spores of bacteria. During the summer such a volume of air often contains as many as 1000, while during the cold months of winter it contains as few as 4 or 5. In rooms with no draft of air, it required 30 to 50 litres of air for a successful inoculation of a sterile culture substance. In Miquel's laboratory 5 litres were found sufficient, while 1 litre of air taken from the subterranean channels in Paris produced the same effect. In a cubic yard of air taken from the wards of a long-used hospital he found as many as 90,000 microbes. This same observer has shown, and others have confirmed it, that sea air contains scarcely any of these microörganisms, and that mountain heights are as nearly free from them for obvious reasons.

In a recent communication to the Royal Society, Dr. Frankland described a new method of examining air for microörganisms. It consists in aspirating a known volume of air through a glass tube containing two sterile plugs of glass wool alone, or of glass wool with fine glass powder, etc., the first plug being more pervious than the second. These plugs are then transferred to sterilized liquid gelatine-peptone, thoroughly agitated with it, and then the gelatine is congealed, so as to form an even film over the inner surface of the flask. The number of colonies which develop are then counted. Klebs has devised a very ingenious but complicated apparatus for filtering a measured quantity of atmospheric air directly upon culture plates, which he used in prosecuting his researches on the microbe of malaria in the vicinity of Rome. Tyndall has shown that floating microbes not only enter our bodies, but are arrested and retained in them, by transmitting a beam of electric light through air before and after it has been respired, which has demonstrated that, however populated before, it is entirely free from organic and inorganic particles after it has once passed through our lungs.

At a recent meeting of the Academy of Sciences of Paris, Strauss ("Sur l'absence de microbes dans l'air expiré," *La Semaine Médicale*, 1887, No. 49) reported a series of bacteriological experiments upon expired air. He confirms the result arrived at by Tyndall, that expired air is optically pure, and adds the further fact that it contains no microbes. Gaucher, Charrin, and Karth examined the expired air in phthisical patients and failed to find bacilli. The bronchial tubes act as filters, permitting only the passage of air.

Neumann ("Ueber den Keimgehalt der Luft im städt. Krankenhause Moabit," *Vierteljahresschrift f. gerichtliche Medicin*, B. xlv. p. 310, 1886) made 35 experiments after Hesse's method to determine the number of germs which exist in a certain volume of atmospheric air. The cultures were made upon meat infusion-peptone-gelatine. With an apparatus of special construction, from 5 to 20 litres of air were taken from one of the rooms in the Moabit Hospital. The examination of air, taken at different elevations, from 1.40 to 3.20 m., showed that it contained about the same number of microbes. In the morning after the rooms had been swept, from 80 to 140 microbes were found in every 10 litres of air. Four consecutive examinations of the same quantity of air taken at the same height showed a gradual decrease in the number of germs, so that at the last examination, at eight o'clock in the evening, only from 4 to 10 germs were found. Cocci were more abundant than bacilli, and microbes were more frequently met with than spores, a fact which had already been established by Hesse.

Buchner (*Münch. med. Wochenschrift*, 1888, Nos. 15–17) has recently made some interesting experiments on the inhalation of microbes with the inspired air, and has come to conclusions somewhat different from those heretofore held. The animals were exposed to a spray of diluted cultures of different bacteria, in such a way that only the finest mist of the spray reached them. Control experiments were made by feeding, so as to exclude accidental infection through the alimentary canal by swallowed spray. The results showed that this happened very seldom, and also that the amount necessary to cause infection through the intestinal canal is much larger than that for the lungs. The bacteria used were those of anthrax, with their spores, and those of chicken cholera and septicæmia. Irritation of the bronchial mucous membrane retarded the ingress of microbes into the circulation. When large quantities of bacilli were used the irritation was often so great that a hemorrhagic pneumonia resulted. The best results were obtained by using only the spores, after destroying the bacilli by drying. The growth of bacilli was very rapid, and in a short time the microbe could be found in the pulmonary capillaries. The rapidity with which the microbes entered the pulmonary tissue was proof positive that a more direct tolerance was effected than through the lymphatics. The author is of the opinion that the chemical irritation exerted on the walls of the capillaries causes the formation of stomata, through which the bacilli grow. These experiments demonstrate also, that the blood microbes are best adapted to pass through the pulmonary tissues into the circulation. The microbes of anthrax, recurrent fever, and malaria are the most striking examples of the blood parasites so far known. In infection caused by blood microbes, the point of entrance cannot always be found, as they may enter through apparently intact surfaces. Even in the case of microbes not strictly blood microorganisms, it is possible for them to penetrate the pulmonary capillaries and be transported to that part of the body where they afterward initiate morbid conditions. These experiments show conclusively that pathogenic microbes suspended in the atmospheric air can enter the organism through the respiratory organs.

Petri ("Zusammenfassender Bericht über Nachweis und Bestimmung der pflanzlichen Microörganismen in der Luft," *Centralblatt f. Bacteriologie u. Parasitenkunde*, B. ii. Hefte 5 u. 6) claims that the methods so far employed in the detection of pathogenic germs in the air have not succeeded in demonstrating more than the microbes of pus, a circumstance which may be ascribed to the methods employed, as some pathogenic germs do not grow upon gelatin at ordinary room temperature, and that the culture of the microbe sought for may become obscured by cultures of other microbes.

Emmerich ("Ueber den Nachweis von Erysipelkokken in einem Secirsaale," *Deutsche med. Wochenschrift*, 1877, No. 3) by means of an aspiration apparatus obtained air from an old dissecting-room, which for a long time had not contained any erysipelatous material, and cultivated from it the cocci of erysipelas, which on being inoculated into animals produced typical erysipelas. The cocci were also found in shavings of the floor and in the plastering of the walls and ceilings.

Eiselsberg ("Nachweis von Erysipelkokken in der Luft chirurgischer Krankenzimmer," Langenbeck's *Archiv*, B. xxxv. Heft 1) made a series of experiments for the purpose of ascertaining the presence of pathogenic microörganisms in the air of hospital wards occupied by surgical patients. By exposing for a certain length of time glass plates coated with sterilized gelatine or agar-agar in a room occupied by four cases of erysipelas, twice colonies of cocci were found, which, however, grew only to a limited extent. Inoculation upon agar-agar and gelatine as well as upon solid bloodserum showed that the new growth was composed of Fehleisen's streptococcus. Its identity with the microbe of erysipelas was further demonstrated by inoculation experiments which proved successful. In another room containing surgical patients, where all the wounds were aseptic, a growth of the staphylococcus pyogenes aureus developed upon culture plates. From erysipelatous patients during the stage of desquamation, he inoculated sterilized gelatin with epidermic scales, and in four out of five cases the specific microbe of erysipelas was cultivated.

At the last Congress of Italian Surgeons, Durante reported experiments, performed at his suggestion, by his assistant, to determine the presence of germs in the air about the bed of a surgical patient, and in a ward. It was found that the greatest number of germs were present about forty inches above the border of the bed; in all other directions the number of germs greatly decreased. Culture materials placed beneath the bed, and about the bottoms of the neighboring walls, remained frequently sterile. The germs most frequently found were staphylococcus pyogenes aureus, streptococcus of erysipelas, and Fränkel's pneumococcus. (*Deutsche med. Wochenschrift*, May 3, 1888.)

Wehde investigated the infectiousness of air in rooms occupied by phthisical patients by exposing plates smeared with glycerin, and with the dust thus collected he inoculated guinea-pigs; the results were negative. From these experiments he concluded that the air expired by tubercular persons does not contain the bacillus, and that tubercular sputum, as long as it remains moist, does not emit microbes. Similar results have been obtained by Celli and Guarnieri, Sirena, Pernice, and Nicolas. On the other hand, Theodore Williams (*Lancet*, 1883), by resorting to the same method, and exposing the plates to the current of air in one of the ventilating shafts in the Brompton Hospital for five days, demonstrated the presence of tubercle bacilli upon the glycerin plates. The most recent and reliable experiments on the existence of tubercle bacilli in hospitals and sick-rooms have been made by Cornet, in Koch's Hygienic Institute, his method consisting in wiping the walls of wards and sick-rooms at places where the dust had been left undisturbed for a long time, with a sterilized sponge, and squeezing it out in bouillon, so as to liberate any germs that might have been caught. This fluid was then injected into the peritoneal cavity of guinea-pigs. If tubercle bacilli were present, the animal invariably manifested symptoms of peritoneal tuberculosis three to four weeks after inoculation. In 7 hospitals in Berlin, 21 wards occupied by tubercular patients were examined. The sponge was only used on surfaces which had never been exposed to direct contact with the expectorations of the patients. In 15 out of the 21 wards tubercle bacilli were present, and of 94 animals which were inoculated, 20 died of tuberculosis. In the homes of fifty-three private patients, the presence of tubercle bacilli was demonstrated by the same method in 20.

That microbes in a dried state, adherent perhaps to particles of dust, follow the currents of atmospheric air, has been demonstrated by the history of endemics of malaria.

Cadene and Malet ("Etude expérimental de la transmission de la mave par contagion médiate ou par infection," *Revue de Médicale*, t. viii. p. 227) have studied experimentally the transmission of the bacilli of glanders through the atmospheric air, and have come to the positive conclusion that the disease is not contracted in this manner. They exposed animals which are known to be very susceptible to the disease, such as the ass, to the expired air of animals suffering from glanders, but the results were always negative. They next produced a bronchitis in animals by inhalations of bromine before they were exposed to the expired air of animals suffering from glanders, but the results remained the same. The expired air of diseased animals was next mixed with steam, and the latter was condensed, but no bacilli could be found, and inoculation in guinea-pigs only yielded negative results. The same negative results were obtained by passing the expired air over water, as the latter was always found sterile. The air in the stables occupied by the diseased animals was also found free from bacilli. Pure cultures of the bacilli were injected into the trachea of a number of animals, with the result that in only two was the disease produced, while all animals inoculated in the same manner which were suffering from bromine bronchitis contracted the disease and died.

It is also well known that microbes which have been carried upward by currents of air descend by virtue of their own weight. When the air is in a quiescent state the lower strata contain microbes in greater abundance than the upper. Klebs found that in the Campagna districts, near Rome, an elevation of three metres afforded perfect immunity against malaria. Soyka made some observations on currents of air as a vehicle for the diffusion of microbes in Klebs's laboratory at Prague. He came to the conclusion that very slight currents convey microbes from the margins of drying fluids, while Naegeli asserted that a current of considerable force is necessary to effect such transportation, and that the lifting up of the particles into the air by the current was greatly influenced by the force of adhesion which exists between the particles and the surface to which they adhere.

The great media for the diffusion of pathogenic microbes over the surface of the globe are the air and water. Some of the microorganisms, as the pus-microbes, appear to be almost omnipresent, while others are diffused over a more limited area, their existence being dependent upon certain conditions of the soil or temperature. Water as a medium of diffusion and vehicle for the entrance into the organism of pathogenic microbes, is of greater interest to the physician than the surgeon. The superficial layer of the soil contains most of the disease germs and spores, as they are deposited upon it from the air, and carried into it by water which contains them, the soil in the latter instance serving the purpose of a filtering substance. To the surgeon, direct infection with microbes of the soil has awakened new interest by Nicolaier's discovery that the bacillus of tetanus has here its natural habitat, and from the well-known circumstance that the bacillus of anthrax is known to multiply in the soil of pastures which have been inhabited by animals suffering from this disease.

CHAPTER IV.

LOCALIZATION OF MICROÖRGANISMS.

EVERY surgeon has had frequent opportunities to observe cases in which a slight subcutaneous injury was followed by a destructive inflammation, an inflammation not produced by the trauma, but by the localization of pathogenic microbes in the tissues altered by the injury. Thus Chaveau (*Comptes rendus*, t. 76, p. 1092) has shown experimentally that subcutaneous crushing of tissue furnishes a favorable condition for the localization of pathogenic microörganisms carried to the part by the circulating blood. When he injected a putrid fluid directly into the circulation of young rams shortly before subcutaneously crushing the testicle, the injured organ always became the seat of septic gangrene, while without such injection the crushed testicle disappeared by necrobiosis and absorption. Gangrene only occurred if the putrid fluid contained bacteria; it did not take place if the fluid was carefully filtered. Extensive subcutaneous injuries, as severe contusions, rupture of tendons or muscles, and crushing of bone, are not followed by suppuration, unless the injured tissues become subsequently the seat of infection with pus-microbes. A patient may have been the subject of tubercular infection for an indefinite period of time, and yet may present the appearances of ordinary health until some slight injury determines localization of the bacillus of tuberculosis in the part injured, an occurrence which is followed by a localized tuberculosis from which, later, regional and general dissemination takes place, to which the patient finally succumbs, unless the tubercular focus is removed by an early operation. These facts suggest very strongly that, in the hypothetical cases, suppuration and tuberculosis would not have taken place in the part injured without the injury, and that the injury certainly would not have produced suppuration or tuberculosis unless the respective patients had been infected previously with specific pathogenic microörganisms. The injury in these cases creates a so-called *locus minoris resistentiæ*, which may signify one of two things: 1. Diminution or suspension of resistance on the part of the injured tissues to the action of pathogenic microörganisms, which was present in the part at the time of injury; or, 2. The injury so alters the tissues that pathogenic germs, which were present in the circulation without having given rise to symptoms, become arrested, and find at the same time the necessary nutrient medium for their multiplication. These questions have been made the subject of patient experimental inquiry by a number of investigators, the most important results of which will now be mentioned.

1. BACILLI OF ANTHRAX.—Huber ("Experimentelle Untersuchungen über Localisation von Krankheitsstoffen," Virchow's *Archiv*, B. vi., 1886) calls attention to the circumstance that a number of infectious diseases, such as tubercular affections of the bones and joints, certain cases of pyæmia, osteomyelitis, and diffuse suppurative inflammation in different tissues, are liable to follow injuries unattended by infection from without. Again, we have local infective processes which occur as complications during the course of typhus and typhoid fevers, syphilis, and other infective diseases. Through these extraneous influences, or accidental conditions, certain parts of the body are rendered more susceptible to the development of serious local disturbances under the action of infective germs than the remainder of the body; in other words, a local predisposition, a *locus minoris resistentiæ*, is created. The experiments of the author were made on rabbits, in which, by the application of croton oil to the ear, he produced a tissue lesion by the inflammation which followed. Only one ear was thus treated, the other being left in a normal condition in order to compare the results of the subsequent action of pathogenic microbes upon normal and inflamed tissues. The virus selected was a pure culture of the bacilli of anthrax, and the reliability of the culture was always tested before inoculations were made. The inoculation was always made at the root of the tail, as far as possible from the seat of inflammation produced by the croton oil. According to the stage of inflammation in which the action of the bacilli of anthrax was intended to be studied, the croton oil was applied either before or after the inoculation. Immediately after death the ears were removed and carefully preserved for subsequent examination, at the same time serum and blood were separately taken in sterilized glass-tubes from the inflamed ear under strict antiseptic precautions. For microscopical examination the specimens were carefully stained, while cultivations were made by inoculating sterilized nutrient fluids, as at that time the method of cultivation on solid nutrient media inaugurated by Koch was not known. These experiments enabled the author to state that in all stages of the inflammatory process the bacilli were never found outside the walls of the capillary bloodvessels in the tissue placed in a condition of inflammation by the croton oil. Their number within the bloodvessels depended upon the conditions of the inflamed bloodvessels. During the first stage of inflammation, marked by œdema without

suppuration, more bacilli were found within the inflamed vessels than in the corresponding vessels of the opposite ear. The bacilli were found equally distributed in the arteries, veins, and capillaries, but were never found anywhere in the para-vascular tissues. The first stage of inflammation lasted on an average seven hours and a half, after which suppuration initiated the second stage. During this stage, the bacilli disappeared within the bloodvessels; this could be more readily seen forty-eight hours after the croton oil was applied. During the third stage, when granulations commenced to form, a complete change was again observed in the bacteriological condition of the inflamed part. The height of this stage is reached on the tenth day. During this stage the bacilli reappear in the inflamed tissue, where they can be seen in considerable number, especially in the interior of new capillary bloodvessels. During the process of cicatrization the number of bacilli in a corresponding area of both ears was about the same. From these observations Huber concluded that the bacillus of anthrax finds, in a soil prepared by inflammation induced with croton oil, a *locus minoris resistentiæ*, which presents more favorable conditions for its growth than tissues in other parts of the body. Suppuration appeared to neutralize the anthracic process by the destructive effect of the pus ptomaïnes upon the bacilli. The conclusions which he has drawn from his experiment may be summarized as follows: Localization of preëxisting microörganisms in tissues prepared by injury or disease takes place, provided that the necessary conditions for their growth are present.

In looking over different pathological conditions we frequently meet with a so-called *locus minoris resistentiæ;* at any rate, if we search only for that which should mean what has been described above, it is not difficult to conceive how slight injuries, wounds, contusions, etc., should in this manner give rise to serious affections. But not only do direct tissue lesions, as hemorrhages, necrosis, hyperæmia, fractures, etc., act in this manner, but a variety of pathological conditions of a general nature may serve the same purpose as imperfect digestion, enfeebled circulation and respiration, and exposure to cold. All these ill-defined conditions belong here, and through their instrumentalities the localization of infective germs is favored. In mixed infections it can be said that certain infective microbes prepare the tissues for the reception and growth of other pathogenic germs.

Muskatblüth (" Neue Versuche über Infection von den Lungen aus," *Centralblatt f. Bacteriologie u. Parasitenkunde*, B. 1, No. 2) has made some very interesting and practical experiments on infection through the respiratory passages. The bacillus of anthrax was used. The infection was made in two ways: 1. Direct injection of a pure culture into the trachea. 2. Injection through a tracheal tube. In the latter case the injection was not made until the tracheal wound had completely healed around the canula, and the tube was left *in situ* until the death of the animal. The tracheal wound in these cases never became infected. Direct injections into the trachea always produced positive results, the animals dying of anthrax in from forty to forty-eight hours. As in these instances the tracheal wound always became infected, it became necessary to kill the animals soon after the injection in order to determine the effect of the bacilli on the pulmonary tissues. On microscopical examination of the specimens of animals killed sixteen hours after direct injection, free bacilli were found in the wound, the walls of the alveoli, in the juice canals and only a few rods in the interior of the alveoli themselves and the lumen of the bronchial tubes. The bacilli were most abundant in the so-called dust cells, derivatives of the alveolar epithelia. No bacilli could be seen in the peri-bronchial and peri-vascular lymph spaces. On examination of the bronchial glands, a very peculiar arrangement of the bacilli was found. Each section contained countless bacilli which occupied almost exclusively the lymph channels of these organs; in the bloodvessels they were scanty. The bacilli and threads surrounded the periphery of the cortical follicles in the shape of a very dense zone, the density of which diminished toward the centre of the follicle and followed the medullary strings. In animals killed seventeen hours after direct injection, small pieces of the lung or liver planted on a proper culture-soil were soon surrounded by a copious growth of bacilli. In animals injected through a canula after complete cicatrization of the tracheotomy wound, killed after forty hours, the bacilli were found most numerous in the pulmonary bloodvessels, while the dust cells contained only a single rod, if any. Most all of the bacilli had left the lymphatic glands, and had passed into the general circulation. From these experiments the author concluded that the bacilli of anthrax can enter the circulation through the mucous membrane of the bronchial tubes, and that the juice canals, lymphatic channels of the bronchial tubes, lungs, and the bronchial glands are the channels through which infection takes place. The carriers, which are directly concerned in the transportation of the bacilli from the mucous surface into the lymphatic system, are yet to be discovered. It seemed strange to the author that no bacilli could be found in leucocytes, but always only in epithelial cells. Final localization of the bacilli of anthrax which have entered the circulation through the lungs takes place in distant organs through the medium of the general circulation by implantation upon the endothelial lining of the capillary ves-

PLATE I.

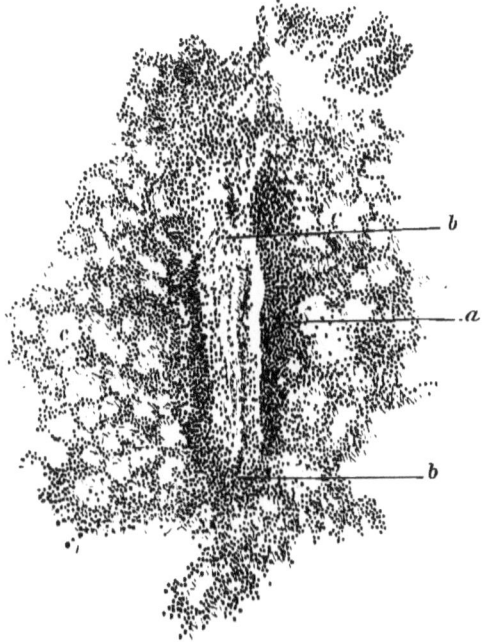

Parasitic interstitial pneumonia, from the lungs of a guinea-pig infected with anthrax. Hartnach. Obj. 4. O. 2. Picro-carmine-gentian.
 a. Periarterial growth.
 b. Wall of artery.
 c. Thrombus with bacilli.

sels. Other experimenters affirm that if the bacilli of anthrax are injected in moderate quantities into the circulation of animals, they disappear soon from the blood without having produced any pathogenic effects; but, if in animals thus injected a contusion is produced in some part of the body, the bacilli pass out of the vessels into the connective tissue along with the blood, grow there, and soon cause the formation of the characteristic inflammatory product, the diffusion of the disease, and its fatal termination.

2. BACILLI OF TUBERCULOSIS.—Volkmann ("Chirurgische Erfahrungen über die Tuberculose," *Verh. der Deutschen Gesellschaft f. Chirurgie*, 1885), from an extensive clinical experience, came to the conclusion a long time ago that a severe trauma seldom, if ever, gives rise to tuberculosis at the seat of injury, but that where local tuberculosis is caused by an injury the trauma was always slight, sometimes almost insignificant. He maintains that the active tissue-changes which follow a severe injury during the reparative process counteract the growth and propagation of the bacillus. Luecke attributes to exposure to cold an important rôle in the causation of tubercular and other infective forms of inflammation, as he asserts that the sudden diminution of blood-supply to the cutaneous surface causes internal congestions which favor the localization of pathogenic germs in some one of the congested parts, otherwise predisposed to the specific inflammation.

Schueller (*Experimentelle und histologische Untersuchungen über Entstehung der skrophulösen u. tuberkulösen Gelenkleiden*, Stuttgart, 1880) studied the localization of the tubercular virus experimentally in the same manner as others have studied the localization of pus-microbes. He inoculated animals with the products of tubercular inflammation, subsequently produced contusions and sprains of joints, and observed that localization usually occurred at the seat of injury. If the tubercular virus was introduced by inhalation, the same typical lesions occurred in the injured joints as when infection was made more directly. In all cases the products of the local lesion corresponded with the character of the material introduced through some remote point. Surgeons are well aware of the danger of general infection following an injury to a part or an organ the seat of a local tuberculosis, more particularly in cases of tubercular disease of joints. Cases like the following are not of rare occurrence.

Szuman ("Brisement forcé eines skrofulös entzündeten Kniegelenkes und konsecutiver acuter allgemeiner Miliar-tuberculose," *Centralblatt f. Chirurgie*, 1885, No. 29) relates the case of a small, well-nourished child, the subject of tuberculosis of one of the knee-joints. The joint was punctured with a fine trocar, and although no pus escaped, it was washed out through the canula with a solution of carbolic acid. The swelling diminished, but the joint became fixed by false ankylosis at nearly a right angle. Rest in bed and extension with weight and pulley had no effect. Under an anæsthetic *brisement forcé* was made, and a plaster-of-Paris bandage applied. The operation was followed by a continued fever, the temperature at times reaching 40° C. (104° F.), rapid emaciation, cough, expectoration, and râles over both lungs, purulent otitis media, and death in a few weeks. The necropsy revealed acute miliary tuberculosis of lungs and several other organs. The joint was the seat of a caseous osteo-tuberculosis.

Verchère ("Coxalgie ancienne, Redressement et immobilisation dans un bonne position, Méningite tuberculeux, Tuberculose miliare generalissée, Mort," *Progrès Méd.*, 1886, No. 24). In a boy sixteen years old, who had suffered for one and three-quarters of a year from coxitis, reduction of a semiflexed leg was made under ether. But little force was used, and the limb was immobilized. Patient was emaciated, but otherwise in fair health. The next morning the temperature was 40° C. (104° F.), and symptoms of meningitis were present. These subsided after ten days, but a moderate fever persisted with gradual decline of health. After two months, pulmonary signs could be detected; four weeks afterward, death. Post-mortem showed general tuberculosis and tubercular meningitis. The author attributes the general dissemination to the *redressement*.

In these cases, as in so many others of a similar character which have been reported by different surgeons, the violence used in straightening the limb caused laceration of some of the small vessels in the tubercular focus, which opened a direct road for the bacilli into the circulation, which by their rapid diffusion through the systemic circulation, caused acute miliary tuberculosis. Imperfect operations for local tuberculosis have been followed by the same disastrous consequences. The veins, in the indurated tissues within and around the tubercular product, when opened by the knife, scissors, or a sharp spoon, do not contract, and cells containing the bacilli, as well as free bacilli, find ready access into the open vessels, and after they have entered the circulation miliary tuberculosis is produced by their localization in distant organs. In operation wounds, after incomplete removal of a tubercular focus, it is, therefore, advisable to prevent the traumatic diffusion of the bacilli by searing their surface with the actual cautery, which not only prevents the entrance of bacilli into the circulation, but, at the same time, adds to the prospects of a permanent result by destroying infected tissues.

Wartmann ("Die Bedeutung der Resection tuberculös erkrankter Gelenke für die Generalisation der Tuberculose," *Centralblatt f.*

Chirurgie, 1887 No. 2), has collected 74 excisions of tuberculous joints made at the Canton Spital in St. Gall, for the purpose of ascertaining the immediate effects of the operation in causing generalization of the tuberculosis. Of this number, 11 died, and in 2 of them death was caused by miliary tuberculosis soon after the operation, the disease being produced by inoculation from the wound surfaces. He next collected 837 resections from other sources, with 225 deaths; of these, 26 cases of acute miliary tuberculosis could be traced to inoculation during the operation.

3. PUS-MICROBES.—When by mechanical, chemical, or other injury, the vitality of tissue cells is lowered, a door is opened for the admission of such organisms which, having once penetrated the tissues or the blood, find a suitable soil for multiplication, and these may prove, by their mere numbers, or by their effects, injurious or fatal to the whole body. Nothing can demonstrate this better than the experiments recently made by Orth and Wyssokowitsch, who found that staphylococci could be injected into the blood of a rabbit without apparent injury to it, but if before the injection a slight mechanical injury was inflicted on the valve of the heart, typical endocarditis was at once produced.

To Rosenbach ("Bemerkungen zur Lehre von der Endocarditis, mit besonderer Berücksichtigung der experimentellen Ergebnisse," *Deutsche med. Wochenschrift*, B. xiii. Nos. 32–33, 1887) belongs the credit of having first studied the influence of trauma in the localization of microbes upon the valves of the heart in cases of artificially-produced ulcerative endocarditis. He found in his experiments and in post-mortem examination in cases of ulcerative endocarditis, microbic emboli in the valves and in the infarcts of other organs, and classifies this affection with pyæmia. The more frequent occurrence of endocarditis in the left side of the heart, he explains by assuming that the microbe finds a better soil in the arterial blood, as when the affection occurs in the fœtus during intra-uterine life, when the blood in both sides of the heart is of about the same composition, the valves on both are affected with the same frequency.

Wyssokowitsch ("Beiträge zur Lehre von der Endocarditis." Virchow's *Archiv*, B. ciii. Heft 6) has determined with great accuracy the influence of traumatism in causing localization of pathogenic germs, and has found that, by introducing a rod into the jugular vein, he was able to cause laceration of the valves, and that then, on subsequent injection of a culture of staphylococci into the blood, ulcerative endocarditis developed at the seat of injury. The effects of the traumatic lesion in this instance are no doubt chiefly due to the production of a *locus minoris resistentiæ* in the endothelial and connective-tissue cells, as the result of the incipient inflammation induced by it. Similar results were obtained by Fränkel and Sänger ("Untersuchungen über die Aetiologie der Endocarditis," *Centralblatt f. klinische Medicin*, 1886, p. 577).

Rinne ("Der Eiterungsprocess und seine Metastasen," *Archiv f. klin. Chirurgie*, B. 39, p. 19) injected pure cultures of the different kinds of pus-microbes directly into the circulation in animals, and found that, as a rule, no harm resulted. In rabbits he injected from two to three Pravaz's syringefuls of unfiltered and filtered suspension of pure cultures in distilled water, and after repeating the injections several times inflicted all kinds of subcutaneous lesions without causing serious disturbances. Only in a few instances were pyæmic metastases observed, and this occurred usually only in cases where undiluted gelatin cultures were used. In several dogs he made subcutaneous fractures and then injected fluid cultures of pus-microbes in large doses into the peritoneal cavity, but no suppuration occurred at the seat of injury. In six rabbits he fractured the femur subcutaneously, and then injected pure cultures into the jugular or one of the auricular veins, but in only one of them did the seat of fracture suppurate, and in this case no other metastases were discovered. In two experiments where he injected osteomyelitic pus diluted with distilled water suppuration was produced at the seat of fracture and at the same time abscesses were found in the heart-muscle and the kidneys at the autopsy.

Rosenbach ("Beiträge zur Kenntniss der Osteomyelitis," *Deutsche Zeitschrift f. Chirurgie*, 1878, p. 369) ascertained that acute suppurative osteomyelitis in animals could only be produced experimentally by injecting pus directly into the circulation and by injuring the medullary tissue a few days before, or after, the inoculation. Kocher ("Die acute Osteomyelitis mit besonderer Rücksicht auf ihre Ursachen," *Deutsche Zeitschrift f. Chirurgie*, 1879, p. 87), Becker ("Ueber die Osteomyelitis erzeugenden Mikroörganismen," *Deutsche med. Wochenschrift*, 1883, No. 46), and Krause ("Ueber einen bei der acuten infectiösen Osteomyelitis beim Menschen vorkommenden Micrococcus," *Fortschritt. der Medicin*, 1884, B. ii. No. 7) repeated the experiments of Rosenbach and came essentially to the same conclusions. Both Kocher and Rosenbach look upon the altered circulation in the injured tissues as the condition which determined localization, while they admit that the immediate tissue lesions, hemorrhage and necrosis, might have the same effect. Upon the same hypothesis Kocher ("Zur Pathologie u. Therapie des Kropfes," *Deutsche Zeitschrift f. Chirurgie*, 1878, B. x. p. 189) explained the occurrence of traumatic suppurative strumitis in a hyperplastic struma.

Ribbert ("Die Schicksale der Osteomyelitiskokken im Organismus," *Berl. klin. Woch.*, 1884, No. 51) used in his experiments an emulsion of a pure culture of osteomyelitis cocci which was injected directly into the circulation. A few days after the injection the microbes had disappeared almostly completely from the circulation. At this stage they could only be found within the interior of the white blood-corpuscles and never free in the plasma of the blood. After twenty-four hours they could be detected in almost every organ of the body. In the liver they were most abundant, but, almost without exception, only in the interior of white blood-corpuscles. In the lungs they were often so numerous as to obstruct completely some of the capillary vessels; when less abundant they were in the interior of white blood-corpuscles which were aggregated in groups. In the spleen and lymphatic glands they were not numerous, but in the kidneys they were again more abundant, especially in the glomeruli, and also in some of the loops of the tubuli uriniferi. Later, they disappeared from all the organs with the exception of the kidneys. In all probability they are not destroyed in the organism, as they thrive in the lungs very well, as has been shown by injections into the trachea, but they are conveyed along the bloodvessels to some excretory organ. In the kidney localization takes place for several reasons. Embolic obstruction is the most important factor. The anatomical structure of the bloodvessels of the glomeruli explains their arrest in this part of the kidney, as in the lungs and liver, on account of a more free anastomosis of the bloodvessels, the cocci are more easily dislodged. In the kidney, the growth of the embolic masses in the vessels of the glomeruli can be easily followed through the several stages. But the process is limited to a few places, as many of the masses are removed by the circulation in the same manner as in the liver and lungs. Abscesses in different parts of the body following injections of pus-microbes into the circulation are produced by embolism. Suppuration is very likely to take place in the connective tissue, as here the circulation is comparatively slow and consequently the removal of cocci attended with difficulties. The microörganisms of osteomyelitis are eliminated through the kidney the same as the microbes of other infective diseases, and consequently localization is very likely to take place in this organ. Six hours after injection large colonies of cocci could be seen in the straight and convoluted tubules. Finally, it is well known that in the living body localization of microbes takes place in the tissues which are the seat of a trauma, and the influence of this exciting cause has been abundantly demonstrated by different experimenters who have produced acute osteomyelitis in animals infected with pus-microbes, by producing fracture of a bone or contusion of the medullary tissue. A number of experiments made, among others by

Ribbert, on the production of myo- and endo-carditis in rabbits, have shown that abscesses can be produced in other organs if the pyogenic microbes are attached to foreign bodies which cannot pass through the pulmonary filtrum. Thus Ribbert was able to produce myocarditis by using a cultivation of staphylococcus pyogenes aureus on potato, if he took the precaution in removing the culture from the surface of the potato to scrape off also the superficial layer of the potato itself. The particles of potato in these experiments determined suppuration by causing localization of the microbes, as the foreign bodies were too large to pass through the capillary vessels and were not capable of removal by absorption.

A subcutaneous fracture occasionally becomes the seat of an acute osteomyelitis.

Steinthal relates two such cases ("Ueber Vereiterung subkutaner Frakturen," *Deutsche med. Wochenschrift*, No. 21, 1887). One case occurred in a man twenty-eight years of age, who, having fallen from a tree, sustained a fracture of the neck of the right femur, also of the trochanter major, and a Colles's fracture of the right forearm. Suppuration began in the hip-joint together with inflammation of the knee-joint. In spite of early and free incisions the patient died, thrombosis of the left femoral vein, and atrophy with nervous disturbances in the right ulnar region being present. Post-mortem revealed lobular pneumonia in the left lower lobe. The second case was that of a woman thirty-four years old in whom, under chloroform narcosis, reduction of a dislocation of the left hip-joint was attempted. During the manipulation the head of the femur was broken off. After twelve days an abscess had formed at the seat of fracture, which was incised and drained. The patient recovered. Steinthal believes that suppuration in these cases was the result of localization of pus-microbes in the tissues necrosed by the injury.

During my recent visit in Zürich (*Four Months Among the Surgeons of Europe*, p. 128, Chicago, 1887) I saw a very interesting case of this kind in Krönlein's wards. In this instance infection probably took place through suppurating wounds distant from the fractures. The patient was a young man who had sustained several subcutaneous fractures from a fall, and at the same time a lacerated wound of the groin. The case progressed favorably until the wound commenced to suppurate, when he was suddenly attacked by osteomyelitis of the fractured bones, which necessitated numerous incisions for the liberation of pus at the points of fracture. There can be no question that, in this case, the pus-microbes entered the circulation at the primary site of suppuration and were arrested at the places of fracture, where they found favorable conditions for multiplication and caused a suppurative inflammation in the medullary tissue. Practically this case should teach us that in a patient who has sustained a simple fracture it is exceedingly important to guard against suppuration in any part of the body, for fear that from a distant purulent focus pus-microbes may enter the circulation and cause a suppurative osteomyelitis of the broken bone, in the same manner as has been done by experiments on animals. At the same time and place I examined a case of suppurative strumitis due to a similar cause. The patient was a man of about forty years of age, who had been operated upon for empyema by rib resection some time ago. He had had a large goitre since childhood. Improvement after the operation progressed uninterruptedly until the empyema was nearly well, when fever set in and the right side of the struma became painful and tender. After a week fluctuation was well marked, and a free incision was made, which gave exit to a large quantity of fetid pus. The fever subsided promptly, and the case again progressed favorably until, a week or two later, the opposite side of the struma was attacked in a similar manner. I was present when this side was incised. A large amount of the same green, fetid pus escaped. The strumitis was unquestionably of embolic origin, the pus-microbes which gained entrance into the circulation from the pleural cavity found in the struma conditions which favored their localization and growth, to be followed by a suppurative inflammation of the swelling.

Lebert observed secondary or metastatic strumitis develop six times in connection with typhus, once with puerperal fever, three times after pneumonia, and once after bronchitis. Kocher ("Zur Pathologie und Therapie des Kropfes," *Deutsche Zeitschrift f. Chirurgie*, B. x.) met with it in cases of typhus, septic endometritis, and after attacks of acute gastro intestinal catarrh. He calls special attention to the fact that the strumitis almost without exception occurred toward the close or after the subsidence of the primary disease. Kocher believes that degenerative changes or injuries of the struma serve as predisposing causes for the arrest of pathogenic microbes which reach the organ through the bloodvessels.

Localization of Microbes in Antecedent Pathological Products.

Antecedent pathological products may serve the same purpose in the body as a trauma in the determination of localization of pathogenic microbes. Suppuration in a tumor, or a hyperplastic gland with an intact cutaneous covering, indicates that in the tumor or swelling pus microbes have been arrested, and that they have met with a soil adapted to their multiplication and the exercise of their pathogenic properties. The atypical vascularization in tumors and the stenosis of the lumen of bloodvessels in inflammatory swellings cannot fail in furnishing conditions which determine filtration of germ-containing blood. If the antecedent pathological product is the result of a previous infection and serves as a medium for localization of another kind of pathogenic germs, we speak of the combined process due to the presence of two varieties of micro-organisms as a mixed infection. The first positive proof of the existence of such secondary processes was furnished by Brieger and Ehrlich. ("Ueber das Auftreten des malignen Oedems bei Typhus abdominalis," *Berl. klin. Wochenschrift*, 1882, No. 44.) These observers saw a malignant œdema develop at the point where musk was injected hypodermatically in a severe case of typhoid fever. They found that in such cases a predisposition is produced by an existing disease to the growth and multiplication of microörganisms which may have been previously present in the organism without producing any pathological lesions. Koch, in his article on "Etiology of Tuberculosis," alludes to the occurrence of mixed infections, as he has seen at the same time bacilli and micrococci present in tubercular products Further, he has observed the bacillus of anthrax side by side in the same tissues and has seen micrococci in the tissues of patients suffering from typhus fever. In reference to the occurrence of micrococci in tubercular deposits in the lungs and spleen, he explained their presence by assuming that they entered the circulation through ulcerations of the tongue, and that they became arrested in the capillary vessels which had lost their normal resisting power by the tubercular process.

Samter ("Mischinfection von Tuberkelbacillen u. Pneumoniekokken," *Berl. klin. Wochenschrift*, 1884, No. 25) discovered in the pneumonic sputa of a man sixty-five years of age, who previously had suffered from a latent bronchial catarrh, besides cocci of pneumonia numerous bacilli of tuberculosis. The necropsy revealed old tubercular deposits in the right, and more recent deposits in the left lung, and around the latter pneumonic consolidation of the parenchyma of the lung. He believed that the pneumonic foci furnished a favorable soil for the localization and growth of the bacillus of tuberculosis. Heubner and Bahrdt ("Zur Kenntniss der Gelenkeiterung bei Scharlach," *Berl. klin. Wochenschrift*, 1884, No. 44) have described the post-mortem appearances of a boy fourteen years of age who had suffered from multiple suppurative synovitis after an attack of scarlatina. The metastatic suppurative process could be traced directly to a circumscribed purulent inflammation of the right tonsil which implicated an adjacent vein, which became the seat of a purulent thrombo-phlebitis, and the point of distribution of infective emboli; chain cocci were found in the pus of the joints and even in the patient's blood.

Fränkel and Freudenberg ("Ueber Infection bei Scharlach,"

Centralblatt f. klinische Medicin, 1885) cultivated from the internal organs of three patients who had died of scarlatina the streptococcus pyogenes and looked upon the presence of this microbe as an evidence that a secondary infection had taken place through the diseased mucous membrane of the pharynx. The important question presents itself whether, in cases of mixed infection, the two kinds of microbes enter the organism at the same time, or whether the primary infection prepares the way for the entrance of the microbes which produce the secondary infection. A third possibility might be maintained, according to which the secondary infection is a purely accidental occurrence, as was claimed by Brieger and Ehrlich. Pus-microbes being present at all times and everywhere, and perhaps gaining entrance into the body more easily than others, it is easy to understand why secondary infection by them is most frequently observed.

Rosenbach frequently found in the products of suppurative inflammation and septic processes more than one variety of pus-microbes. He frequently met with both kinds of staphylococci in the same pus, or with one form of staphylococcus and the streptococcus pyogenes.

Löffler ("Untersuchungen über die Bedeutung der Mikroörganismen für die Entstehung der Diphtherie," *Mittheilungen aus dem Reichs-Gesundheits-Amte*, 1884, Band ii.) cultivated from the membranes of a case of scarlatina-diphtheria cocci which, when injected into the circulation of animals, produced multiple suppurative synovitis.

Huber ("Experimentelle Untersuchungen über Localisation von Krankheitsstoffen," Virchow's *Archiv*, Band cvi.) attributes the occurrence of suppuration and gangrene in croupous pneumonia, phlegmonous inflammation and suppuration in erysipelas, and suppuration in tubercular processes, to secondary infection, in most instances with pus-microbes. Schnitzler ("Combination von Syphilis und Tuberculose des Kehlkopfes"), after having observed and carefully studied a number of cases, has come to the conclusion that syphilitic ulceration of the larynx may pass into tubercular, as the syphilitic ulcer furnishes a good culture-soil for the bacillus of tuberculosis.

Bumm (*Le Bulletin Médical*, December 25, 1887), in a communication to the Medical Society of Munich, discusses a theory under the name of mixed infection, which he describes as the penetration into the organism of several species of bacteria. In some the secondary infection is purely accidental, as for example, a tuberculous patient can be attacked with erysipelas, a lying-in woman suffering from gonorrhœa may become the subject of septic infection. Another and practically more important variety of mixed infection he speaks of, where a more direct relation exists between the different microbes, in the sense that the one precedes the other and prepares the soil for the growth of the latter. These forms are characterized by being constantly associated with certain definite microbes. The pneumococcus may prepare the soil for fructification of the bacillus of tuberculosis or the microbes of suppuration in individuals that otherwise would have been immune to the action of these germs. The gonococcus can also modify the mucous membrane of the genito-urinary tract, chiefly in women, in such manner as to render easy the invasion of other pathogenic microbes. Gonorrhœal infection of the vulvo-vaginal glands furnishes a good illustration: as long as the infection remains purely gonorrhœal the acute purulent stage is succeeded by a chronic stage which may last for some months, the swelling gradually subsides and subsequently atrophy and sclerosis of the gland follow. If, however, a purulent infection is added to the gonorrhœal the gland soon becomes enlarged and tender, and suppuration follows. In the abscess and its vicinity no gonococci can be found; the pus only contains the pyogenic staphylococcus which has exterminated the gonococcus. Cystitis which accompanies gonorrhœa is, again, a variety of mixed infection. The stratified epithelium of the bladder is impenetrable to the gonococcus. According to Bumm, the cystitis is due to another species of microbe resembling the gonococcus, but differing from it by taking a different staining. The gonococcus expends its action in the superficial layers of the mucous membrane exclusively. Suppurative parametritis following gonorrhœa is analogous to a gonorrhœic bubo, which is always caused by a secondary infection with pus-microbes.

A valuable contribution to our knowledge of mixed infection has recently been made by Babes (*Bacteriologische Untersuchungen über septische Processe des Kindesalters*, Leipzig, 1889), of Bucharest. His investigations consist of a series of bacteriological studies of the dead bodies of children. Within a few hours after death tissue was taken from different organs, with which sterilized culture material was inoculated, the strictest antiseptic precautions being exercised throughout. In acute infectious diseases, such as diphtheria and scarlatina, he found the spleen, kidneys, liver, lungs, and blood infected with numerous colonies of streptococci, putrefactive bacteria, capsule cocci, more rarely staphylococci and various bacilli. Of special interest are his researches on the manner of localization and extension of the secondary infection after different primary diseases. In eight cadavers he found one or more species of bacteria in the internal organs. In a case of septic omphalitis he found the bacillus of green pus. In six cases of different kinds the pus streptococcus grew upon the culture substances, and only in one was the yellow staphylococcus found, and in five cases various putrefactive bacilli were cultivated. In some instances he was able to demonstrate the point at which the different secondary invasions had taken place. Thus in a case of sepsis after scarlatina in which streptococci were found in every part of the body, a pure streptococcus pneumoniæ was found in the lower part of the left lung, while a number of foci in the upper part of the right lung contained only bacilli.

Rinne (*Verhandlungen Deutscher Naturforscher und Aerzte*, in Wiesbaden, 1887) made some experiments which convinced him that a trauma which results in an ordinary inflammation does not furnish a *locus minoris resistentiæ* for the localization of pathogenic germs, as has been described above. He claims that the living body possesses the capacity to eliminate large numbers of living pus-microbes when these are in tissues where they are not exposed to direct contact with oxygen. He has found that cocci do not collect at points the seat of subcutaneous injury, and that they do not migrate into sterile abscesses from distant parts, and not even after intravenous or intraperitoneal injection. Even an injection of a pure culture of pus-microbes into a part injured subcutaneously, or into young or mature cicatricial tissue around encapsulated foreign substances, produced no suppuration. From these experiments he was led to believe that tissue lesions which produce an inflammatory reaction do not predispose to metastatic suppuration. On the other hand, it was also demonstrated experimentally that tissue lesions become a *locus minoris resistentiæ* after the tissues have become permeated by the chemical products of the microörganisms. The *locus minoris resistentiæ* for the pus-microbes is a tissue which, by chemical or by chemical and mechanical lesions, has lost its normal resistance to the microbes, which is not the case if a trauma is followed by an active reparative process.

The same author ("Der Eiterungs-process und seine Metastasen," *Archiv f. klin. Chirurgie*, Band xxxix. Heft 2) has recently made numerous experiments on rats, rabbits, guinea-pigs, and dogs, to show that subcutaneous mechanical lesions never suppurate, even after inoculation with larger doses of pure cultures of pus-microbes. The subcutaneous lesions were made with a tenotome under strict antiseptic precautions and the puncture sealed with collodium. The cultures, usually diluted with sterilized water, were injected either subcutaneously into the peritoneal cavity, directly into the circulation, or into the injured tissues, but metastatic suppuration was never produced. Circumscribed suppuration was produced around woollen threads impregnated with a pure culture and introduced into the tissues. He believes that in these cases the pus-microbes in the meshes of the threads produced ptomaïnes, which chemically injured the tissues before they could be removed by the tissues.

When chemical abscesses were produced by the injection of croton oil, nitrate of silver, ammonia or cadaverin, and the animal was inoculated with pus-microbes in another part of the body, the chemical pus was always found sterile. The same was observed if a phlegmonous inflammation was first produced and one of the chemical irritants was injected later in some distant part of the body.

When a chemical substance which produces pus is injected with pus-microbes into the tissues, the former prepares the soil for the latter, but does not determine localization of microbes introduced into the circulation. The cause of this is probably the active cell-proliferation produced by the chemical substance.

Aschoff (*Ueber die Einwirkung des Staphylococcus pyogenes aureus auf Entzündetes Gewebe;* Dissertation, Bonn, 1889) found that the action of pathogenic microbes upon tissues in a state of active cell-proliferation induced by chemical irritants is intensified. A pure culture of the yellow coccus was injected into inflammatory swellings, caused by the subcutaneous use of tincture of iodine. The necrosis and exudation were much more marked than when the same injection was made into healthy tissues, while the regenerative processes were also correspondingly retarded and the microbes manifested greater activity and power of resistance.

Orloff ("Materialien zur Frage über die Eintrittswege der Mikroben in den Thierischen Organismus," *Centralblatt f. Bacteriologie u. Parasitenkunde*, B. iii. No. 15) made many interesting experiments to ascertain if pathogenic microörganisms can enter the body through healthy intact, as well as through irritated diseased mucous membranes. He injected a pure culture of the staphylococcus pyogenes aureus into the trachea, duodenum, and ileum of animals, or administered the same culture by feeding. The experiments were made on rabbits and guinea-pigs. In some of the animals the bronchial mucous membrane was altered by application of croton oil, solution of nitrate of silver, and in two instances by injuring it mechanically with a catheter through a laryngotomy wound before the culture was injected. The same conditions were produced in the stomach and intestines by similar means prior to the administration of the culture. In 12 experiments the staphylococcus produced no symptoms when introduced into the healthy stomach, although a pure culture was fed for from 1 to 14 days. In the post-mortems made 12 to 72 hours after the feeding of the microbes, all the internal organs were found sterile, and only the lower portion of the colon contained staphylococcus colonies. All examinations of the blood during life yielded negative results. In 7 experiments in which the stomach was irritated by chemicals before the feeding of the culture was commenced, 6 animals remained well, and in the one that died the microbes had not passed beyond the prima via, as the blood and internal organs were found sterile. Of 12 animals in which the pure culture was injected into the healthy bronchial tubes, 9 were killed one-half of an hour to 11 days after the injection, and 3 died. In the animals which were killed no trace of microörganisms could be found in the blood, they had not passed beyond the lung tissue. In the 3 fatal cases the animals died 17, 25, and 70 hours after the injection, and the microbes could be found in nearly all of the internal organs and the blood. In the 2 cases in which the trachea had been altered by traumatic irritation, the animals were killed 24 and 25 hours after inoculation, and on examination it was seen that the infection had remained local. In 6 experiments in which the bronchial tubes were irritated with a solution of nitrate of silver 6 hours before inoculation, the animals died 20, 24, 40, and 48 hours after the injection, having before death shown evidences of serious pulmonary trouble. The post-mortem showed in all of them pulmonary œdema and pleuritis, but staphylococci were found in only 2 of them in the effusion. In 6 animals fed on the culture, in 4 on a healthy stomach and in 2 after an artificial gastro-intestinal catarrh had been produced, he made a subcutaneous fracture, and suppuration followed at the seat of the fracture. He believes that in all cases in which the microbes entered the circulation the entrance was effected through some perhaps inappreciable lesion.

Gussenbauer (*Deutsche Chirurgie von Billroth u. Luecke*, Lief. 4, p. 126) describes one of those cases which are not of infrequent occurrence where recovery after an attack of suppurative lymphangitis was followed by a phlegmonous inflammation in the axilla. The patient was twenty-five years old, in good health otherwise, when he contracted after a slight injury of the hand a lymphangitis and lymphadenitis of the arm and axilla attended by circumscribed gangrene of the skin and flexor tendons. The lymphatic glands did not suppurate but remained slightly enlarged. Suppuration continued for eight weeks when recovery was complete and the man remained in good health for eight months. At this time he suffered from an acute abscess in the axilla of the same side which could have been only caused by the pus-microbes which had remained in the enlarged lymphatic glands since the first attack. Smirnoff has recently published an interesting dissertation (St. Petersburg, 1889), in which he describes his examinations of synovial fluid removed from joints. He has found that not infrequently in the course of an infectious disorder the pathogenic microbes may be detected in the synovia. He examined 51 cases including erysipelas, pneumonia, abscess, phthisis, typhoid fever, diphtheria, and gonorrhœa, while in some the result was negative, in others not only was the specific microbe present, but other microörganisms were also detected, showing that secondary infection had occurred. In some he found the gonococcus or pneumococcus with the staphylococcus or streptococcus. Transportation of the microbes, from the primary focus to the joints, occurred either by way of the bloodvessels or lymphatics. The frequency of articular infection seems to be governed by the size and form of the bacteria. The small round or oval cocci, such as the pneumococcus, gonococcus, and pus-microbes, obtain more ready entrance than the bacilli.

In many instances of recurring suppuration years after the primary injury or disease, we have reason to believe that the microbes were introduced with a foreign body, or became encapsulated in the granulation tissue during the healing process, and remained there in a latent condition until by some accidental cause the surrounding tissues had undergone changes favorable for their growth. I have seen numerous cases of secondary osteomyelitis ten to twenty-five years after the primary attack, which occurred in the sclerosed bone, usually near one of the epiphyseal extremities. In all of these cases the disease was deeply located, the primary starting-point being in the same locality as in the first attack. A patient who has once suffered from osteomyelitis during childhood, is always prone to suffer from recurring attacks, and the disease, without exception, selects the old site. The difficulties which deep cavities in bone present to the final process of definitive healing, offer an explanation why microbes are more liable to become buried and permanently retained, than in suppuration in soft parts.

Nepveu (*Revue de Chirurgie*, 1885, p. 353) reported two cases of gunshot injury of the extremities in which, fourteen years after the definitive healing, suppuration occurred at the former site of injury.

Numerous instances are on record in which foreign bodies have remained in the tissues encapsulated for an indefinite period of time without giving rise to any symptoms until the microörganisms which were introduced with them exercised their pathogenic qualities, because the surrounding tissues had undergone changes favorable to such a process. A very interesting case of this kind is reported by Rinne ("Der Eiterungs-process und seine Metastasen," *Archiv. f. klin. Chirurgie*, Band xxxix. p. 70). In 1871, a man, thirty-four years of age received a gunshot wound through the left shoulder-joint. Langenbeck made primary resection of the joint, and the wound suppurated for a number of weeks profusely but healed completely at the end of three months. The patient remained in perfect health for eleven years, when he was suddenly taken with symptoms of acute osteomyelitis. When admitted into Friedrich-shain Hospital a few days later a free incision were made over the resected end of the humerus and a large quantity of fetid pus escaped. Free drainage and disinfection had no effect in diminishing the severity of the general symptoms. Through another incision, made a few days later, small particles of clothing were removed. As the symptoms became more and more threatening amputation was made through the shoulder-joint four weeks after the commencement of the attack. Patient died a few days later after well-marked symptoms of pyæmia had manifested themselves for several days.

The examination of the amputated arm showed old foci of inspissated pus in the medullary cavity of the humerus, which leaves no doubt that the second attack of osteomyelitis was caused by microbes which had remained in a latent condition for eleven years.

Gussenbauer (*Deutsche Chirurgie von Billroth u. Luecke*, Lief. 4, S. 125) relates two such cases. In one a bullet perforated through the scapula from behind and lodged in the deep tissues. The shoulder-joint suppurated and had to be resected. The bullet could not be found. The wound finally healed and the patient remained well for twenty-three years, when suppuration occurred at the site of operation; fragments of cloth were removed through an incision and the man recovered.

At a recent meeting of the French Congress of Surgeons, Socin reported a case that had come under his observation, in which a bullet remained in the tissues, perfectly encapsulated, for sixty years, when finally the pus-microbes which entered the organism with it, and had remained all this time in a latent condition, caused the formation of an abscess.

CHAPTER V.

ELIMINATION OF PATHOGENIC MICROÖRGANSIMS.

The probable existence of pathogenic microbes in the healthy body, and the spontaneous subsidence of many infective processes, make it important to consider the ways and means by which pathogenic microörganisms are rendered harmless in the living body, or are removed by elimination through some of the secretory organs. It is not at all improbable that the localization which bacteria effect may come to play an important part in the study of micro-parasitic pathology. Whatever be the explanation, there is no doubt that the microörganisms hitherto found have various relations with the tissues.

In all infective processes in which life is not destroyed, and the products of inflammation do not find their way to the surface spontaneously or by treatment, the microbes are removed with the excretions as dead foreign bodies, or are eliminated through some of the excretory organs in an active state.

1. *Phagocytosis.*

Metschnikoff introduced the term "phagocytosis" to designate the destruction of microbes within the organism by leucocytes and the fixed tissue-cells.

Wyssokowitsch (" Ueber die Schicksale der ins Blut injicirten Microörganismen im Körper der Warmblüter," *Zeitschrift f. Hygiene,* Band i. S. 1–45) has studied experimentally the destiny of microörganisms injected into the blood of warm-blooded animals, and found that they were not eliminated. He demonstrated by microscopical examination of the blood and by cultivation experiments, that the microbes disappear entirely from the blood or diminish greatly in number soon after intravenous injection. The saprophytes disappear soonest. The toxic bacteria are slowest in disappearing and seldom leave the blood entirely. After a brief period, when their number is at a minimum, they again reappear in the blood and multiply with great rapidity until they kill the animal. Examination of the urine during life and soon after death showed that elimination of bacteria through the kidneys did not take place; that, on the other hand, the urine contained bacteria in cases in which extravasation of blood into the parenchyma of the kidneys had taken place. During the first few hours after intravenous injection, and before any organic changes had taken place in the kidneys, the urine contained no bacteria. It was also shown that no elimination took place through the intestinal canal, except in cases in which the mucous membrane was the seat of hemorrhagic extravasations, or other serious local lesions. In two animals infected during lactation, the milk was sterile, showing that no elimination took place through the mammary gland. Microscopical examination revealed an increase of white blood-corpuscles. Inclusion and death of bacteria in blood-corpuscles were not observed. The bacteria injected into the circulation soon became deposited in certain organs, notably the spleen, liver, and medullary tissue of bone, and consequently separated from the circulating blood. Non-pathogenic microbes disappear entirely and permanently from the blood soon after injection. The spores retain their vitality in the organism for an unusually long time; for instance, active spores of ordinary mould for 7 days, and the spores of the bacillus subtilis for 62 to 78 days.

Ribbert (*Der Untergang pathogener Schimmelpilze im Körper*, Bonn, 1887) has also made extensive investigations concerning the fate of pathogenic microbes in the organism when introduced by intravenous injection. In rabbits he injected into the circulation such small doses of a pure culture of aspergillus flavescens that the animals did not die in consequence of the infection. On examining the different organs at variable intervals after the injection it was easy to determine in what manner the germs were destroyed or eliminated. In the experiments made for the special purpose of ascertaining the fate of microbes in the liver, the injection was made directly into one of the mesenteric veins. In all other cases the injection was made into one of the veins of the external ear. In the liver, it was seen that when the microbes reached the capillary vessels and the terminal branches of the portal vein they were surrounded by a dense cluster of leucocytes. As long as they remained within these surroundings, their multiplication appeared to be retarded. Mitosis was not observed in the liver cells which were compressed by the inflammatory exudation, but occasionally it occurred in some of the epithelial cells lining adjacent bile-ducts. The microbes were seen either disappearing directly in the surrounding leucocytes, or some of the liver cells were transformed into giant cells, into the protoplasm of which they found their way, and they were destroyed. The giant cells either undergo fatty degeneration and disappear with their contents by absorption, or they are again transformed into normal parenchyma cells. When larger quantities of the same microbe were injected no such aggregation of leucocytes around them occurred in the vessels and tissues of the liver, and multiplication at once took place, and after mural implantation of the microbes had taken place perforation of the vessel wall was effected by means of ray-like projections which developed from the microbes, and which were seen to insinuate themselves between the cells. If the animal survived the injection for a few days the rays became narrower and the granular detritus of the microbe was removed by giant cells. In studying the fate of the same microbes in the lung, the injections were made either into a vein, the trachea, or directly into the parenchyma. The microbes surrounded by leucocytes were seen not only in the vessels but also in the alveoli. Limitation of the infection and the final destruction of the microbes were here also accomplished by the leucocytes and giant cells. Injection of larger quantities was likewise followed by rapid permeation of the vessel wall by the microbes and their early appearance in the paravascular tissues. In the kidneys the growth of the microbe was more rapid than in the liver and lung, the rays and filiform projections formed earlier and developed to an unusual extent. The accumulation of leucocytes, on the other hand, was retarded and less marked, so that the microbes were only surrounded by them after the thread-like projections had formed. The leucocytes containing spores were not only found in the interstitial tissue, but also in the tubuli uriniferi. The same microbes injected into the anterior chamber of the eye produced upon the iris a thin layer of fibrin, in which the leucocytes gathered around the microbes in the form of minute nodules and the microbes were soon lost in their interior. Only a slight increase of growth was observed in this locality. The fixed cells of the iris take no part in the formation of the nodules, and the work of destruction of the microbes is accomplished exclusively by the leucocytes. Culture experiments with small particles of organs containing the microbe showed that their power to infect is sometimes lost after 24 hours and always after 8 to 14 days. The retardation and arrest of germ-growth by the leucocytes and giant cells are first effected mechanically, but to a greater extent by the abstraction of oxygen. It is also possible that the diffusion of the ptomaïnes of the microbes is also limited by the grouping of cells around centres of infection. He does not believe that the leucocytes digest the microbes, as was claimed by Metschnikoff. Fibrin-production could be detected by Weigert's method around the periphery of the nodules. Experiments with aspergillus fumigatus yielded the same results. From his own work Ribbert has come to the conclusion that retardation of germ-growth by leucocytes in infections with schizomycetes is an established fact in pathology. The retarding influence of the organism to germ-growth differs according to the kind of germs, as the germs by the production of noxious substances affect the envelopes of cells, and most of the pathogenic microbes assimilate nutritive material from adjacent tissues more readily than ordinary mould.

Microbes which diffuse themselves rapidly and which do not form colonies cannot become encapsulated by leucocytes. By counting the blood-corpuscles, it was found that after injections of small doses of aspergillus fluorescens ten times more leucocytes were found than in normal blood. Ehrlich's method of staining proved that the leucocytes which surrounded the spores as well as the new corpuscles in the blood were polynuclear and represented the pyelogenous leucocytes. In animals which recovered after infection reinoculation was followed by more prompt encapsulation of the spores by the leucocytes.

Fig. 1.

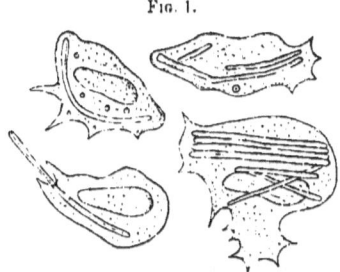

Leucocytes of frog with anthrax bacilli. (METSCHNIKOFF.)

Metschnikoff (" Ueber den Kampf der Zellen gegen Erysipel Kokken," Virchow's *Archiv*, B. 101, p. 209), who was the first to describe the struggle of the cells of the organism against invading microbes, gives the name of phagocytes to the cells which possess the property of absorbing and digesting microbes (Fig. 1). He made his researches on the daphnia, which are found invaded by a parasite of the higher animals. The spores of this parasite penetrate with the food into the intestine, from which they escape into the cavity of the body of the daphnia. As soon as they have done so, a struggle commences between them and the white corpuscles, which, isolated or in groups, absorb the spore and destroy and transform it into shapeless granules. The daphnia is thus saved. When the victory is not accomplished the spores are not arrested and destroyed by the leucocytes, and they invade the tissues of the animal and kill it. Metschnikoff affirms that the same struggle takes place in the higher vertebrates and in man between the microbes and the cell-elements, but in these the process is more complicated, as there are two kinds of phagocytes. The leucocytes he calls "microphagi," and to the permanent tissue-cells that are capable of absorbing minute solid bodies, and are provided only with one large nucleus, such as the connective-tissue cells, epithelial cells of the pulmonary alveoli, he applies the term "macrophagi." In infectious diseases, in which the phagocytes do not protect the system from progressive and rapid infection, death is inevitable. In anthrax the microphagi are powerless to absorb the bacteridia. The case is different, however, when, instead of a strong virus, an attenuated culture is employed. Metschnikoff employed in his experiments a watery solution of vesuvin, which does not color living microörganisms, but stains dead ones brown. In this way he saw most of the rod-shaped microörganisms, encased in the protoplasm of the microphagi, assume a brown color, whereas the cells remained unaffected and in a living state.

Hess ("Weitere Untersuchungen zur Phagocytenlehre," Virchow's *Archiv*, B. cx. p 313) has examined the influence exerted by leucocytes in retarding the pathogenic action of the staphylococcus pyogenes aureus. He injected a minute quantity of a pure culture of this microbe into the substance of the cornea in rabbits. The eyes were extirpated, hardened, and examined microscopically 24 hours to 8 days after inoculation. During the first 24 hours the cocci multiplied very rapidly at the point of infection; 36 hours after inoculation leucocytes were seen enclosing the infected area, and all of them contained microbes. As the suppurative process advanced the intra-cellular cocci increased in number. Gradually the cocci disappeared in the cells, and after the sixth day often none could be found. The free cocci in the tissues formed a very narrow zone around the margins of the ulcer, while the microbe-carrying leucocytes were found in the tissues at a greater distance. In two cases which terminated fatally, an extensive hypopyum formed and no phagocytosis had occurred, although many leucocytes were present. In one experiment on a cat, the ulcer resulting from the inoculation was very small, and in this instance phagocytosis was more marked than in the rabbits. In the destruction of the cocci in the ulcer, the cells springing from the conjunctival sac played also an important part, and the secretions from the conjunctiva contained soon after the inoculation many cells with microbes in their protoplasm. Warm fomentations at first appeared to favor the increase of microbes, but later, by stimulating the action of the phagocytes, they retarded it.

Christmas-Dircking-Holmfeld ("Ueber Immunität und Phagocytosis," *Fortschritte der Medicin*, No. 13, 1887) repeated the experiments of Metschnikoff, and claimed that in animals immune to anthrax, or rendered artificially immune, the bacillus when introduced into the tissues is devoured by the leucocytes. He experimented with a very virulent culture and cultures attenuated by age, or by Koch's method of attenuation. These experiments were made on mice, rabbits, and young and old rats. He soon found out that the results were greatly modified by the degree of susceptibility of the animal to the anthrax bacillus. In animals very susceptible to the disease, as in mice and rabbits, the inoculation of a very active culture produced little or no inflammatory reaction at the point of inoculation. In animals with a greater degree of immunity, as young rats, the injection produced suppuration at the point of inoculation, which assumed the character of an abscess, as the susceptibility was diminished as in old rats. The injection of a mitigated culture produced suppuration even in rabbits. The intensity of the inflammatory reaction following the inoculation with the bacilli of anthrax stood in an inverse ratio to the degree of the susceptibility of the infected animal. The bacilli were never found in the interior of the pus-corpuscles, and it followed as a natural conclusion that death and disintegration were effected by the pus serum. That the pus-corpuscles did not destroy the bacilli was also proved by another experiment. A pure culture of anthrax bacilli was put in hermetically-sealed glass tubes containing pus, which when kept at the temperature of the body killed the bacilli in from two to three days. Metschnikoff asserts that the latter observation does not prove that the pus-corpuscles did not act the part of phagocytes, as under the conditions described they might have retained the properties of such.

Wehr ("Ueber den Untergang des Staphylococcus pyogenes aureus in den durch ihn hervor-gerufenen Entzündungs-processen in der Lunge," Dissertation, Bonn, 1887) injected pure cultures of the staphylococcus pyogenes aureus into the trachea of rabbits and favored the descent of microbes into the finer divisions of the bronchial tubes by standing the animals for some time upon the hind legs. Soon after the injection he found cocci not only in the interior of leucocytes which had grouped around the seat of injection, but also in the interior of epithelial cells. He assigns to epithelial cells phagocytic action. During the first week after injection he found cocci without exception in the epithelial cells. In none of his experiments could the microbes be found in the blood or in any of the internal organs.

The doctrine of phagocytosis has recently received a substantial support from the researches of Nuttal and Buchner ("Uber die bakterienödtende Wirkung des Zellenfreien Blutserums," *Centralblatt f. Bakteriologie u. Parasitenkunde*, B. v. No. 25). These researches show that not only the cellular elements of the blood are antagonistic to pathogenic bacteria, but that the serum possesses a similar destructive power. Their experiments with different microbes showed that both defibrinated and freshly-drawn blood manifest a decidedly deadly action upon them for more than four hours after it has been drawn from the body. For example, the number of anthrax bacilli in a given quantity of material was reduced in two hours from 4800 to 56, by being mixed in a test-tube with defibrinated blood; and three hours later only 3 living bacilli remained. The destructive action of the blood on putrefactive bacteria is, however, much less marked, and against some of them, at least, the blood manifested little or no germicidal action.

Ribbert ("Ueber den Verlauf der durch Staphylococcus aureus in der Haut von Kauinchen hervorgerufenen Entzündung," *Deutsche Wochenschrift*, No. 6, 1889) studied experimentally the phagocytic action of leucocytes in the tissues infected with pus-microbes. He made the skin of rabbits over a limited area aseptic, and made inoculations with pus-microbes by introducing a pure culture of the yellow coccus through punctures made with a cataract-needle. In cases where the small wound healed rapidly he found on examining the tissues that even during the course of the first day all of the microbes had reached the interior of the leucocytes and fixed connective-tissue cells and showed signs of destructive changes. Later the cocci disappeared completely. Thirty-six hours after inoculation he found distinct mitotic changes in the connective-tissue cells within and in the immediate vicinity of the inflammatory focus and somewhat earlier in the epithelial cells. If suppuration occurred he found a considerable aggregation of leucocytes, which apparently were being destroyed by the cocci, consequently phagocytosis was not present; on the other hand, the cocci underwent destructive changes to the fourth day, and were then taken up by the macrophagi. The death of the cocci under these circumstances was not caused by the phagocytes but was owing to the exclusion of oxygen or other nutrient material by the wall of cells in their vicinity.

Ruffer ("Notes on the Destruction of Microörganisms by Amœboid Cells," *British Medical Journal*, August 30, 1890) is a firm believer in the phagocytic action of leucocytes and other amœboid cells. He first studied sections of the Peyer's patches of rabbits, removed with antiseptic precautions, plunged at once into absolute alcohol, and then stained with carmine and gentian-violet. He often found, crowded in between the layer of epithelial cells in the inner surface of the mucous membrane, leucocytes holding microörganisms in their protoplasm. In the submucous tissue he also found lymphoid cells enlarged to the size of the so-called epithelioid cells. He also found leucocytes in the interior of the macrophagi. He asserts that the leucocytes wander out to the surface of the mucous membrane, seize the microörganisms, and bring them back to be destroyed by the large phagocytes. Even the large cells in some lymphoid structures may wander to the surface and absorb and destroy, microbes. He is of the opinion that the action of the microörganisms taking place in the normal lymphoid tissues of the alimentary tract resembles, in all particulars, the destructive process following on the inoculation of pathogenic organisms into resistant animals.

The author asserts, from further experiments, that the large epithelioid cells of the spleen, lymphatic glands, and those of the lungs, are really macrophagi developed from the lymphoid cells.

Lubarsch ("Ueber Abschwächung der Milzbrand-bacillen im Froschkörper," *Fortschritte der Medicin*, No. 4, 1888), in his experiments on frogs with the anthrax bacillus, infected the animals by inserting into the lymph sac portions of the internal organs of animals which had died of anthrax. His observations led him to the conclusion that the phagocytes in reality do devour the microbes, and that disintegration of the intra-cellular microbes is an active process on the part of the cells.

2. *Elimination of Microbes through the Kidneys and other Organs.*

The rapidity with which some microbes disappear from the blood is very remarkable; it is in many cases a matter of minutes, certainly of an hour or two, and this disappearance from the blood must be due to an active process on the part of the constituents of the blood on them. Mere unsuitability of soil is not sufficient to account for the rapidity of the phenomenon. That they are eliminated through the kidneys is shown by various observations, and this is an important point to remember, as probably explaining certain cases of pyelitis occurring in patients who have never had any instrument passed, and in whom urethra and bladder are perfectly normal. The salivary glands, more especially the parotid, occasionally take part in the excretion of pus-microbes, thus offering an explanation of the not infrequent occurrence of abscesses in the parotid gland after suppurations elsewhere.

Rosenstein ("Vorkommen der Tuberkel-Bacillen im Harn," *Centralblatt f. d. med. Wissensch.*, 1883, No. 5) and Babes (*Fortschritte der Medicin*, B. i. p. 4) found bacilli in the urine in patients suffering from tuberculosis of the genito-urinary tract.

Fardel ("Les bacilles dans la tuberculose miliaire. Tuberculose glomerulaire du rein," *Archiv de Physiologie*, 1886) mentioned the condition of the kidneys in cases of miliary tuberculosis as an evidence that the bacilli are diffused through the bloodvessels. He found in a thrombosed capillary vessel in a glomerulus, in which no inflammatory changes had as yet occurred, numerous bacilli. They were also found in the epithelial cells of a convoluted capillary vessel. The author was also able to demonstrate that the deposits in the kidney occurred around the capillary vessels in the glomeruli, often including the latter completely. He believes that the bacilli migrate from the glomeruli into the surrounding tissue, when they give rise to miliary nodules. In cases of tuberculosis of the kidneys the number and arrangement of the bacilli, as they are found in the urine, are characteristic of this disease.

Monpurgo (Schmidt's *Jahrbücher*, B. ccxii. p. 128) found in the urine of a woman suffering from renal tuberculosis numerous bacilli arranged in groups resembling the letter "S," an appearance which is only observed in pure cultures. Koch described this bacteriological condition in a case of miliary tuberculosis. For such a culture to form in the genito-urinary apparatus, it is necessary that the bacilli should be located in a place where they are not washed away by the urine, and where they find favorable soil for their growth, conditions which are only furnished in the kidney.

Neumann ("Ueber die diagnostische Bedeutung der Bakteriologischen Untersuchungen bei inneren Krankheiten, *Berl. klin. Wochenschrift*, Nos. 7, 8, 9, 1888) found the specific microbes in the urine in cases of typhus, septicæmia, and pyæmia. In a case of acute endocarditis and acute osteomyelitis, he cultivated from the urine the staphylococcus pyogenes aureus. He believes that the microörganisms which circulate in the blood localize in the capillary vessels of the kidney, where they often cause minute, multiple lesions without implication of the entire parenchyma of the organ. Through the altered tissues some of the microbes enter the tubuli uriniferi and are eliminated with the urine.

Seitz found the bacillus of typhus in the urine in 2 out of 7 cases, Konjajeff in 3 out of 20 cases, and Hueppe only once in 16 cases, Neumann in 8 out of 48 cases. The last observer ("Ueber Typhus Bacillen im Urin," *Berl. klin. Wochenschrift*, February 10, 1890) in some instances found them so numerous that under the microscope the urine appeared like a fluid culture. In these cases the bacillus multiplies in the bladder. In two cases he also found the streptococcus pyogenes, an occurrence which he considered as an evidence of the existence of complications.

Philipowicz ("Ueber das Auftreten pathogener Microörganismen im Harne," *Wiener med. Blätter*, 1885, No. 22) found the bacillus of tuberculosis not only in three cases of tubercular pyelonephritis, but also in cases of acute miliary tuberculosis. If the organisms were not present in sufficient number for detection by the microscope, their presence in the urine could be proved by the injection of the urine into the peritoneal cavity of guinea-pigs. He also found bacilli in the urine in cases of glanders. In mice which had died of anthrax, the urine contained the bacilli in large numbers. In patients who had succumbed to ulcerative endocarditis, pus-microbes were also found in the urine.

Schweiger ("Ueber das Durchgehen von Bacillen durch die Nieren," Virchow's *Archiv*, B. c. Heft 2) has shown conclusively by his careful clinical observations that the urine from scarlatinal patients is contagious; for varicella, typhus recurrens, and malaria the same holds true. In typhus Gaffky has found bacilli in the vessels of the kidneys. As in most infective diseases the kidneys show textural changes, it was natural to conclude that the renal lesions were caused by microbes on their way out of the body. Schweiger looks upon all kidney lesions found in the course of infective diseases as of bacillary origin. To prove that microbes pass through the kidneys, he cultivated a bacillus which Reimann had discovered in the pus of ozæna. This bacillus is stained an intense green color in a culture of gelatin and agar after twenty-four hours. The cultures of this green bacillus were suspended in a sterilized physiological solution of salt, and injected directly into the circulation. The experiments were made on a dog, cat, and rabbit. The bacillus did not pass directly through the kidneys, but a certain length of time intervened between the injection and its appearance in the urine, as though somewhere an obstacle to its free passage had been met with. At first only isolated bacilli were found in the urine, but later in large numbers. In one instance he extirpated one kidney, and two days later, during the first stage of compensatory hypertrophy of the remaining organ, he injected a culture directly into the carotid artery. The animal died suddenly two and a half hours after the injection in an attack of convulsions. Under strict antiseptic precautions the urine was removed from the bladder, and with it a culture of agar-agar was inoculated. The next day the culture showed a beautiful growth of the same bacillus. The author believes that the kidney, the seat of increased vascular pressure, furnished a favorable condition for the rapid passage of the microbe. He found the microbes most frequently in the glomeruli, and in the space between these and Bowman's capsule; and again, quite abundant in the bloodvessels and in the lumen of the first portion of the convoluted tubuli uriniferi, and only rarely in the perivascular connective tissue. Only once a bacillus was found between two epithelial cells of the convoluted tubules. In the cells themselves no bacilli were found. Upon these observations he bases his advice to favor elimination of microbes through the kidneys in all infective processes by administering large quantities of water, and even diuretics.

Escherich ("Bacteriologische Untersuchungen über Frauenmilch," *Fortschritte der Medicin*, B. iii. p. 231) examined the milk for bacteria in patients suffering from puerperal infection, and, without exception, found that it contained the staphylococcus pyogenes aureus or albus. After having satisfied himself by the examination of twenty-five healthy cases that normal milk contained no microörganisms, he examined the milk of such patients whose bodily temperature was increased by puerperal processes, or lactation, who, in fact, presented evidences of septic conditions, and here he found regularly micrococci present in the milk, principally the staphylococcus aureus and albus. In puerperal women, who had fever from other causes, such as pulmonary tuberculosis, otitis media, etc., no microörganisms were found in the milk. He believes that the microbes are introduced into the milk through the blood, which they enter in puerperal septicæmia, through wounds or abrasions of the genital tract.

Bumm ("Zur Aetiologie der puerperalen Mastitis," *Archiv f. Gynäkologie*, B. xxiv. p. 262) cultivated from a case of puerperal mastitis a diplococcus which resembled the gonococcus very much. He injected the culture under his own skin, and produced an abscess. Karlinski injected a pure culture of staphylococcus pyogenes aureus into a vein of a puerperal rabbit on the third day after labor. In twenty-four hours the animal's milk proved to contain numerous staphylococci. On the fourteenth day the animal died of pyæmia.

Bollinger ("Ueber Tuberkelbacillen im Euter einer tuberkulösen Kuh, und über die Virulenz des Secretes einer derartig erkrankten Milchdrüse." *Bayr. Ärztl. Intelligenzblatt*, 1883, No. 16) found tubercle bacilli in the parenchyma of the udder, as well as in the secretion in the milk ducts in a cow suffering from tuberculosis, which, on being inoculated in guinea-pigs, produced typical tuberculosis.

Hirschberger (*Archiv f. klin. Medicin*, June, 1889) has made some very interesting investigations in reference to the presence of tubercle bacilli in the milk of tuberculous cows. The material was furnished by the abattoir of Munich, and the examinations were made in Bollinger's laboratory. The whole udders of cows that were known to be tuberculous were sent to the laboratory, where, under the strictest antiseptic precautions, the milk ducts were opened and the milk removed was injected into the peritoneal cavity of guinea-pigs. Twenty inoculation experiments were made in as many animals, and in eleven, or in fifty-five per cent., the result was positive. In most of those cases the disease appeared as a diffuse miliary tuberculosis of the peritoneum, omentum, spleen and liver; the milk of cows suffering from advanced tuberculosis proved the most virulent.

Ernst (*Medical News*, Sept. 28, 1889) examined 114 samples of milk obtained from 36 cows suffering with tuberculosis of some organ other than the udder; 17 samples were found to contain the specific bacillus. These 17 specimens came from 10 cows.

Inoculation with the infected milk produced the disease in 50 per cent. of the cases treated. Feeding experiments were also made, with the result of inducing the disease in a number of calves and young pigs. These experiments furnish positive proof that the bacillus of tuberculosis is eliminated through the mammary secretion even in cases where the gland is not the seat of any tubercular lesion.

Ribbert found, twenty-four hours after injecting osteomyelitic cocci directly into the circulation of animals, the microbes in all of the internal organs; later, only in the kidney. Lübbert found the staphylococci in osteomyelitis frequently in the urine. Lebedoff saw the streptococcus of erysipelas in the skin and umbilical cord of a child born eight days after its mother had recovered from an attack of erysipelas.

Clinical observation as well as experimental research has shown that in all localized infective processes the leucocytes act as an advance-guard in protecting the tissues against the ingress of the microbes by mechanically obstructing the way; later, the active granulation tissue performs the same function. As inclusion of the microbe in the cell protoplasm may have a great deal to do with the destruction of the microbes, it is correct to speak figuratively of a struggle of cells against microbes. When the microbes have become disseminated throughout the organism by the circulating blood, they are also brought to the excretory organs, through which many of them are eliminated without having lost their virulence during the passage through the body.

CHAPTER VI.

ANTAGONISM AMONG MICROÖRGANISMS.

ONE of the most recent achievements in bacteriology is the discovery of the antagonism which exists among certain pathogenic microörganisms. That such antagonism exists has been demonstrated by cultivation, and inoculation experiments.

Pawlowsky (Virchow's *Archiv*, B. cviii.) has furnished strong experimental evidence of the antagonism which exists between the pneumococcus of Friedländer and the bacillus of anthrax; as, in eight rabbits infected with a fatal dose of anthrax, all of the animals were saved by a subsequent injection of a pure culture of the pneumococcus. The same author also ascertained that the micrococcus prodigiosus is also antagonistic to the coccus of erysipelas. Thus, ten rabbits were first inoculated with anthrax bacilli, and then cultivations of the micrococcus prodigiosus were injected subcutaneously into each animal on two occasions, two and twenty-four hours after injection; of these ten animals, eight recovered. He also found that subcutaneous injection of anthrax bacilli, and cultivations of pneumonococci were fatal to rabbits; and that subcutaneous injection of cultivations of anthrax bacilli and staphylococcus pyogenes aureus was not followed by the death of the animal; four rabbits treated with a culture of the staphylococcus aureus recovered; of seven anthracic rabbits treated by subcutaneous injection of a culture of streptococcus erysipelatosus, two died.

Emmerich (*Fortschritte der Medicin*, B. v. 1887) has also studied experimentally the antagonism among pathogenic microbes in the living organism. His experiments on rabbits have shown the value of the erysipelas cocci as a protective and curative agent in anthrax in these animals. In one series of experiments the rabbits were first inoculated with a large quantity of a reliable culture of the streptococcus of erysipelas, and then, two to fourteen days later, the animals were again inoculated with a pure culture of the bacillus of anthrax. Of fifteen animals treated in this way, seven recovered, while all the control animals infected with anthrax, but not protected by the microbe of erysipelas, died; of the seven animals which died after inoculation of both microbes, some succumbed to the anthrax bacillus, and some to the microbe of erysipelas. Therapeutic inoculations with cultures of the streptococcus erysipelatosus in animals suffering from anthrax were not as successful. In later experiments made in the same direction in conjunction with Mattei, it was ascertained that when the cocci of erysipelas were injected both into the circulation and the subcutaneous tissues of rabbits twenty-four hours before the infection with anthrax, the bacilli, even when administered in large quantities, were destroyed in from twelve to seventeen hours, and could no longer be found either at the seat of infection or in the blood and internal organs, whether by microscopical examination, or by cultivation experiments.

Neumann ("Ueber den Einfluss des Erysipelas auf den Verlauf der constitutionellen Syphilis," *Allg. Wiener med. Zeitung*, 1888, No. 4) communicates two observations of his own in which erysipelas exerted a decided beneficial effect on syphilis.

The first patient was a woman, fifty-six years old, who within two and a half months passed through three attacks of facial erysipelas, during which the cutaneous gummata in that part of the face affected with erysipelas disappeared completely. The second case was a man, twenty-six years old, who contracted a hard chancre six weeks before he was attacked by facial erysipelas. During the acute attack the primary sore improved and the secondary symptoms did not appear until seventy-three days after the infection, and then in only a mild form, so that the erysipelas had the effect of postponing the secondary symptoms.

The experimental work of Watson Cheyne (*London Medical Record*, 1887) on the antagonistic action of the streptococcus erysipelatosus upon the bacillus of anthrax is of the greatest practical and scientific importance. He experimented on rabbits ; (*a*) by inoculating them with the microbes of erysipelas, and two to fourteen days later with anthrax ; (*b*) by simultaneous intravenous, or subcutaneous inoculation of both ; (*c*) by inoculating the virus of erysipelas, after anthrax had been artificially produced. Of the first series of fifteen animals, seven recovered, while all the control animals died. Of the second series of sixteen animals, only two recovered. In these sets of experiments, by injection of a half million of anthrax bacilli, all the control animals died, whilst all those that received intravenous injection of a pure culture of the microbes of erysipelas, though they were never ill, recovered.

Schwimmer ("Ueber den Heilwerth des Erysipels bei verschiedenen Krankheitsformen," *Wiener med. Presse*, 1888, Nos. 14, 15, 16) has studied with special care the antagonistic properties of the streptococcus of erysipelas in different infectious diseases. In syphilis, the lesions occurring in the erysipelatous area healed more promptly. In a case of obstinate orchitis and epididymitis, on both sides, absorption took place during an attack of facial erysipelas; no material improvement was observed except that in one case an ulcer healed more rapidly. The author saw tubercular glands in the neck disappear during attacks of erysipelas of the face and neck.

Bruns (*Beiträge zur klin. Chirurgie,* B. iii. Heft 3) has collected twenty-two cases of malignant tumors including one of his own, of melano-sarcoma of the breast, in which a final cure followed an attack of erysipelas. Out of 5 sarcomata, 3 were permanently cured, while the other 2 were diminished in size, but soon returned to their former size. The effect of the erysipelatous invasion proved negative in 6 cases, in which the diagnosis between carcinoma and sarcoma could not be positively made, as also in 3 cases of ulcerative epithelioma. It is stated that in cicatricial keloid and lymphomata the attack of erysipelas proved curative.

Garrè (" Ueber Antagonisten unter den Backterien," *Correspondenzblatt f. Schweizeraerzte,* B. x.) has studied the antagonisms among bacteria on artificial culture-soils. He has made many careful experiments to determine the growth of a culture of germs on different nutrient media by removal of the entire culture with a minute spade, and inoculation of the same soil with another microbe. From the results thus far obtained, he has ascertained that some microbes affect the soil favorably for the growth of other varieties, while others render it sterile. For example, a culture-medium impregnated with the ptomaïnes of the bacillus fluorescens putridus remains perfectly sterile when inoculated with pus-microbes. These investigations have an important practical bearing, as future research may not only show the way to secure immunity from infection by pathogenic microbes by prophylactic inoculations with harmless microbes, but may likewise establish a system of rational treatment by inoculations of cultures of antagonistic bacteria for therapeutic purposes.

CHAPTER VII.

INFLAMMATION.

OUR ideas of the nature of inflammation have been materially changed by the knowledge we have obtained from bacteriological investigations which have been made during the last fifteen years. Inflammation is no longer viewed as a disease. Many heretofore obscure inflammatory lesions are now known to have been caused by definite, specific microbes. Modern pathology has established the fact that the condition called inflammation is a restorative process, which has for its object the repair of injured tissues, or the neutralization or removal of the primary microbic cause. From a scientific and practical standpoint, all inflammatory affections can be divided into two classes: 1. Simple, or plastic inflammation. 2. Infective, or destructive inflammation.

1. *Simple or Plastic Inflammation.*

A simple, or plastic, inflammation is a regenerative process, induced by a trauma, or disease, in which the tissues are in an aseptic condition, and the products of tissue-proliferation are transformed into normal permanent tissue. During the first stage of inflammation the tissues are weakened, so that they cannot resist in any way the entrance of pathogenic microörganisms, should any reach the tissues by direct contact, or through the circulation. In the second stage, the weak tissue has become removed, and its place is occupied by vigorous granulation tissues, which possess great power in resisting the attack of microbes. In the third stage, the granulation tissue is becoming converted into mature active tissue, Prolonged vascular engorgement and the first stage of inflammation frequently furnish the conditions which determine transformation of a simple into a septic inflammation. This is frequently observed in cases of acute intestinal obstruction, in which the vascular engorgement and incipient inflammation of the intestinal tunics so alter the tissues of the bowel that pathogenic microbes pass from the intestinal canal into the peritoneal cavity, where they set up a septic inflammation. As long as no septic infection takes place the new embryonal cells undergo transformation into tissues which correspond to the anatomical seat of the lesion. As no toxic ptomaïnes are formed in simple inflammation, sepsis and febrile disturbances are either absent entirely or, when present in a slight degree, are caused by the absorption from the tissues of the products of tissue-wear, or the introduction of fibrin ferments derived from extravasated blood. It is only a matter of time until the process still known as simple inflammation will be no longer considered and classified with inflammatory processes, but will be assigned a separate place in pathology, as it resembles in every respect the physiological processes as we observe them during the growth and regeneration of the body.

2. *Infective Inflammation.*

The characteristic features of this form of inflammation are, that it is caused by the presence of specific microbes, and that the products of the inflammatory process do not undergo transformation into tissue of a higher type.

Thoma ("Ueber die Entzündung," *Berl. klin. Wochenschrift*, Nos. 6, 7) regards suppuration as a qualitative alteration of inflammation. He defines it as an exudation plus fermentation, in which the latter is not essentially caused by microörganisms, as it may also be caused by the tissues, as they also are known to possess zymotic properties. Caseation is also looked upon as the result of fermentation, but not in the sense described by Weigert, but more allied to digestive and putrefactive processes.

The new tissue is destroyed either rapidly or slowly, by the action of the ptomaïnes; the cells undergo retrograde metamorphosis, and are absorbed after the primary cause has ceased to act, or they are so rapidly destroyed that their interstitial removal is no longer possible and the product of the inflammation is removed spontaneously, or by the intervention of art, or, finally, life is destroyed, either by disturbances due to mechanical causes or by ptomaïne-intoxication. The intensity of the inflammation depends as well upon the nature of the microbes as their quantity. The microbes of suppuration, gonorrhœa, and erysipelas always cause an acute inflammation, while tubercular processes and allied affections are noted for their chronicity. Glanders and actinomycosis are chronic or subacute affections. In the same organ, the primary seat of the inflammatory process will be modified by the kind of microbe which has caused the infection. Thus, in suppurative mastitis, the abscesses which are caused by the staphylococci always begin in the deeper part of the organ, and spread toward the surface, while in infection with streptococci of the same part, the inflammation starts from some superficial abrasion and first attacks the skin, whence the process extends in a central direction to the deeper portions of the gland, where suppuration takes place (Cheyne). This difference depends on the manner of invasion of the two microbes. The staphylococci enter the organism through the milk ducts, and act from their interior, whereas the streptococci, like the streptococcus of erysipelas, enter the tissues through the lymphatic vessels, and their pathogenic action is primarily observed at the surface. Bumm excised a portion of the wall of a commencing abscess of the breast, and was able to demonstrate the presence of staphylococci in the interior of the acini, and their penetration thence into the interacinous tissue. The phlegmonous inflammation of the breast caused by the streptococci takes place along the course of the lymphatics, and primarily involves the interacinous connective tissue.

During the first stage of microbic inflammation, the increased afflux of blood may be looked upon as an attempt to wash away the microbes. The increased velocity of the blood-current is well calculated to prevent mural implantation of the microbes, and to detach such as have become fastened upon the vessel wall. The next attempt on the part of the tissues is to limit infection by the production of granulation tissue. The tissues which first come in contact with the primary cause of the inflammation are converted into embryonal tissue, which forms a wall of protection for the surrounding tissues around the primary area of infection. In acute suppuration this granulation tissue is transformed into pus, and the process extends until limitation takes place by the primary cause becoming less virulent, when the abscess wall, composed of living granulation tissue, forms a boundary-line to the suppurative process. In chronic infective processes, as tuberculosis and actinomycosis, the granulating stage remains for an indefinite period of time, and in the former affection, under favorable conditions, retrograde degenerative changes are prevented, the embryonal cells are transformed into connective tissue, and a spontaneous cure is the result. During the third stage the microbes have either been removed or, at least, their pathogenic properties no longer exist, and a process of repair is initiated. It has been shown experimentally that microbes enter the organism most rapidly during the first stage of inflammation, as studied in tissues the seat of a simple inflammation produced by the action of chemical irritants, and subsequently infected by the introduction of pathogenic microbes. It has been found, as regards pus-microbes, that if they are circulating in the blood, the induction of a severe inflammatory action does not lead to their localization in the part, while if the inflammation is chronic, or the trauma less intense, they become arrested in the inflammatory depot and set up suppuration. Thus, Rinne concluded from his experiments on suppuration, quoted elsewhere, that a violent inflammatory action did not produce a *locus minoris resistentiæ*, but that the slightest injury caused by the chemical products of the bacteria themselves sufficiently weakened the part to enable the organisms to grow in it. Acute osteomyelitis and local tuberculosis follow not severe, but, as practical experience has shown, trifling injuries. In phthisical patients a severe wound or fracture is not followed by local tuberculosis, but slight injuries determine localization. Chronic infective inflammations, as tuberculosis, syphilis, and actinomycosis, often lay the foundation for acute suppurative inflammation, as the foci of granulation tissue determine localization of floating pus-microbes, and the embryonal tissue furnishes the most favorable local conditions for their pathogenic action. Acute septic peritonitis has been made the subject of careful experimentation with special reference to its etiology.

Pawlowsky (Beiträge zur Aetiologie und Entstehungsweise der akuten Peritonitis," *Centralblatt f. Chirurgie*, 1887, No. 48) made 10 series, with 101 experiments. The chemical irritants, or cultures, were introduced through the canula of a small trocar under strict antiseptic precautions, and the small wound carefully closed with iodoform collodium. The first series consisted of experiments with croton oil on three dogs and nine rabbits. The amount of croton oil injected varied from six drops to one-tenth of a drop. The smallest doses produced no symptoms. Large doses produced a severe, acute hemorrhagic peritonitis, the intensity of which was proportionate to the amount of the irritant injected. The peritoneal effusion, under the microscope, was seen to contain red and white blood-corpuscles. Inoculations of different nutrient media with the fluid yielded negative results. In the next series of experiments, an aqueous solution of trypsin and pancreatin was injected for the purpose of determining whether the digestive ferments in the event of intestinal perforation could produce peritonitis. The experiments established that trypsin acts as a powerful irritant upon the peritoneum. Injection of one-half gramme of trypsin dissolved in sterilized water caused in rabbits a severe hemorrhagic peritonitis, with a copious exudation and death in from four to four and one-half hours. In doses of from one-fourth to one-tenth of a gramme hemorrhagic peritonitis was also produced, but death did not occur until twenty to twenty-four hours after the injection. One-twentieth of a gramme produced no symptoms. Nutrient media inoculated with the products of inflammation remained sterile. Next, the peritoneal cavity was infected with plate cultures of different microbes suspended in sterilized water. The first experiments were made with non-pathogenic germs. Four rabbits and one dog were injected with large quantities of a micrococcus which was obtained from a plate culture inoculated with pus; the micrococcus was exactly similar to the staphylococcus albus, for which it was first mistaken; later, it was shown that it was not a pus-microbe, as it did not liquefy gelatin. All of the animals recovered. Two rabbits inoculated with an entire culture of yellow sarcinæ upon agar, mixed with one-tenth of a drop of croton oil, also recovered. The experiments with pathogenic microbes always produced positive results. Three series, with three separate microörganisms, were carried out. The staphylococcus pyogenes aureus grown from osteomyelitic pus was first used. In seventeen out of forty-one experiments, this microbe alone was used; in eleven it was mixed with croton oil, in six with trypsin, and in seven with agar-agar. From injections with the staphylococcus alone, the following results were obtained:

1. Large quantities produced fibrinous suppurative peritonitis.
2. In quantities of two plate cultures, four rabbits succumbed to suppurative peritonitis.
3. Half of this quantity produced the same results.
4. In yet smaller quantities, still the same result, and peritonitis only failed to develop when a very minute quantity of the culture was used.

In all cases in which peritonitis was produced, inoculations of the products of inflammation upon nutrient media yielded positive results. In hardened specimens of the peritoneum, stained with various coloring agents, the microörganisms could be seen in the lymph spaces. The suppurative nature of the peritonitis thus induced became more apparent the longer life was prolonged. An entire agar culture of the bacillus pyocyaneus caused death in from twenty-four to forty-eight hours. The autopsy revealed a fibrinous hemorrhagic peritonitis. The exudation consisted largely of red corpuscles and a large number of bacilli. Pure cultures of the bacillus could be obtained by inoculating the fluid upon agar-agar. One-fifth of this quantity proved harmless.

In hardened sections, the bacilli were found in the lymph spaces of the central tendon of the diaphragm, the parietal peritoneum, visceral peritoneum, and in the capsule of the spleen and liver, also in the uriniferous tubules and Malpighian bodies of the kidney. The next series of experiments was made to ascertain what caused the inflammation in cases of perforative peritonitis. The fresh intestinal contents of a healthy animal, just killed, were divided into three parts, one of which was at once injected into several rabbits without filtration, in doses of one syringeful. The second portion was filtered, and of the filtrate two and a half to three and a half syringefuls were injected into each rabbit; the third portion was sterilized according to Tyndall's directions for eight days, and then one syringeful was injected into each animal. The results were as follows : Four rabbits died of fibrinous suppurative peritonitis from the injection with the first portion. Four rabbits injected with the filtered feces, recovered, as did one rabbit inoculated with the sterilized portion. At the autopsy, particles of the intestinal contents were found in the peritoneal cavity covered with fibrin, and microscopically peculiar, short bacilli. This microbe he called bacillus peritonitidus ex-intestinalis cuniculi, and describes the cultures upon agar-agar plates as shining, grayish-white, oil paint-like colonies. It does not liquefy gelatin. The bacillus is non-motile. With cultures of this bacillus he made nine experiments on rabbits and two on dogs. Each animal, which received an entire agar culture, died of hemorrhagic peritonitis in from twenty to twenty-four hours. Smaller quantities produced death from the same cause in from twenty-four to seventy-two hours. Still smaller doses produced a suppurative peritonitis, and death after a number of days. Of the two dogs, each injected with an agar culture, one died after twenty-four hours of incipient hemorrhagic peritonitis, the other recovered after an illness of several days' duration. In the fatal cases, the bacilli were found in different internal organs, and could again be reproduced by inoculations with the infected tissues upon nutrient media. He believes that this bacillus is the essential cause of perforative peritonitis. He also asserts that the fibrinous form of peritonitis is the least dangerous, as the layers of fibrin tend to limit the ingress of microbes into the organism. The fibrinous purulent variety is the next formidable form, while in the most rapidly fatal cases of septic peritonitis the local lesion is not characterized by any macroscopical tissue-changes.

Alexander Fränkel (*Wiener klin. Wochenschrift*, 1888, Nos. 30–32) testifies to the harmlessness of pure cultures of microbes when injected into the peritoneal cavity of rabbits. Fehleisen (*Archiv f. klin. Chirurgie*, B. xxxvi. p. 978) injected pus from abscesses containing the staphylococcus pyogenes aureus in doses of from 4 to 8 c.cm. into the peritoneal cavity of rabbits without producing peritonitis in every instance, or even as a rule, although in some instances the animals died from injection of a much smaller quantity. Fehleisen is of the opinion that the number of microbes in the pus does not determine its virulence.

Orth (" Experimentelles über Peritonitis," *Berl. klin. Wochenschrift*, October 28, 1889) agrees with Grawitz that when a pure culture of pus-microbes is injected into a healthy peritoneal cavity no suppuration is produced. But his experiments proved what is of the greatest practical interest, that if the peritoneum is wounded under antiseptic precautions peritonitis is invariably produced, if somewhere else in the body suppuration existed at the same time. If, for instance, an abscess in the subcutaneous tissue was produced, and then the intestine was temporarily rendered impermeable, the animal died without exception of peritonitis. The same result followed if the pus-microbes were injected directly into the circulation, but not if they were introduced through the alimentary canal. Laruelle ("Etude bactériologique sur les péritonites par perforation," *La Cellule*, T. V. Louvain, 1889) produced peritonitis artificially in dogs and rabbits, and from his observations came to the conclusion that the localization of pus-microbes in the peritoneal cavity is greatly favored by the action of chemical substances. He believes that peritonitis is not produced by the specific microbe described by Pawlowski. He claims that the microörganism discovered by Pawlowski is the bacillus coli communis.

Weichselbaum ("Der Diplococcus pneumoniæ als Ursache der primären, akuten Peritonitis," *Centralblatt f. Bakteriologie*, B. v. No. 2) found in three cases of acute peritonitis the diplococcus of pneumonia and no pus-microbes. In one case the peritonitis was complicated with pneumonia, in the second case it was followed by double pleuritis, and in the third case the disease was doubtless primary.

Wegner (*Archiv f. klin. Chirurgie*, 1877) has shown by his experiments that a great variety of fluids free from septic germs, such as water, bile, urine, blood, etc., can be injected into the peritoneal cavity of rabbits without any serious results following. Even large quantities of unfiltered air proved innocuous when introduced in the same manner. Putrescible fluids when injected in small quantities were rapidly absorbed without producing peritonitis, but when the quantity injected was large and insufflation of air, unfiltered, was practised at the same time, putrefaction and death from septic intoxication occurred.

Grawitz (Virchow's *Archiv*, B. cviii.) proved that saprophytic bacteria, when injected into a normal peritoneal cavity, were promptly destroyed and absorbed. In cases in which the injection was made into a peritoneal cavity which had previously undergone alterations by injury or disease, or in which the quantity of fluid was too great for rapid absorption, symptoms of intoxication, as described by Wegner, resulted, but these symptoms were unaccompanied by suppurative peritonitis. A healthy peritoneal cavity has also been found capable of disposing of a limited quantity of pure cultivations of pus-microbes, the germs being removed by absorption and destroyed in the circulation. But when pyogenic organisms are introduced into an abdominal cavity, where the absorptive powers of the peritoneum have been diminished or suspended by antecedent pathological conditions, suppurative peritonitis is the usual result. When pus-microbes are introduced in large quantities, even into a healthy peritoneal cavity, the preformed ptomaïnes, by their chemical actions, so alter the tissues that the process of absorption is impaired, and suppurative peritonitis again results in consequence. The greatest clinical difference between simple peritonitis produced by a trauma or chemical irritants, and septic peritonitis, consists in the course and extent of the inflammation. Simple inflammation produced by aseptic causes remains limited to the seat of the trauma, and does not extend much beyond the surface area to which the irritant is applied; while septic peritonitis is always characterized by its progressive character, as the cause upon which it depends is multiplied within the peritoneal cavity. The same can be said of the two kinds of inflammation in any other tissue or part of the body. The same conditions which were found to favor the development of septic peritonitis, such as trauma and the presence of fluids, are equally potent in determining localization of microbes in other parts of the body, and intensify their pathogenic action by creating conditions which prevent their absorption and destruction, and, on the other hand, they furnish a nutrient medium in which the microbes find a proper soil for their multiplication.

Rinne (" Der Eiterungs process und seine Metastasen," *Archiv f. klin. Chirurgie*, B. xxxix p. 13) is of the opinion that on account of the rapidity with which absorption takes place in the peritoneal cavity that the peritoneum, when in a normal condition, is almost immune to infection with pus-microbes. He injected from 30 to 35 c.c. of an aqueous suspension of a pure culture of pus-microbes into the peritoneal cavity of healthy animals, and was never able in this manner to produce a peritonitis. He had no better success with injections of a mixture of a gelatin culture of staphylococcus aureus and a turbid culture of the same coccus in bouillon. He also injected from day to day a boiled putrid solution to which was added a culture of the staphylococcus aureus without any inflammation following. The experiments, as a rule, were made on dogs, although in several instances rabbits, guinea-pigs, and white rats were used. He believes that the difference in the results obtained by him and Grawitz, as compared with Pawlowski's, consist in the nature of the abdominal wound. Pawlowski made an incision down to the muscles and then perforated the abdominal wall with a blunt trocar, while he and Grawitz used a sharp, hollow needle for making the injection. To prove that his injections entered the peritoneal cavity he added coal-dust, which he found in the peritoneal cavity in making, subsequently, the autopsy.

CHAPTER VIII.

SUPPURATION.

THE wonderful results which were obtained by the antiseptic treatment of wounds made it exceedingly probable that all wound-infective diseases were caused by living microörganisms. The probability was increased when Koch, in 1879, showed the direct connection existing between certain traumatic infective diseases in animals and the never-absent definite microörganisms. It requires no longer any arguments to show, at this time, that all wound-infective diseases, among them particularly suppuration, are, without exception, caused by the introduction into the tissues of the organism of specific pathogenic microbes. This part of the work has been prepared with special reference to the etiology of acute suppuration, as chronic suppuration is intimately associated with that of surgical tuberculosis and other forms of infective diseases, which usually pursue a chronic course, and differs greatly in its pathology from the other.

Etiologically, most of the purulent processes constitute more of a unity than was formerly believed, and the clinical varieties are mostly determined by the intensity of the infection and by the manner of localization. The most conclusive evidence of the correctness of this assertion is furnished by the fact that the same streptococcus which produces a simple abscess is likewise the most frequent cause of progressive gangrene and of that most grave form of suppuration—pyæmia.

HISTORY.—As in the case of nearly all infective diseases, years before the specific microörganisms of suppuration were discovered, living organisms were found and described in pus, and were believed to be the cause of the suppuration. In 1865, Klebs detected in the tubuli uriniferi in cases of pyelo-nephritis following suppurative cystitis, between the pus-cells small, round cocci, which he believed produced the infection. In 1872, the same author (Schusswunden, Leipzig, 1872) published the result of his researches during the Franco-Prussian War on septic wound diseases. In this work he again referred to the organism which he had previously described, and showed that it existed in the tissues and organs the seat of suppurative inflammation before pus had formed.

He also showed how these organisms enter the circulation and are the direct cause of pathological changes in distant organs. Even at that time he placed great stress on the fact that, as long as the cocci remained only in the tissues at the point of infection they caused only local inflammatory conditions or necrosis, but as soon as they entered the circulation fever and other symptoms of general septic infection followed.

It was not until 1881 that Ogston ("Report upon Microörganisms in Surgical Diseases," *British Medical Journal*, March, 1881, p. 369) announced his great discovery, which has since revolutionized the study of acute suppuration. This patient investigator examined the pus of 69 abscesses for microörganisms, and found in 17 of them a chain coccus (streptococcus), in 31 cocci which arranged themselves in groups which resembled a bunch of grapes (staphylococcus), and in 16 both of these forms were present. In cold abscesses neither of these microörganisms was found. He also found that these two forms of microbes differed in their action on the tissues, as the streptococcus, following the lymphatic channels, was seen to be the cause of diffuse suppurative processes, while the staphylococcus was found only in abscesses which were circumscribed. Rosenbach took up the work where Ogston left it, and, as the fruit of a number of years of patient study and research, published his classical work in 1884 (*Microörganismen bei den Wund-infections Krankheiten des Menschen*, Wiesbaden, 1884). This work must serve as a basis for all future research on suppurative inflammation. Rosenbach availed himself of the advantages offered by an improved technique in bacteriological research, and cultivated the pus-microbes upon solid nutrient media, and pointed out the difference in macroscopical appearances of the cultures of the different kinds of pus-microbes which enabled him to differentiate between them by the naked-eye appearances of the cultures upon the nutrient media. He discovered the staphylococcus pyogenes aureus, the micrococcus pyogenes tenuis, and three bacilli saprogenes.

Passet should be mentioned next after Rosenbach, in the long list of distinguished men who have made the etiology of suppuration a special study. Passet ("Ueber Microörganismen der eitrigen Zellgewebs-entzündung des Menschen," *Fortschritte der Medicin*, 1885, Nos. 2, 3) discovered and described the staphylococcus citreus, cereus albus and flavus, and from a peri-rectal abscess he cultivated the bacillus pyogenes fœtidus. The streptococcus which he found, he claimed was different from the one described by Rosenbach, as it resembled more closely the streptococcus of erysipelas.

Description of the Different Kinds of Pus-microbes.

1. STAPHYLOCOCCUS PYOGENES AUREUS is the pus-microbe most frequently present in acute abscesses. Cocci singly, or aggregated in masses. It grows readily upon gelatin, agar-agar, coagulated blood-serum, and potato. It possesses the property of liquefying gelatin. It grows readily at the ordinary temperature, but more rapidly when the temperature is not less than 30° C. (86° F.), and does not exceed that of the normal temperature of the body. It peptonizes albumen and coagulates milk. The culture grows in the track of the needle and upon the surface of the nutrient medium. The culture presents a gold-yellow appearance.

Lübbert ("Biologische Spaltpilzuntersuchung. Der Staphylococcus pyogenes aureus und der Osteomyelitis-coccus." Würzburg, 1886) has made an extended and thorough study of this microbe. It was examined in reference to its behavior to light, temperature, and the various culture substances. Kreatin was found to be its simplest nitrogenous nutrient material. Carbonic acid arrests its growth, while oxygen not only favored its growth, but also accelerated the production of the orange-yellow pigment material which distinguishes it from the other staphylococci. Milk rendered sour by the addition of a pure culture of this microbe, and tested with the salts of barium, showed the presence of methyl-alcohol, lactic and butyric acids. It was ascertained that it produced carbonic acid and absorbed oxygen. Corrosive sublimate, 1 : 81400, and thymol, 1 : 1100, arrested its growth. A dried culture, exposed for an hour to a temperature of 86° C. (176° F.), was rendered completely sterile. Its pathogenic properties were found greatest in cultures made directly from the diseased tissues, while its virulence diminished through successive cultures. A culture, attenuated in this manner, again regained its virulence by being passed through the animal body. Large quantities of this microbe injected into the subcutaneous tissue of rabbits produced a general infection, the kidneys and muscles of the heart being the organs first attacked. If injected into the knee-joint of a dog suppuration occurred, followed by disintegration of the joint. It is identical with the organism which has been described in acute osteomyelitis, and at first supposed to be the specific organism of that disease.

2. STAPHYLOCOCCUS PYOGENES ALBUS.—Same as aureus, but produces no pigment. Both Passet and Klebs have observed in the white culture of this coccus small yellow dots, which, when isolated, lost their color. These authors, therefore, consider the yellow and white staphylococcus as varieties of the same kind. Its pathogenic properties, both in man and animals, are somewhat less

intense than those of the aureus. The cultures of both the yellow and white staphylococcus upon gelatine present an irregular surface, and the margins are dotted with numerous minute globular projections. Both of these microbes liquefy gelatin, but agar-agar and coagulated blood-serum are not similarly affected. The cultures, especially if kept moist, retain their virulence for a very long time. Rosenbach found a culture upon serum active after the lapse of two years.

3. STAPHYLOCOCCUS PYOGENES CITREUS.—Like the former, liquefies gelatin. Cocci singly, or in pairs, or zoöglœa. If cultivated on nutrient gelatin, or agar-agar, a sulphur or lemon-yellow growth develops after twenty-four hours, which, at that time, resembles the aureus, but later does not change into an orange-yellow color. In both, the development of pigment only takes place where the colonies are in contact with the air. According to Passet, its pathogenic properties are somewhat less than those of the aureus and albus. This latter statement has received the confirmation of Cheyne. When inoculated under the skin of mice, guinea-pigs, or rabbits, an abscess forms after a few days, from which a fresh culture of the microbe can be obtained.

4. STAPHYLOCOCCUS CEREUS ALBUS.—Cocci also obtainable from pus, but distinguished by forming on nutrient gelatin a white, slightly shining layer, like drops of white wax, with somewhat thickened, irregular edge. The needle puncture develops into a grayish-white, granular thread. In plate cultivations, on the first day, white points are observed, which spread themselves out on the surface to spots of one-half a millimetre in diameter; when cultivated on blood-serum, a grayish-white, slightly shining streak develops; and on potato the cocci form a layer which is similarly colored. This microbe is not pathogenic in rabbits.

5. STAPHYLOCOCCUS CEREUS FLAVUS.—If cultivated on gelatin, the growth, which is at first white, becomes a citron-yellow color, somewhat darker than staphylococcus pyogenes citreus. Both varieties of staphylococcus cereus are found in pus and cultures in colonies. Inoculations in rabbits have proved harmless.

6. STAPHYLOCOCCUS FLAVESCENS.—This organism was found in an abscess by Babes, and occupies an intermediate position between the staphylococcus aureus and albus. On gelatin, the growth forms a colorless layer, and causes liquefaction. It is fatal to mice, sometimes causing abscesses, and, in large doses, septi-cæmia.

7. MICROCOCCUS PYOGENES TENUIS.—Rosenbach found this microörganism in a large abscess which had given rise to no general symptoms. It is of rare occurrence. On agar-agar it forms an exceedingly delicate, almost invisible, white film. The

PLATE II.

Staphylococci. (After Rosenbach.) 962 diam.
a. From a culture twenty-four days old.
b. From a culture two months old.

individual cocci are irregular in shape, and larger than the staphylococci. In all cases in which this microbe is the sole cause of the suppuration, the process appears to have been not attended by any very severe inflammatory symptoms, and little or no general febrile disturbances. Rosenbach made no experiments to test its pathogenic properties in animals.

This microbe was never found by anyone else but Rosenbach until February, 1888, when Raskina (*Transactions of Russian Medical Association*, 1889, p. 327) isolated it from the pus and organs in a case of scarlatina complicated with pyæmia, which proved fatal on the eighteenth day of the commencement of the primary disease. At the necropsy multiple miliary abscesses were found in the kidneys at the junction of cortex and medullary portion. From these the micrococcus pyogenes tenuis was obtained in a state of pure cultivation, and from the parenchymatous portion of the kidney a diplococcus of unknown species was cultivated. Inoculations of rabbits with a pure culture of the micrococcus gave negative results, even though the coccus was present in the blood twenty-four hours after inoculation, hence it is problematical as to its being a pyogenic microbe. Like the staphylococcus cereus, it probably belongs to the so-called metabiotic microbes of Garrè, occurring secondarily after suppuration has been established by genuine pyogenic microbes.

8. STREPTOCOCCUS PYOGENES.—Cocci singly, or arranged in chains often of great length, Fig. 2 (Rosenbach). Cultures grow very slowly on ordinary nutrient media at the summer temperature, but with greater rapidity at the temperature of the body. Cultivated in a streak on the surface of gelatin on a glass plate, this microbe forms at first whitish, somewhat transparent rounded spots, of the size of small grains of sand. On nutrient agar-agar it grows most energetically at a temperature of 35°–37° C. (95°–98.6° F.). Even if the inoculation is made with a needle in a continuous line, the culture appears in small dots. In its further growth, the culture is elevated in the centre, and presents a pale brownish color, while the periphery is flattened, except at the extreme margin, which is again raised, and often with a spotted appearance. Still later, the periphery develops successive layers or terraces. The growth is so slow that in two or three weeks the maximum width of the culture-streak is about two or three millimetres. In a vacuum, peptonization of albumen and beef takes place rapidly. In the subcutaneous tissue of rabbits in small quantities they cause

Fig. 2.

Streptococcus pyogenes.
(ROSENBACH.)

a transient redness; when larger quantities are used some authors claim they produce small circumscribed abscesses. If a pure culture is injected into a serous cavity, it causes, first, inflammation, and, later, effusion, which is again absorbed.

9. BACILLUS PYOGENES FŒTIDUS.—Passet found this organism (Fig. 3, Passet) in the pus of a peri-rectal abscess This bacillus grows on gelatin, forming a delicate white or grayish layer on the surface, but causes no liquefaction. When cultivated on agar-agar and potato it has the appearance of a light brown glistening layer, which emits a very offensive odor. In milk this smell is not produced. This organism is not pathogenic in rabbits. In mice traces of the culture do no harm; the injection of several drops causes septicæmia. Injection of about ten minims of the cultivation into guinea-pigs causes an abscess in which the bacilli alone are found; direct injection into the circulation causes sepsis.

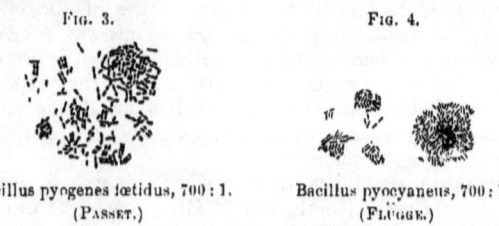

FIG. 3. FIG. 4.

Bacillus pyogenes fœtidus, 700 : 1. Bacillus pyocyaneus, 700 : 1.
(PASSET.) (FLÜGGE.)

10. BACILLUS PYOCYANEUS.—Ernst found in blue pus two kinds of bacilli, Fig. 4 (Flügge), which he designated as bacillus pyocyaneus. Ledderhose ("Ueber den blauen Eiter," *Deutsche Zeitschrift f. Chirurgie*, B. xviii. Heft 3), by extensive cultivations of these bacilli, obtained a considerable quantity of pyocyanin and by chemical analysis determined its formula to be $C_{14}H_{14}N_2C$. In doses of one gramme as muriate of pyocyanin injected into the circulation of different animals, he observed no toxic symptoms. When a pure culture of the bacilli was injected, he observed inflammation and suppuration, and attributes this result, not to the presence of pyocyanin, but to other as yet unknown phlogistic and pyogenic substances.

Experimental and Clinical Evidences which Prove that Pus-microbes are the Only and Essential Cause of Suppuration.

Rosenbach found that in dogs and rabbits a small quantity of a pure culture of the staphylococcus pyogenes aureus injected under the skin produced a most violent suppurative inflammation; cultures of the staphylococcus pyogenes albus produced the same effect. Cultures of the streptococcus pyogenes produced only slight inflammation in rabbits, while they proved very fatal in mice.

Passet ("Ueber die Aetiologie und Therapie der Impetigo, des Furunkels und der Sykosis," *Monatshefte f. prakt. Dermatologie*, B. vi. No. 10, 1887) took a pure culture of the staphylococcus pyogenes aureus the size of a pea grown upon potato and mixed it with 5 c.c. of distilled water. Of this fluid he injected under the skin of a mouse 0.1 c.c.; the animal recovered. Another mouse was treated in the same manner, but 0.4 c.c. of gelatin, liquefied by the same microbe, was used, and this animal died in eighteen hours. Cocci were found in the blood. In rabbits and dogs, a subcutaneous injection of 1 c.c. liquefied gelatin-culture of the aureus usually produced an abscess at the point of inoculation. If the dose was increased to 5 c.c. of such a culture, the animals died in from eighteen to twenty hours, at the same time a local inflammation was found at the point of injection. In all of these cases the blood contained numerous cocci. Of the culture of the streptococcus it was found necessary to inject a considerable quantity in order to produce suppuration. Liquefied gelatin-cultures of the staphylococcus pyogenes aureus and albus, in doses of 1 c.c., injected into the abdominal cavities of rabbits, were well tolerated, and death was only produced when the dose was increased to from 4 to 6 c.c. Injection of cultures of the streptococcus pyogenes into the peritoneal cavity was even better tolerated, and usually had to be repeated several times to produce death from peritonitis. A needle dipped into a culture of pus-microbes he could insert into joints without causing suppuration; but the injection of from 0.3 to 0.5 c.c. of a mixture of pus-microbes and water into the hip-joint of rabbits produced suppurative arthritis, rupture of the capsule, and diffuse para-articular phlegmonous inflammation and suppuration, and often death of the animal. Injection of one or two drops of a liquefied gelatin-culture of the staphylococcus aureus, or albus, into a vein of a rabbit did not produce any serious disturbance, but if the dose was increased to from 0.5 to 1, it, as a rule, caused a fatal disease. In such cases, multiple suppurating foci were found in the kidneys, liver, spleen, and lungs, with pleuritic and peritoneal effusions, pericarditis and myocarditis, also serous and purulent effusions into joints and muscular abscesses. A pure culture on potato scraped off and injected into the circulation produced mycotic endocarditis. Injection of a pure culture of streptococcus pyogenes was more frequently followed by joint complications and peritoneal effusion. The effect of the introduction of pus-microbes in man is the same as in animals. Garrè ("Zur Aetiologie der acut eitrigen Entzündungen," *Fortschritte der Medicin*, 1885, No. 6) made a superficial abrasion on one of his fingers and applied a pure culture of the staphylococcus pyogenes aureus; the only symptom observed was a slight redness eighteen to twenty-four hours after the inoculation. He then made three small incisions and inoculated himself with a larger quantity of the culture, which was followed by a slight sub-epidermal suppuration. The same author made numerous cultivations from pus of different forms of acute abscesses, and always succeeded in demonstrating the presence of one or more of the pus-microbes. From the product of acute, diffuse phlegmonous inflammation the culture consisted usually of streptococci.

Fehleisen repeated the same experiments with cultures of the different kinds of pus-microbes, and, if he succeeded in causing suppuration, this was always very slight. He also found minute doses administered subcutaneously harmless, while a larger quantity of pus-microbes suspended in water, almost without exception, caused abscesses, and, in animals, very large doses produced death from sepsis before suppuration could take place.

Bockhardt introduced a trace of the mixed cultivation of staphylococcus aureus and albus into the cutis of his left forefinger; after forty-eight hours a small abscess had formed, and was opened, and the pus contained the staphylococcus pyogenes aureus.

He also made on himself endermic inoculations with a mixture of staphylococcus pyogenes aureus and albus. Fourteen hours after injection there were at the seat of inoculation, in a space about the size of a five-shilling piece, twenty-five impetigo pustules, varying from the size of a pin's head to that of a lentil, a few, but not the majority, of these being traversed by hair. They contained the cocci employed. Bumm injected pure cultivations of staphylococcus pyogenes aureus into the subcutaneous tissue of his own arm and into the arms of two other persons. In each instance an abscess developed, which varied from the size of a pigeon's egg to that of a man's fist, according to the time which elapsed before they were opened, and these abscesses contained the same species of microbes as were injected.

Fehleisen ("Zur Aetiologie der Eiterung, Arbeiten aus der Chirurgischen Klinik der Königl. Universität," Berlin, Dritter Theil., 1887), who believes that the ptomaïnes produced by the pus-microbes are the direct cause of suppuration, has also made numerous experiments on animals by injecting pure cultures, or pus. In order to ascertain whether the presence of atmospheric air had anything to do with the formation of pus, he made subcutaneous sections of muscles and tendons with a tenotomy knife which had been dipped in a mixture of water and pus-microbes, and, as suppuration did not occur, he smeared the blade of the tenotome with a pure culture of pus-microbes and kept the instrument under a glass globe for six hours for the liquid to dry on it, and repeated the same operation, and still no suppuration followed. In two dogs of the same age, he dissected up a triangular flap of skin over the abdomen with the base toward Poupart's ligament, and in one of the animals infected the wound with a pure culture of pus-microbes before suturing, while in the other the wound was closed without such infection. The non-infected wound healed by primary intention, while in the other animal a large abscess formed. He also exposed the femoral vessels in animals, and inoculated the surface over the vessels with pus-microbes, with the result of causing only an abscess, but never progressive septic infection. In rabbits and dogs, the injection of one drop of pus into the connective tissue produced no reaction—absorption taking place. In both of these animals the injection of from 1 to 2 c.c. of pus under the skin always developed large abscesses. In a rabbit, 0.4 c c. of streptococci-containing pus caused death in forty hours; suppuration at the point of injection. Streptococcus culture was obtained from the blood taken from the liver. Wegner ascertained that, in order to produce suppurative peritonitis in rabbits, it is necessary to inject 6 c.c. of pus, and in dogs 25 c.c. Fehleisen injected into the peritoneal cavity of rabbits 8 c.c. of streptococci-containing pus, with the result that the animals were almost immediately seized with toxic symptoms, followed by death in twelve hours. After death, double pleuritis and peritonitis were found. Streptococci could be cultivated from the blood. In two rabbits, 8 c.c. of tubercular pus were injected into the peritoneal cavity without producing any immediate symptoms until tuberculosis developed. In rabbits, staphylococcus-containing pus produced no constant symptoms. Some died after an injection of from 1 to 2 c.c., others recovered after injection of from 4 to 5 c.c. From these experiments, he came to the conclusion that the virulence of pus does not depend upon the number of microbes it contains, but is proportionate to the intensity of the inflammation in the individual from which it was taken.

Zuckermann (*Aus dem Laboratorium f. Chirurgie, Pathologie u. Therapie der k. Universität zu Kasan.*), as the result of his own investigations and experiments, formulates his own ideas in reference to the causes of suppuration in the following conclusions:

1. Mechanical, chemical, or thermal influences, if microbes are excluded, do not produce suppuration.

2. If suppuration followed any of these causes, it was not without the admission of microbes.

3. Chemically-pure substances may be mycotically impure; even

some disinfecting agents are, it appears, not always free from microbes.

4. As causes of suppuration the following microbes can be enumerated: staphylococcus pyogenes aureus, albus, and citreus; streptococcus pyogenes, and, in foul abscesses, also the bacillus pyogenes fœtidus.

5. Inoculations with staphylococcus and streptococcus have, when injected in large quantities, produced local suppuration, or death by general infection.

6. The pus-microbes must, in the face of the frequent occurrence of suppuration, have a wide diffusion through nature.

7. Pus-microbes can enter the organism through the respiratory passages, the intestinal canal, and skin. Entrance is most frequently effected through the skin.

8. The staphylococcus and streptococcus are found most frequently in pus.

The same writer to show the frequency of occurrence of the different kinds of pus-microbes, has tabulated from different sources 495 abscesses, and he states that the staphylococcus was present in 71 per cent., the streptococcus in 16 per cent., these two microbes together, in 5.5 per cent., and the remaining pyogenic microbes only exceptionally.

Tricomi ("Referat," *Berl. klin. Wochenschrift*, Jan. 23, 1888) made a bacteriological examination of 80 acute abscesses, 8 phlegmonous inflammations, and 5 furuncles, and never failed in finding the microbes of suppuration. He makes no distinction between the streptococcus and staphylococcus and includes both forms under the term micrococcus pyogenes. His own experiments on mice, rabbits, and guinea-pigs have led him to the opinion that pus-microbes have a specific destructive action on the connective tissue, and he asserts that their pathogenic action is limited to this tissue. The results of his observations on animals he applies to human pathology, and attempts to establish the fact that, in the formation of abscesses in the internal organs, destructive processes must precede and prepare the soil for the microbes.

He does not believe that indifferent and chemical substances, mycotically pure, can cause suppuration.

Watson Cheyne, in his admirable lectures on suppuration, puts himself on record as believing that no aseptic substances or chemical irritants ever cause suppuration. He affirms that the product of inflammation which accumulates at the point of implantation or infection in such cases is not true pus, but a putty-like substance which is more consistent than pus from the absence of ptomaïnes. He asserts that for the formation of a true abscess we require the peptonizing ferment produced by the microörganisms, or, at any rate, of a chemical substance which prevents coagulation of the exuded fluid. He describes a minute abscess as containing a colony of microbes which, when fixed in the tissues, cause a limited coagulation-necrosis of the tissues in immediate contact with the microbes by the action of the ptomaïnes; a few hours later a zone of leucocytes aggregate around the dead tissues. The products of inflammation also remain fluid, probably also from the peptonizing effect of the microbes, and thus an abscess is formed.

The above observations are conclusive in showing that pus-microbes can be cultivated from the pus of every acute abscess, and that, in man and animals, the injection of a sufficient quantity of a pure culture into the tissues is followed by suppuration; and thus far, positive proof has been furnished of the direct etiological relationship which exists between pus-microbes and suppuration. Quite recently a number of pathologists have gone one step further, and claimed that pus-microbes are not the direct cause of suppuration, but that their presence is essential for the production of ptomaïnes, to which they attribute pyogenic properties. We will, therefore, consider

The Relation of Ptomaïnes to Suppuration.

Grawitz and de Bary ("Ueber die Ursachen der subcutanen Entzündung und Eiterung," Virchow's *Archiv*, B. cviii. S. 68), after detailing the results of their experiments with injections of chemical irritants during their studies on pus-production, give an account of their experiments with the ptomaïnes, of pus-microbes. They claim that these ptomaïnes, like chemical irritants, prepare the soil in the tissues for the growth and reproduction of pus-microbes The action of these substances becomes apparent by injection of sterilized cultures where the only active agents could be the preformed poisons. They injected 4 c.c. of a sterilized culture of the staphylococcus pyogenes aureus under the skin of a dog, and produced suppuration. The contents of the abscess were examined for microbes, but none were found. They claim that the presence of oxygen is of the greatest importance in the production of ptomaïnes.

Grawitz ("Ueber die Bedeutung des Cadaverins f. das Entstehen von Eiterung," Virchow's *Archiv*, B. cx. Heft 1) experimented with a pure preparation of cadaverin, prepared by Brieger from bacteria. This belongs to the class of non-toxic cadaver alkaloids, and is a colorless fluid, the chemical formula of which is identical with pentamethylendiomin; a $2\frac{1}{2}$ per cent. solution of this substance destroyed the staphylococcus pyogenes aureus in an hour, and a small quantity added to a culture of pus-microbes arrested further growth. A solution absolutely free from microbes, injected under the skin of animals, according to strength and quantity used, produced cauterization or inflammation, terminating in suppuration or inflammatory œdema, followed by resolution and absorption. The pus produced by cadaverin contained no bacteria as long as the skin remained intact. The injection of a mixture of a solution of cadaverin and pus-microbes caused a progressive phlegmonous inflammation.

Scheuerlen ("Weitere Untersuchungen über die Entstehung der Eiterung, ihr Verhältniss zu den Ptomainen und zur Blutgerinnung," *Mittheilungen aus der Chirurgischen Klinik der Königlichen Universität*. Berlin, Dritter Theil. 1887) has on a previous occasion shown that croton oil, turpentine, and other irritants cannot produce suppuration. In all cases in which pus was produced by any of these agents it contained bacteria. He, therefore, takes it for granted that pus can only be produced by microörganisms. How this is done is difficult to prove. The action of pus-microbes upon the tissues must be either physical or chemical. The action of ptomaïnes on the living tissues was studied by Panum as early as 1856, and later by Bergmann and his scholars. Recently this subject has been studied in a systematic manner by Brieger. Scheuerlen was the first to study their local effects. He introduced into the subcutaneous connective tissue of rabbits aseptic glass capsules containing sterilized putrid infusion of meat; the wounds were treated under the strictest antiseptic precautions, and healed by primary union. After the wound was healed, he broke off subcutaneously both ends of the glass capsule, so as to bring the fluid it contained into contact with the tissues. Three to six weeks after implantation of the capsule the parts were incised and examined. The ends of the capsule were always found to contain a few drops of thin yellow pus, which, under the microscope, showed all the characteristic appearances of this fluid. The surrounding tissues were not affected. Cultivation experiments yielded negative results. When the capsules were broken, the ptomaïnes came in contact with the leucocytes which had accumulated around the foreign body, and by their local toxic effects transformed them into pus corpuscles, while no evidences of local infection were present, which shows conclusively that the transformation of the leucocytes into pus corpuscles was accomplished by the ptomaïnes. In about twenty experiments the pus was found only inside the capsule. In the cases in which the tube had been implanted for eight weeks, the conditions remained the same. Weigert has repeatedly shown that the difference between a purulent and fibrinous exudation can be readily demonstrated, as the former does not coagulate, although white corpuscles and plasma may be present. Klemperer believes that this is due to a previous destruction of fibrinogen by the microörganisms. The putrid meat infusion used by Scheuerlen caused limited suppuration, and on that account it must also have possessed the property to prevent coagulation. To prove this, he made the following experiment: The abdomen of a rabbit was opened while the animal was under the influence of chloroform, and blood was drawn directly from the aorta into a glass tube containing putrid extract of meat. As the fluids gradually became mixed, the blood assumed a brownish-red color; coagulation did not take place after hours and days, while in the control experiments, with solution of salt, the blood coagulated firmly after the lapse of a few minutes. Microscopical examination of the mixture of blood and putrid fluid showed no fibrin and no rouleaux of red blood-corpuscles. Both the red and white corpuscles were contracted and corrugated. He next made thirty cultures of the staphylococcus pyogenes aureus on agar-agar gelatin, and the same number of cultures of the albus, and after completion of their growth, fourteen days later, he sterilized them with boiling water, and after shaking the fluid removed the cultures and boiled them for a few minutes, and finally filtered them; and thus obtained about 150 c.c. of a light yellow fluid. This was reduced to 8 c.c. by boiling; when used, the fluid was again filtered. The filtrate was put in capsules and introduced into the subcutaneous tissues of animals in the same manner as in the preceding experiments. The suppuration which followed was again found to be limited to the inside of the broken glass capsules, the same as in the experiments with the sterilized putrid meat infusion. The cadaverin and putrescin, two ptomaïnes prepared by Brieger, were next experimented with in the same manner. In preventing coagulation the results were even more marked than with the former substances. Both also produced the same localized suppuration in the interior of the broken glass capsule. Cadaverin has lately been produced chemically by Ladenberg from trimethylen-cyanuret as pentamethylendiomin.

These experiments leave no doubt that ptomaïnes exert a chemical influence on leucocytes and embryonal tissue, which transforms these tissues into pus corpuscles. The suppuration which is thus produced, however, never extends beyond the tissues which are brought in direct contact with them, and, therefore, always remains circumscribed. In this respect the experiments just cited do not correspond with suppuration as we meet it at the bedside, as here from the same causes, and apparently under the same conditions, the process presents the greatest variations in reference to its intensity and extent. In one case the suppuration remains circumscribed, resulting in a furuncle; in others the regional infection is more extensive, and a diffuse phlegmonous inflammation is the result; while in a third class, the local infection leads to general invasion, and the patient dies of sepsis or pyæmia. The clinical forms are noted for the progressive character of the infection, which is due to the multiplication of microörganisms within the body, and the production of ptomaïnes proportionate in amount to the number of microbes present. Practically, the matter remains the same as before it was known that ptomaïnes could cause suppuration, as pus-microbes must be introduced into the organism before ptomaïnes can be produced, and for the practical surgeon it is immaterial to know whether suppuration is the direct or indirect result of the presence of pus-microbes. Scientifically, however, this question has an important bearing, and has again awakened interest in the question:

Can Suppuration be Produced by Chemical Irritants?

Grawitz and de Bary produced by subcutaneous injections of turpentine, variable results in different animals. Injections of turpentine, with or without pus-microbes, produced inflammation, but no suppuration in rabbits and guinea-pigs; while in dogs the same injections invariably caused suppuration. Croton oil was found to possess no influence in retarding the growth of pus-microbes. Injection experiments demonstrated that this substance in small quantities in the connective tissue of rabbits caused a serous or fibrinous exudation, while larger doses acted as a caustic, and were only occasionally followed by suppuration. Injections of a mixture of pus-microbes and croton oil always caused suppuration. They maintained that certain chemical substances, used in a definite degree of concentration, injected into the subcutaneous tissue of animals, prepared the tissues for the growth of pus-microbes.

In a later series of experiments on the production of suppuration Grawitz ("Beiträg zur Theorie der Eiterung," Virchow's *Archiv*, B. cxvi. S. 116) obtained similar results, and still maintains that aseptic turpentine, when introduced in sufficient quantity into the tissues, causes suppuration. Inoculations of different nutrient media with such pus showed that it was sterile. He maintains that turpentine does not destroy pus-microbes. He also found that pus produced by turpentine injections had a distinctive effect on pus-microbes. This action of sterile pus he attributes not to the presence of ptomaïnes but to the action of its albuminous constituents. His experiments led also to the important observation that when gelatin cultures are over-saturated with albumin or peptone, pus-microbes will no longer grow upon them.

Councilman ("Zur Aetiologie der Eiterung," Virchow's *Archiv*, B. xcii. S. 217–230) introduced turpentine and croton oil in aseptic glass capsules into the connective tissue of animals, and after the wound had healed and the capsules had become encysted, ruptured them subcutaneously. He observed that these substances caused a circumscribed suppuration.

Uskoff ("Giebt es eine Eiterung, unabhängig von niederen Organismen?" Virchow's *Archiv*, B. lxxxv., 1881) found by his experiments that a considerable quantity of indifferent substances, such as milk, olive oil, etc., if injected subcutaneously in animals, either at once, or by repeating the injection from time to time, caused suppuration, and that turpentine administered in this manner always acted as a pyogenic agent.

Orthmann (" Ueber die Ursachen der Eiterbildung," Virchow's *Archiv*, B. xc. S. 544–554), under Rosenbach's supervision, repeated Uskoff's experiments, and by resorting to most strict antiseptic precautions could not verify the correctness of his conclusions in regard to the pyogenic properties of indifferent substances. His experiments with croton oil, turpentine, and metallic mercury resulted in inflammation and suppuration. Cultivation experiments with pus thus produced showed that it was sterile.

Scheuerlen (" Die Entstehung und Erzeugung der Eiterung durch chemische Reizmittel," 1887) made a series of very carefully conducted experiments, and came to the conclusion that aseptic substances never produced suppuration. He modified Councilman's method by substituting capillary glass tubes for the glass capsules, which were rendered perfectly aseptic, and were then charged with from one to four drops of turpentine, after which they were sealed in the flame of a spirit lamp. They were inserted into the tissues through a long, hollow needle, in order to avoid the necessity of making an incision. After the glass was put in the proper place the needle was withdrawn, and the puncture closed with iodoform collodion. After the puncture was healed the glass tube was broken underneath the skin. He extended his investigations to other irritating substances aside from turpentine, but in no instance was the experiment followed by suppuration.

Quite recently, Grawitz and de Bary (" Ueber die Ursachen der subcutanen Entzündung und Eiterung," Virchow's *Archiv*, B. cviii. S. 67–103) again upheld the theory that suppuration can be produced by irritating substances independently of microörganisms. Among the many important conclusions drawn from their numerous experiments may be mentioned the following:

1. Weak solutions of nitrate of silver, 0.5 per cent., if administered subcutaneously, are absorbed; strong solutions, 5 per cent.

in dogs produce suppuration, in guinea-pigs only inflammatory swelling.

2. To a number of chemical agents, such as concentrated solution of salt, acids, etc., a considerable quantity of a culture of staphylococcus can be added without suppuration following the injection.

3. In rabbits and guinea-pigs, even a large quantity of turpentine can be injected without causing suppuration. In dogs, turpentine injection into the subcutaneous tissue acts as a pyogenic substance *par excellence*. These experimenters affirm that the microbes of suppuration alone cannot produce pus in the subcutaneous tissue in dogs and rabbits. On the other hand, they claim that chemical substances, mycotically pure, can, under certain conditions, produce suppuration, and must, when used in proper doses and concentration in the right kind of animals, produce suppuration without fail.

Nathan (*Archiv f. klin. Chir.*, Bd. xxxvii. S. 875) makes an interesting contribution on the etiology of suppuration in view of the fact that Grawitz and de Bary assert that ammonia, silver nitrate, and turpentine injected subcutaneously into the tissues of dogs produce a suppuration in which no microbes can be found. He used dogs in his experiments; the point of injection was shaved and carefully disinfected with sublimate 1 : 1000; the substance used was, after careful sterilization, injected by means of a needle thoroughly purified by heat. During the whole experiment the field of operation was irrigated with sublimate solution. It was found that abscesses did at times, but not invariably, appear as a result of these injections. Plate cultures always showed that these abscesses contained microörganisms, though both cover-glass preparations and test-tube cultures failed to demonstrate them. The development of microörganisms was explained by the theory that by constant licking the dogs infected the puncture.

Janowski ("Ueber die Ursachen der acuten Eiterung," Ziegler's *Beiträge zur path. Anatomie*, Bd. vi. Heft 3) experimented with sterilized oil of turpentine on dogs and rabbits, producing by subcutaneous injections suppuration in the former animal in a few days, while in the rabbits the suppurative process was delayed four to six weeks. With caustic ammonia he obtained only negative results, which led him to assert that when this substance caused suppuration in the hands of other experimenters, it must have been contaminated with pus-producing microbes.

P. Kaufmann ("Ueber den Einfluss des Digitoxins auf die Enstehung Eitriger Phlegmone," *Archiv f. Exp. Pathologie*, Bd. xxv. S. 397) has shown experimentally that digitoxin when injected subcutaneously into the tissues of animals produces suppuration independently of pus-microbes. In dogs this result followed injections in doses of from one-half to one-third of a milligramme dissolved in alcohol.

From a clinical aspect, the absence of pus-microbes in non-purulent inflammatory products speaks strongly in favor of a microbic cause of suppuration.

Ruiys ("Ueber die Ursachen der Eiterung," *Deutsche med. Wochenschrift*, 1885, No. 48) made some exceedingly interesting experiments on the pyogenic action of different substances, selecting the anterior chamber of the eye as the seat for injection. The results of his experiments were such that he claimed, in a most positive and emphatic manner, that suppuration never takes place without micoörganisms.

Biondi expressed himself to the same effect.

If we think for a moment how difficult it is in experimenting on animals with indifferent substances and chemical irritants to procure for the seat of injection a perfectly aseptic condition, it is not difficult to conceive that opinions still differ in regard to the immediate cause of suppuration. At the same time, Watson Cheyne has shown most conclusively in his article on "Suppuration and Septic Diseases," to which frequent allusion has been made, that the number of bacteria introduced greatly modifies not only the intensity of symptoms, but also the character of the disease. His experiments were made with cultivations of Hauser's proteus vulgaris. He estimated that $\frac{1}{16}$th c. c. of an undiluted cultivation of this microbe contains 225,000,000 of bacteria, and when this quantity was injected into the muscular tissue of a rabbit it produced speedy death. $\frac{1}{40}$th c. c. administered in the same manner caused an extensive abscess at the point of injection, and death of the animal in six or eight weeks. Doses of less than $\frac{1}{300}$th c. c. produced no effect—in fact, doses of less than $\frac{1}{12}$th to $\frac{1}{120}$th c. c., or, in other words, fewer than about 18,000,000 bacteria, seldom caused any result. The same observer found that in the case of staphylococcus pyogenes aureus that it was necessary to inject something like 1,000,000,000 cocci into the muscles of rabbits, in order to cause a rapidly fatal result, while 250,000,000 produced a small abscess. In the case of the tetanus bacillus, death did not occur in rabbits when fewer than 1000 bacilli were introduced. He believes that the preformed ptomaïnes in these cases alter the result. It is therefore quite possible that, in the experiments in which injection of pus-microbes did not produce suppuration, an insufficient number of cocci were injected to produce the desired result, and that where inert substances and chemical irritants caused suppuration the injected material was contaminated, or that infection at the point of injection occurred through the wound or subsequently through the circulation. The latter mode of infection should always be borne in mind where the presence of an aseptic body in the living tissues has apparently been the cause of suppuration. The tissues altered by the action of chemical irritants constitute a foreign substance which may determine localization of floating microbes, while, at the same time, the chemical alterations which they have caused in the tissues have prepared a favorable soil for their reproduction. Practically, in man, suppuration without microörganisms is only possible on the surface of the body, as the products of a suppurative inflammation in any of the internal organs always show the presence of pyogenic microbes which can be cultivated, and it is only rational to conclude that the inflammation and subsequent suppuration were caused by pyogenic microörganisms or their products, the ptomaïnes.

Pus-microbes in Different Suppurative Affections.

Direct infection with pus-microbes can only take place through the cutaneous and other accessible surfaces or through wounds. Suppuration in the interior of the body, in the absence of a wound or other recognizable infection-atrium, must be considered in the light of an auto-infection with pyogenic microbes.

1. SUPPURATIVE AFFECTIONS OF THE SKIN.—Longard ("Ueber Folliculitis abscedens infantum," *Archiv f. Kinderheilkunde*, Bd. viii. Heft 5, 1887) has made a careful microscopico-bacteriological examination of nine cases of furunculosis in young children. In four of these cases he found the staphylococcus pyogenes albus alone, in five cases in combination with the staphylococcus pyogenes aureus; the identity of these microbes with those described by Rosenbach was demonstrated by cultivations and experiments on rabbits. The microbes were not found in the fecal discharges of the patients, but were found, in small numbers, in the diapers of healthy unclean children, as well as in the diapers of those suffering from folliculitis. He believes that the pus-microbes were the direct cause of the affection, and that the infection took place through the sweat-glands, as the microbes were found in abundance upon the inner surface of the membrana propria of these appendages of the skin. As soon as they reached the subcutaneous connective tissue they produced suppurative inflammation.

Experiments on dogs and rabbits, by cutaneous inoculations with pus-microbes cultivated from the furuncles, produced a slight swelling and redness, and in some instances the formation of small pustules. The result was always the same whether the pus was taken from the cultivation grown from a furuncle, a suppurating wound that healed without fever, or from a pyæmic patient. The cutaneous inoculation experiments of Garrè, Bockhart, and Bumm upon themselves have been previously referred to, and they prove that many of the circumscribed suppurative affections of the skin are caused by direct inoculation with pus-microbes which enter the connective tissue, either through a slight abrasion or through the glands of the skin.

2. SEROUS CAVITIES.—A most interesting investigation of the conditions under which infection of the peritoneum can take place has been made by Grawitz ("Statistischer und experimentell pathologischer Beitrag," *Charité Annalen*, Bd. xi. S. 770). From his own experiments he came to the conclusion that the injection into the healthy peritoneal cavity of schizomycetes, pyogenic as well as non-pyogenic, produced no unfavorable results. Peritonitis was caused only when the microbes were mixed with a caustic fluid, or when the peritoneal cavity contained a fluid which could serve the purpose of a nutrient medium; also, when the amount of fluid injected exceeded the absorbing capacity of the peritoneum; or, finally, if the peritoneum was injured at the same time, or the abdomen was opened by a penetrating wound. Strange as it may appear, he claims that the injection into the peritoneal cavity of intestinal contents caused no peritonitis as long as the punctured wound remained aseptic.

Leyden ("Ueber spontane peritonitis," *Deutsche med. Wochenschrift*, 1884, p. 212) demonstrated the presence of streptococci in the exudations of cases apparently spontaneous peritonitis.

Garrè ("Bacteriologische Untersuchungen von serösen Trans und Exsudaten und Atheromen," *Corresbl. f. Schweizerärzte*, 1886, No. 17) examined carefully for the presence of microörganisms in hydrocele fluid, serous, peritoneal, and pleural effusions, and the contents of joints the seat of serous synovitis, etc., by means of microscopical examination and cultivation experiments, always with negative results.

The same author ("Bacteriologische Untersuchungen des Bruchwassers eingeklemmter Hernien," *Fortschritte der Medicin*, B. x. S. 486–490), in order to test the observations of Nepveau, according to whom the transuded fluid in a hernial sac constantly contains bacteria, examined eight cases of incarcerated hernia. Although the microscope was used together with delicate reagents for microörganisms and cultivations in gelatin, bacteria were found in only a few instances. The fluid in the sac of non-incarcerated herniæ examined by these methods was found absolutely sterile. Analysis of all the cases examined demonstrated that the length of time the incarceration has existed has no significance as regards the presence of bacteria; since, in recent cases, positive results were obtained, while in a case in which incarceration had existed for eight days the result was negative. Furthermore, the odor of the fluid is no indication of the presence of bacteria, as these may be absent even in fluid having a fecal odor. In cases in which the bowel becomes gangrenous, bacteria pass through its walls and death is caused by septic peritonitis.

Fränkel ("Ueber puerperale Peritonitis," *Deutsche med. Wochenschrift*, 1884, p. 212) has found the streptococcus pyogenes in a great variety of puerperal diseases; especially in cases in which the local affection implicated the lymphatic vessels. In such cases, the microbes found entrance into the pelvic tissues from abrasions or ulcers in the vagina, and by extension of the inflammatory process the broad ligaments and the peritoneum are successively reached; after the peritoneum has once been reached, rapid diffusion takes place, and finally the diaphragm and pleura are implicated in the same process, and the microbes reach the blood and cause sepsis and pyæmia.

Weischelbaum (*Centralblatt f. Chirurgie*, August 17, 1889) has shown that peritonitis is not always caused by pus-microbes as has been heretofore believed; he has found the diplococcus of pneumonia unaccompanied by any other microörganism in three cases of peritonitis. In one case peritonitis and acute pneumonia existed at the same time; in the other, double pleuritis followed the peritonitis; but in the last case the peritonitis was undoubtedly primary and in the absence of any other microbes in the products of the inflammation must have been caused by the diplococcus of Friedländer.

Orth (*British Medical Journal*, March 1, 1890) has shown that the pathogenic properties of pus-microbes are strongly modified by certain preëxisting pathological conditions, although large doses of pure cultures of the staphylococcus and streptococcus injected into the peritoneal cavity of rats failed to cause any lesion of the peritoneum, the same microbes in the same doses caused fatal results when mixed with material which could only be absorbed slowly. Preëxisting disease of the peritoneum favored the action of the microbes, in ascitic animals a very small quantity of a culture of staphylococcus caused septic peritonitis. The same result followed when any intra-abdominal structure was wounded. These experiments show the great danger which may follow infection of the peritoneum after laparotomy, especially if fluids or solid particles are allowed to remain after the operation.

Fränkel (*Berl. klin. Wochenschrift*, May 14, 1888) made a bacteriological study of twelve cases of empyema. In three cases in which no special cause could be traced the pus contained exclusively the streptococcus pyogenes. In three cases the pus contained only pneumococci. Other authors have found, in such cases, also other pus-microbes. Fränkel thinks that when this is the case, they have been deposited in consequence of a secondary invasion. The presence of streptococci in the pus from a suppurating pleural cavity presents nothing characteristic, as the microbe is also found in cases in which the empyema is secondary to pneumonia and tuberculosis. On the other hand, he assigns to the pneumococcus in pus removed from a pleural cavity a diagnostic significance, as it proves, beyond all doubt, that the suppurative pleuritis occurred in the course of a pneumonia as a secondary affection, consequently its presence in the pus is positive proof that a pneumonia exists, or has existed, even if the clinical and physical symptoms were not sufficiently clear to indicate its existence. In four cases the empyema had a tubercular origin, in two of which pneumothorax existed at the same time. The presence of the bacillus of tuberculosis in the pus is not easy to demonstrate, but the absence of this microbe is no sign that the disease is not tubercular, as inoculations with such pus in animals almost constantly produce typical tuberculosis. In the pus of tubercular pyo-pneumothorax, if microörganisms are present, the bacillus of tuberculosis can be found, and the pus shows no tendency to undergo putrefactive changes, in contradistinction to empyema occurring in non-tubercular subjects, in whom spontaneous discharge through the bronchial tubes takes place.

In the discussion on Fränkel's paper, Senator maintained that putrefaction is prevented by the parenchyma of the lungs acting as a filter, preventing ingress of bacteria with the inspired air, and by the presence of a large amount of carbonic acid gas in the air of the cavity, as it is well known that microbes do not thrive as well in such an atmosphere as in ordinary air. Fränkel believes that the absence of putrefaction in such cavities is due to the fact that few, if any organisms, except tubercle bacilli, are present in pus, and these do not cause putrefaction. It seems that tubercular pus does not furnish a favorable soil for the growth of other germs.

Ehrlich (*Berl. klin. Wochenschrift*, May 14, 1888) has made a bacteriological study of the pus in nineteen cases of empyema; in only seven of these could the bacillus of tuberculosis be found; in the remaining twelve this microbe could not be found, and upon this negative ground the existence of tuberculosis was excluded. Further observation in these cases after operation corroborated the diagnosis. He asserts, therefore, that in the purulent pleuritic exudation in tubercular patients, in empyema, and pyo-pneumothorax, the presence of the specific microbic cause can always be demonstrated. In a case of pneumo-hydrothorax he failed to find the bacillus until the effusion had undergone transformation into pus, when its presence could be readily demonstrated. Some of the pus-corpuscles contained as many as twenty bacilli. He places the greatest importance on a bacteriological examination of the pus as a means of differential diagnosis between suppurative and tubercular empyema.

3. PURULENT ARTHRITIS.—Reference has already been made to the bacteriological researches of Garrè, who never found bacteria of any kind in the serous effusions of joints. Suppurative synovitis, in an intact joint, is always caused by localization of pus-microbes in the synovial membrane, where their presence excites a purulent inflammation. In this manner the metastatic suppurative synovitis, as it occurs in pyæmia, in some cases of gonorrhœa, and in some of the general infective diseases, is caused. In animals susceptible to purulent infection, the injection into a joint of a pure culture of pus-microbes is usually followed by destructive purulent inflammation, and, not infrequently, by the formation of extensive para-articular abscesses.

Hoffa ("Bacteriologische Mittheilungen aus dem Laboratorium der chirurgischen Klinik des Prof. Maas," Würzburg, *Fortschritte der Medicin*, B. iv. S. 75), Kranzfeld ("Zur Aetiologie der acuten Eiterungen," St. Petersburg, 1886, *Centralblatt f. Chirurgie*, 1886, p. 529), and Krause ("Ueber acute eitrige Synovitis bei kleinen Kindern und über den bei dieser Affection vorkommenden Kettencoccus," *Berl. klin. Wochenschrift*, 1884, No. 43) have studied with special care the bacteriological origin of suppurating joints in small children a streptococcus, the identity of which with the one described by Rosenbach was proved by cultivation experiments. In one case the same microbe was also found in the products of a purulent meningitis, which followed in the course of the joint disease. The same streptococcus was found by Heubner and Bahrdt ("Zur Kenntniss der Gelenkeiterung bei Scharlach," *Berl. klin. Wochenschrift*, 1884, No. 44) in pus from a suppurating joint, and in the diphtheritic membranes of a scarlet fever patient. Clinical experience and experimental research appear to prove that purulent synovitis occurring independently of osteomyelitis is, in the majority of cases, caused by the streptococcus pyogenes.

4. ACUTE SUPPURATIVE OSTEOMYELITIS.—Acute suppurative inflammation in bone, when it occurs independently of an external wound, and consequently of direct infection, furnishes one of the most interesting, and, thanks to the patient and persevering investigations of a number of the foremost pathologists, one of the best known forms of purulent infection. For years it has been contended by some who made the etiology of acute osteomyelitis the subject of experimentation, that it is caused by a specific microbe not found in other forms of suppuration. Convincing evidence, however, has been accumulating for a number of years which seems to leave no further doubt that the ordinary microbes of suppuration are the cause of this form of suppurative inflammation, and that the

gravity of the symptoms which attend the disease, as compared with other suppurative processes, is owing to the anatomical location and structure of the inflamed tissues, rather than to any difference in the microbic cause.

Rosenbach ("Vorläufige Mittheilung über die acute Osteomyelitis beim Menschen erzeugenden Microörganismen," *Centralblatt f. Chirurgie*, 1884, p. 65), as early as 1881, cultivated the staphylococcus from osteomyelitic pus. In one case the yellow and the white staphylococcus were found combined, in another case the staphylococcus albus alone, while in a third case the staphylococcus aureus and the streptococcus pyogenes were found present together. Rosenbach produced the same result in his experiments by injection of a pure cultivation of pus-microbes from a furuncle of the lip, as Struck did with cultivations from the pus of osteomyelitis, and with osteomyelitic pus injected into the subcutaneous connective tissue he produced an ordinary abscess. Recurrent attacks of osteomyelitis years after the primary disease, he explains by assuming that after the first attack some of the microbes are left in the tissues, and remain in a latent condition until at some subsequent time local conditions are created which enable them again to display their pathogenic properties. Ogston found the staphylococcus in the pus of a case of acute osteomyelitis.

Struck ("Ueber eine im Kaiserlichen Gesundheitsamt ausgeführte Arbeit, welche zur Entdeckung des die acute infectiöse Osteomyelitis erzeugenden Microörganismus gefühl that," *Deutsche med. Wochenschrift*, 1883, No. 46) obtained from the pus of an acute case of osteomyelitis upon gelatin, an orange-yellow culture; the identity of this cultivation with the staphylococcus pyogenes aureus was soon generally recognized. By injecting a pure culture into the circulation of animals which had been subjected a few days before to injury of bone, as contusion or fracture, he produced a suppurative inflammation at the seat of trauma.

Even before the microbic cause of acute osteomyelitis was understood, Kocher ("Die acute Osteomyelitis mit besonderer Rücksicht auf ihre Ursachen," *Deutsche Zeitschrift f. Chirurgie*, B. xi. S. 87) believed that the infection, in some cases at least, occurred through the intestinal canal, and made some experiments to prove this point. In dogs he produced subcutaneous fractures, and then fed them large quantities of putrid material, and, in some cases, succeeded in producing suppuration at the seat of injury. In his clinical experience he also observed that in many cases of acute suppurative osteomyelitis the premonitory symptoms pointed to the gastro-intestinal canal as the *portio invasionis*.

Krause ("Ueber einen bei der acuten infectiösen Osteomyelitis des Menschen vorkommenden Micrococcus," *Fortschritte der Medicin*,

1884, Nos. 7, 8) cultivated from osteomyelitic pus the staphylococcus pyogenes aureus and albus, which he also found in the effusion in joints, when this occurred as a complication of the disease. Injection of a pure culture of these cocci into the peritoneal cavity of animals caused suppurative peritonitis. Intravenous injections, with or without previous fracture, were followed most frequently by suppuration in joints and muscles. If a bone was fractured subcutaneously before the injection, he frequently observed suppuration at the seat of fracture, and from the pus the staphylococcus could again be cultivated. Foci in the kidneys were always present in all of these experiments.

Garrè ("Zur Aetiologie acut eitriger Entzündungen," *Fortschritte der Medicin*, 1885, p. 165) corroborated by his own experimental work the observations made by Rosenbach, and, in addition, he showed that in acute suppurative osteomyelitis the staphylococcus is also present in the blood.

Müller (" Die acute Osteomyelitis der Gelenkgebiete," *Deutsche Zeitschrift f. Chirurgie*, B. xxi. Hefte 5 u. 6) succeeded in cultivating the staphylococcus pyogenes aureus from the yellow granulations in cases of acute epiphysary osteomyelitis.

Rodet's (" Etude experimentelle sur l' osteomyelite infectieuse," *Compt. rend.*, 1884, No. 14) researches deserve special mention, as he succeeded in producing suppurative osteomyelitis in animals without inflicting a trauma before or after the infection. This result could only be obtained by resorting to intravenous injections. The purulent inflammation, which was generally circumscribed, was usually located near the epiphysis; it seldom extended over a considerable portion of the shaft. In many cases epiphyseolysis occurred, and very frequently a suppurative arthritis of the adjacent joint. In the most acute cases, the animals died within twenty-four hours without any appreciable changes in the bones. The detection of the microbes in the blood was the most difficult; they were found most readily in the kidneys, in which often multiple abscesses were found. Subcutaneous inoculation resulted in local suppuration; osteomyelitis could not be produced in this manner. Young animals were more susceptible to inoculations. According to Rodet, the osteomyelitic cultures lose their virulence after thirty to forty days. In one of his experiments, which he details very minutely, he employed a culture of the thirteenth generation, and produced epiphyseal osteomyelitis of both femora, of one tibia, and of one humerus. As the result of his observations, he locates as the primary seat of osteomyelitis of the long bones the medulla in close proximity to the epiphyseal line. When separation of the epiphysis was observed, the pathological fracture always occurred on the side of the diaphysis.

Ribbert ("Die Schicksale der Osteomyelitis-coccen im Organismus," *Deutsche med. Wochenschrift*, 1884, No. 24) made investigations to ascertain the extent of diffusion of the osteomyelitic cocci in the organism. Twenty-four hours after direct injection into the circulation he found them in all the organs, later only in the kidneys. In regard to their localization, the following conditions must be taken into consideration:

a. Embolic obstruction of capillary vessels.
b. Elimination of pus-microbes through the kidneys.
c. The influence of traumatism.

Lübbert (*Biologische Spaltpilzuntersuchung. Der Staphylococcus pyogenes aureus und der Osteomyelitis-coccus.* Würzburg, 1886) has studied the effect of the staphylococcus pyogenes aureus, the microbe most frequently found in osteomyelitis in the different tissues. From his experiments he came to the conclusion that the intensity of its action varies greatly without a sufficient cause for it being known. Inoculations with it of superficial abrasions produced no effect. Subcutaneous inoculations resulted in the formation of abscesses which at times became quite diffuse. Inoculations of granulation surfaces proved harmless. Injections into the pleural and peritoneal cavities were oftenest followed by intense general symptoms. Injections into the trachea through a tracheotomy wound produced suppurative tracheitis and foci in the lungs. Intravascular injections were followed by symptoms indicative of sepsis. Foci were also found in the intestinal mucous membrane. Feeding experiments proved harmless.

Kraske ("Zur Aetiologie und Pathogenese der acuten Osteomyelitis," *Archiv f. klin. Chirurgie*, B. xxxiv. S. 701) has studied from a clinical standpoint, the manner of infection in cases of acute osteomyelitis. In one case he could trace the infection distinctly to a furuncle of the lip; but, as a rule, he thinks that infection takes place through a wound or abrasion of the skin. Infection through the intestinal canal he considers possible, but not proven; more frequently it takes place through the respiratory organs, and in one case he could locate the infection here with certainty. He asserts that recurring attacks should not always be looked upon as the result of former infection, but as a consequence of a new infection. He formulates the result of his clinical studies as follows:

a. The staphylococcus pyogenes aureus can produce osteomyelitis, and, in fact, is most frequently met with in the osteomyelitic products.

b. In a certain number of cases acute osteomyelitis is the result of a mixed infection and is then most prone to pursue a severe course.

c. It is possible that the result of further investigation will show that every microörganism that possesses pyogenic properties is capable of causing a typical osteomyelitis in man.

Rinne ("Der Eiterungsprocess und seine Metastasen," *Archiv für klin. Chirurgie*, B. xxxix. S. 21), who failed in producing metastatic abscesses with pure cultures of pus-microbes, rendered four rabbits pyæmic by injecting osteomyelitic pus directly into the venous circulation. He used pus later from a case of acute osteomyelitis with grave symptoms, and diluted it with distilled water, and of such a mixture he injected a syringeful into one of the auricular veins of four rabbits. One died in twenty-four hours with symptoms of toxæmia, and the autopsy showed nothing but a beginning pneumonia of the left lung. The other three animals died seven to ten days after the injection, and in all of them the necropsy showed suppurating foci in the kidneys and the heart muscle. No abscesses in muscles or suppuration in joints. The plate cultures made from the pus used for the experiments showed the staphylococcus aureus albus and bacillus pyocyaneus. With the exception of the albus, all of the microbes were also cultivated from the pus of the metastatic abscesses in rabbits.

In a later examination (*Ibid.*, p. 271) the same author expresses the opinion that the indirect causes of suppurative osteomyelitis are changes brought about in the medullary tissue by the microbes and their ptomaïnes of general febrile diseases, such as typhus, scarlatina, diphtheria, etc., which prepare the soil for the action of pus-microbes, or the disease is produced by direct extension from a localized suppurative lesion, as a furuncle, through lymphatic vessels or along vessel sheaths or nerve trunk to the medullary tissue.

The structure and location of the capillary vessels in the vicinity of the epiphyseal cartilage in young persons determine the localization of pus-microbes in this part of the long bones, and, almost without exception, the inflammatory process starts from here.

The rapid local diffusion of the process is largely due to the unyielding nature of the tissues around the primary focus, and to the fact that the bloodvessels are directly concerned in the extension of the process by becoming the channels for the dissemination, their contents forming the nutrient medium for the pus-microbes. Thrombo-phlebitis is a constant and early condition in every case of acute osteomyelitis. Coagulated blood is an excellent culture substance for the pus-microbes, and it serves the double purpose of a nutrient substance, and a medium for the local spread of the disease. General dissemination and metastatic foci in distant organs, or in other bones, are often observed because the pus-microbes re-enter the vascular system again, and by so doing cause a coagulation-necrosis of the intima and thrombosis; and subsequently intravascular growth and general dissemination from such centres of germ growth take place.

In some cases even during the earliest stages the general symptoms are out of all proportion to the local lesion, presenting a clinical picture characteristic of intense septic intoxication. It is very possible that the ptomaïnes produced by the pus-microbes in the medullary tissue of bone may be more virulent or produced in larger quantities than in suppurative inflammations in other organs. Again, the ptomaïnes gain here more ready entrance into the circulation, as, at least, part of them are produced within the bloodvessels and the extravascular products are forced rapidly into the circulation on account of the unyielding nature of the tissues around the primary focus of inflammation.

CHAPTER IX.

GANGRENE.

GANGRENE, resulting from mycosis of the tissues, is caused by one of three well-defined conditions:

1. The microbes are so numerous in the capillary vessels that their presence interferes mechanically with the blood supply, and death of the part ensues in consequence of greatly diminished or suspended nutrition.

2. The microbes in the tissues produce ptomaïnes which destroy the tissue by their direct destructive chemical action on the protoplasm of the cells.

3. The specific inflammation caused by the microbic infection is so intense that the inflammatory products in the paravascular tissues accumulate so rapidly, and in such abundance, that nutrition is suspended by impairment or suspension of the arterial blood supply or mechanical interference with the venous return of the blood from the part, or both of these conditions combined.

For these reasons no one variety of microbes can be the sole cause of gangrene. In its different forms, different microbes will be found. In cases of inoculation anthrax, when at the point of inoculation an abundant growth of the bacillus takes place, the connective-tissue spaces and bloodvessels become so blocked with the bacilli that circulation is mechanically arrested, and a circumscribed gangrene is the result. In the progressive gangrene which Koch produced artificially in rabbits by subcutaneous inoculation of putrid fluids, the gangrene always occurred in advance of the line of microbic invasion, and must, on that account, have been caused by the local toxic effects of the ptomaïnes. In phlegmonous inflammation, when the process is very acute and diffuse, gangrene frequently follows as one of the consequences of the inflammation, and the microbe which is found in the gangrenous part is the same as that which caused the inflammanition.

Tricomi ("Il micro-parasitica della gangrena senile," *Rivista internazionale di Medicina e. Chirurgia*, 1886) has found a slender, long bacillus in the blood of patients suffering from senile gangrene. The same bacillus was also found in the secretions of the gangrenous part at the line of demarcation in the lymph spaces and in the subcutaneous connective tissue beyond the seat of gangrene. He cultivated the microbe successfully upon gelatin, agar-agar, blood serum, and potato. The bacillus was readily stained with red and blue aniline dyes, and often a spore could be seen either at one of its extremities or near its centre. Injection of one-half to one gramme of a pure culture grown upon gelatin was injected under the skin over the back of a guinea-pig, house-mouse, or rabbit, and death was produced in from two to three days. A form of gangrene resembling senile gangrene was always found at the point of inoculation, but no changes in the blood or other organs were observed. Gangrene could also be produced in other animals by inoculating the secretions from the gangrenous part. The researches of Arloing and Chaveau are of the greatest importance in the elucidation of the etiology of gangrene as it is sometimes observed in connection with septic infection. These observers have published the results of some recent investigations on the pathology and prophylaxis of gangrenous or gaseous septicæmia of man. It is the prevailing opinion that this disease is a surgical complication, of which the exclusive cause is the introduction of a specific microbe into a wound. In man the microbe exists in the connective tissue which surrounds the wound, and in the contents of bullæ, which may be developed in its neighborhood. The microbe is a short, thick, mobile rod, of homogeneous structure, or else provided with one spore, rarely two, at one of the extremities. When the bacillus appears in the blood at the end of this disease, and sometimes only after death, its size is smaller than in the local lesion, and it may appear in the form of a micrococcus. A number of animals, such as the horse, ass, sheep, pig, dog, cat, guinea-pig, white rat, rabbit, duck, and other fowls, have been successfully inoculated, but the rabbit is not very susceptible to this disease, in this respect presenting a marked contrast with another form of septicæmia. The most prompt way of producing the disease was found to be by subcutaneous injection. The smallest dose capable of causing death when injected into the connective tissue, never proved fatal when injected into the veins or arteries; in the latter case, only a temporary intoxication was produced. If a large dose was injected into a vein, death ensued, with well-marked septic infection of the serous membranes. It was found difficult, if not impossible, to produce the disease by feeding-experiments. Attempts to inoculate a healthy wound involving the skin, connective tissue, and muscles were unsuccessful. On the contrary, the microbes of this disease find a favorable soil in dead tissues not exposed to air. Protection from the disease is afforded in sheep and dogs by the successive injection of two or three moderate doses of the virus into the circulation. Inoculation into the subcutaneous connective tissue of animals thus protected, gave rise to a simple circumscribed phlegmonous inflammation which ran its course in eight days. The pus from such an abscess may contain the specific microbe, and may act as the original virus. Experiments showed that this disease could be transmitted from mother to fœtus. The bacillus could be destroyed by heat alone if exposed to a temperature from 90° to 100° C. (194° to 212° F.) for a quarter of an hour. In the dried state the virus is extremely resistant. However, it can be destroyed after immersion in a hot bath (120° C., 248° F.) in from ten to fifteen minutes. There can be no doubt that the bacillus described here is a saprophyte, and that it is identical with one of the bacilli saprogenes described by Rosenbach.

Brigadier-Surgeon Godwin (*British Medical Journal*, July 23, 1887) reports a case of progressive gangrene with emphysema, starting from an inflamed corn, which extended with such rapidity that a week after it commenced it had extended so high as to necessitate amputation in the thigh. The man suffered from grave septic intoxication, the symptoms of which persisted for a number of days after the amputation, until the ptomaïnes had been eliminated. Culture experiments with the pus showed the presence of streptococcus pyogenes and staphylococcus pyogenes albus.

William Koch ("Milzbrand u. Rauschbrand," *Deutsche Chirurgie*, Lieferung 9) states that in a case of progressive gangrene with emphysema in a young man, he found a bacillus which resembled the bacillus of glanders, and proved its identity by cultivation and inoculation experiments.

Ciarrocchi (Virchow u. Hirsch's *Jahresbericht*, B. xi. S. 642, 1888) describes a case of metastatic gangrene of the skin, caused by infection through a lacerated wound of the last phalanx of the little finger. Ten days after the accident the patient became ill, complaining of chilly sensations, which were followed by fever and thirst. The next day yellowish-white spots appeared on the chest, which were diagnosticated as gangræna cutanea. Suspecting a causal relation between the injured finger and the gangrene of the skin, cultures were made from the necrotic tissue upon gelatin and the following microbes were isolated: staphylococcus pyogenes aureus, staphylococcus cereus albus, and a brown culture in which a short bacillus was found. Inoculation experiments could not be made as the material was lost, but the results of the microscopical examination agreed with those of Demme, which he obtained from gangrenous spots in the skin of five children affected with erythema nodosa. The latter observer also isolated the two varieties of pus-microbes and a short, fine bacillus. This bacillus, when injected under the skin of animals, produced gangrenous spots and a nodular efflorescence.

Jaffe and Leyden (*Deutsches Archiv f. klin. Medicin*, B. ii. S. 488) found leptothrix threads in gangrenous foci of the lung, and attributed to them the cause of the gangrene; but their presence only proved the possibility of the entrance of foreign bodies from the mouth into the lungs, and, while they played no essential part in the causation of the gangrene, the observation shows in what way other and more deleterious microörganisms can enter the lungs with the air through the bronchial tubes.

Bonome (*Deutsche med. Wochenschrift*, 1886, No. 52) examined nine cases of gangrene of the lungs in man, and found in three of them staphylococcus pyogenes aureus alone; in five staphylococcus pyogenes alone, and in one both microbes together. He made a number of experiments for the purpose of producing gangrene in the lungs of rabbits, by injecting pus-microbes into the circulation, but the results were negative if only the cultivations were used. He obtained positive results by mixing with the cultivations particles of the pith of elder, and injecting this mixture into the jugular vein. The result was multiple small abscesses, and gangrene of the lung. Injection of the pith fragments alone caused no effect. The pith fragments in these experiments determined an intense inflammation in the peri-bronchial tissues by causing alteration of tissue, thus preparing the soil for the localization and pathogenic action of the pus-microbes. The gangrene of the lung in man does not follow in consequence of the croupous inflammation caused by the pneumococcus, but its occurrence indicates that secondary infection has taken place with one or more varieties of pus-microbes, or with saprophytes.

In compound fractures of the limbs by direct violence, gangrene, with or without emphysema, is a comparatively frequent complication when infection takes place, as the death of the tissues from the trauma, and the nature of the wound, furnish the most favorable conditions for the growth of pus-microbes and the bacilli of putrefaction.

CHAPTER X.

SEPTICÆMIA.

SEPTIC processes were among the first to excite interest in the part played by microörganisms in disease, and it is due to this fact that so much more has been said and written on septicæmia than on any other microbic disease in surgery. Although some of the best pathologists have been diligently investigating this subject for years, we still remain in the dark concerning its true etiology, and its relation to other infective processes. True sepsis is looked upon as a general infection from some local source, unattended by any gross pathological changes. Some writers have claimed the difference between septicæmia and pyæmia to be a quantitative and not a qualitative one, while others maintained that pyæmia was a specific disease *sui generis*, and that it was in nowise related to sepsis. They resemble each other so far, that both are caused by microorganisms.

HISTORY.—The first reliable investigations into the microbic origin of sepsis were made by Rindfleisch in 1866, and somewhat later by Klebs, Recklinghausen, Waldeyer, and Hueter. Rindfleisch found bacteria in abscesses, while the researches of Klebs (*Beiträge zur pathol. Anatomie der Schusswunden*, Leipzig, 1872) initiated a new era in the etiology of septic diseases. The latter author differentiated between septicæmia and pyæmia, although he claimed that putrid and septic infection were the same. He found in the tissues altered by septic processes, also in the lymph spaces and in the blood, a microbe, a round coccus, isolated and in groups, which he termed *mikrosporon septicum*. Rosenbach (*Microörganismen bei den Wound-infections Krankheiten des Menschen*, Wiesbaden, 1884) in three cases of septicæmia which he subjected to bacteriological examination, found the staphylococcus pyogenes aureus present each time in the pus; in two of these cases he isolated and cultivated from the products of septic inflammation the bacillus saprogenes. In two of the cases no cultivation could be obtained from the blood. In two cases of gangrene, with general septicæmic symptoms, the microbe found was the streptococcus pyogenes. Intoxication symptoms from the introduction of putrid material, he attributed to the presence of one or more varieties of the bacillus saprogenes, which he designates, respectively, Nos. 1, 2, 3. No. 1 he cultivated from putrid blood, and this he believes to be quite harmless, while the remaining two possess pyogenic properties.

No. 1. Large rods, which, cultivated on nutrient agar-agar, form an irregular sinuous streak, with a mucilaginous appearance. They grow also very readily on blood-serum, and all cultivations yield the odor of rotting kitchen-refuse. It is not pathogenic.

No. 2. Rods shorter and thinner than No. 1. They develop very rapidly on agar-agar, forming transparent drops, which become gray. They were isolated from a patient suffering from profusely sweating feet. The cultivations yielded a characteristic odor similar to the last. They are pathogenic.

No. 3. Rods isolated from the putrid marrow of a case of compound fracture, cultivated on agar-agar; an ash-gray, almost liquid culture is developed, with a strong characteristic odor of putrefaction. Injected into the knee-joint or abdomen of a rabbit, they cause suppurative inflammation.

According to Chaveau ("Septicemie gangreneuse," *Publ. de l'Acad. de Méd.*, No. 34, 1884), the microbe of septicæmia is identical with the *vibrion septique* described by Pasteur. Both of these authors claim that this microbe is anaërobic, and Chaveau only succeeded in cultivating it in a vacuum. He made many experiments on guinea-pigs, sheep, and horses, by injecting the liquid contents of bullæ which he found in cases of septic gangrene. In doses of one-fifth of a drop in guinea-pigs, and from two to four drops in horses, it produced rapid death. In all cases the necropsy showed at the point of injection localized œdema and turbid serum in the peritoneal, pleural, and pericardial cavities. In the fluids, the microbe could always be demonstrated under the microscope. The disease could be reproduced in other animals by inoculation with the serous fluid contained in any of the serous cavities. The microbe proved less virulent when injected directly into the circulation. All animals which recovered after intravenous injection were protected against any further subcutaneous inoculations.

Gautier and others have more abundantly proved that not only after death, but even during life, the animal organism, by virtue of its physiological powers, is able to elaborate the alkaloids to which the name of leucomaïnes has been applied, bodies which are, many of them, essentially toxic in their properties, and which resemble so closely the poisonous cadaveric alkaloids to which Selmi first called attention.

Watson Cheyne ("Report on Micrococci in Relation to Wounds, Abscesses, and Septic Processes," *British Medical Journal*, 1884, pp. 553, 559, 645) asserts that the microbes of sepsis only grow *in loco*, and act by producing toxic ptomaïnes, or if they occur in the blood, they do not make emboli; they are not always cocci, sometimes rods.

At the last meeting of the German Medical Congress (*Berl. klin. Wochenschrift*, 1888, No. 18) Jürgensen read a paper on kryptogenetic septico-pyæmia, in which he referred to 100 cases of this disease which had come under his own personal observation, and in which it was impossible to locate the source of infection. The microbes found were either streptococci or staphylococci, or both together, in the same patient. He stated that the streptococcus circulated in the blood, while the staphylococcus produced local processes.

Vidal (*Gaz. hebdom.*, No. 22, 1888) has reported to the Académie de Médecine de Paris the results of his studies of the "*forme septicemique pure*" in puerperal fever, of typhoid type without suppuration. In all of his cases he found the streptococcus pyogenes, and from this, and the result of his culture and inoculation experiments, he comes to the conclusion that it is impossible, in the present state of our knowledge, to distinguish between the various forms of streptococci, and that one and the same form can set up any of the various forms of puerperal infection.

Besser (*St Louis Medical and Surgical Journal*, 1888, No. 2) states that he has examined 22 additional cases of traumatic septicæmia, and found streptococcus in every one of them. During the patient's life he discovered the microbe (*a*) in blood, in 4 of 16 cases examined; (*b*) in pus or fluid discharge from the primary focus, in 17 of 17; (*c*) in urine, in 3 of 4, and (*d*) in sputa, in 3 of 3; while after death the microörganism was present (*a*) in blood, in 7 of 15; (*b*) in organs, in 16 of 18; and (*c*) in pus or uterine discharges, in 12 of 12. In 6 of 22 staphylococci were simultaneously detected, side by side with masses of bacteria of many other species. In 3 cases, however, the streptococcus alone could be found. The author supposes that septicæmia is produced solely by the streptococcus. The microbe itself penetrates into the organism but very seldom.

From the above historical consideration it becomes evident that the essential cause of septicæmia has as yet not been demonstrated. The streptococcus pyogenes has been found most frequently in the products of septic inflammations, but whether any of the pus-microbes alone are capable of producing true sepsis remains to be demonstrated by future research. As the introduction into the circulation of the products of putrefaction is followed by a complexus of symptoms which closely resemble septicæmia, and as different microbes have been cultivated from septic patients, it would seem that this disease can be produced by any of the microbes which, after their introduction into the organism, have the capacity

PLATE III.

Micrococcus growth in the capillaries of the lungs in a case of puerperal sepsis, after septic thrombosis of the internal spermatic vein.
 a. Filling the capillaries.
 b. Rupture of one of them.

to produce a sufficient quantity of poisonous ptomaïnes to give rise to progressive septic intoxication.

ARTIFICIAL SEPTICÆMIA IN ANIMALS.—In the latter part of the seventeenth century, Kircher and Leuwenhoek claimed that putrid substances contained minute microscopical worms which caused the putrefaction. Perty and Naegeli assigned to the minute organisms a vegetable origin instead of animalculæ, as had been previously done, and they were classified under the name of schizomycetes. In 1857, Pasteur made the important discovery that specific agents are the cause of the various forms of fermentation and putrefaction. No discovery, perhaps, attracted such universal and deep attention as Pasteur's theory of fermentation. This theory was strengthened somewhat later by Lemaire's observation, that all fermentative changes in fluids are suspended on the addition to the fluids of phenic acid, from which he concluded that fermentation must be due to living organisms. As for a long time all septic affections were supposed to be caused by a process of fermentation and putrefaction, these theories led to a diligent search for microörganisms in the fluids of septic patients, and to experimentation on animals with putrid substances. Koch, in his great work on wound-infective diseases, published in 1878, described two distinct varieties of septicæmia, one in mice and the other in rabbits. In fifty-four infected mice he found small bacilli in the interior of the white blood-corpuscles, and also in the capillaries after they were set free by the destruction of the blood-corpuscles. They were also found in the serous cavities and in the lymphatic glands and vessels. The bacillus of mouse septicæmia was found very difficult to cultivate, and Koch first succeeded in cultivating it upon a composition of gelatin with the aqueous humor of the eye of the ox. The growth, however, was very feeble, and successive cultivations upon the same soil were uncertain. Later, Loeffler succeeded better with a nutrient medium composed of infusion of meat, to which were added 1 per cent. of peptone and 0.6 per cent. of common salt, the whole rendered faintly alkaline with sodium phosphate. The bacilli appeared upon this clear and transparent soil as opacities upon its surface. In rabbits he found large oval micrococci free in the capillaries. The progressive character of septicæmia was well shown by Koch and Davaine in rabbits, as the latter could cause rapid death by injecting a single drop of a mixture prepared by adding to a quantity of blood of a rabbit which had died inoculated with a twenty-fourth generation, diluted with one trillion times its quantity of pure water.

Darwin believed that septicæmia could be produced in a more and more virulent form with every successive inoculation from animal to animal by the intensity of the virus being increased by passing through different media. Buchner even went so far as to claim that perfectly harmless bacteria might thus be made to assume pathogenic qualities. Koch and his pupils took a most decided stand against such mutability of form, or action of any bacteria.

Coze and Feltz (*Recherches expérimentales sur la présence des infusoires dans les maladies infectieuses*, Strassburg, 1866) produced sepsis in animals by injecting putrid substances directly into the circulation, and could not only demonstrate the presence of bacteria in the blood, but were able to propagate the disease from one animal to another.

Gaffky ("Experimentell erzeugte Septicæmie," etc., *Mittheilungen aus dem Kaiserl. Gesundheitsamte*, B. i. S. 80) investigated Davaine's septicæmia experimentally. He procured the infection by using water from a stagnant river, and, by continually controlling the experiments with the microscope, using Koch's methods, and working only with pure cultures, he was able to prove beyond a doubt that the theories of progressive virulence of bacteria were untenable. He showed that the highest degree of virulence was already attained in the second generation. He also proved that the wrong conclusions were due to impurification in the experiments, and that when the proper precautions are taken in the process of sterilization to prevent the admixture of other microörganisms, the introduction of one kind always produces in the same animal the same definite specific result. The most interesting conclusions to be drawn from the experiments in Koch's laboratory, point to the fact that septicæmia is only a general term which includes a number of morbid processes, and this is well illustrated by the injection into the tissues of the "vibriones septiques" of Pasteur. Surface inoculation with these bacilli produced no effect, their pathogenic influence became only evident after injections into the subcutaneous connective tissue. Gaffky found that this bacillus grows most readily upon potato. Koch applied to the condition produced by this bacillus the term "malignant œdema." A minute quantity of these bacilli, taken from a second potato—that is, a second generation—injected under the skin of a guinea-pig proved fatal on the second day with all the usual signs of malignant œdema and septic intoxication. This bacillus is not only contained in stagnant water, but can also be obtained from garden earth.

Sternberg ("Induced Septicæmia in the Rabbit," *American Journal of the Medical Sciences*, July, 1882) produced marked septicæmia in rabbits by injecting subcutaneously his own saliva in small doses. Injections of 1.25 to 1.75 c.c. with few exceptions produced death, usually within forty-eight hours. The constant and characteristic lesion found was a diffuse cellulitis, or inflammatory œdema, extending in all directions from the point of injection, attendant with an abundant exudation of bloody serum swarming with micrococci. Hemorrhagic extravasations in the connective tissue and in various organs were of frequent occurrence, and changes in the liver and spleen such as are common to quickly fatal septic diseases, were commonly found. The disease could be communicated by dipping a hypodermic needle in the blood of a rabbit just dead as the result of an injection of saliva and inoculating a healthy rabbit; a rapidly-fatal septicæmia was produced. Ogston states that in cases of septicæmia in man, micrococci are present in the blood and are excreted in a living state in the urine. This statement has been confirmed by Eiselsberg, who examined the blood of almost all cases in Billroth's clinic which were suffering from septic fever, and was able to demonstrate the presence of staphylococci and streptococci, most frequently of staphylococcus pyogenes albus, in the blood, and yet apparently no abscesses formed.

Smith (*Annals of Gynecology*, vol. ii. No. 12) isolated and cultivated from two cases of puerperal sepsis a streptococcus which by inoculation and cultivation experiments differed from the streptococcus of Fehleisen and the ordinary streptococcus of suppuration. He made a series of gelatin cultures with blood taken from the heart. After an interval of two or three days many colonies appeared. Rats inoculated with a pure culture died in three or four days, the microbe being found in their blood. Inoculations were also made in the ears of rabbits, and at the end of twenty-four hours a circumscribed redness without tendency to diffusion was apparent, the redness disappearing in two or three days. Another series of cultures and inoculations was made with blood taken from the finger of a woman sick with puerperal fever, and this produced similar results.

Dowdeswell ("Report on Experimental Investigations on the Intimate Nature of the Contagion in certain Acute Infective Diseases," *British Medical Journal*, July 10, 1881) has made numerous experiments to determine the nature of Davaine's septicæmia of rabbits and Pasteur's septicæmia of guinea-pigs. The former were made by injecting subcutaneously five drops of putrid ox-blood. If infection was produced, which was generally the case, death followed within forty hours; characteristic textural changes in organs were not found. One drop of blood from infected animals caused the death of a second animal. Blood from the second animal, diluted from 10 to 100,000 times, produced a fatal effect in from twenty-four to twenty-seven hours. Blood from the sixth generation was diluted ten million times, and still produced a fatal sepsis on being injected subcutaneously; while a drop diluted one hundred million times produced septicæmia in some animals and a localized suppuration in others. The microbe appeared in the blood as small rods usually arranged in pairs. Pasteur's septicæmia of guinea-pigs was induced by injecting into the peritoneal cavity a few drops of putrid blood pure, or mixed with diluted ammonia. Serum, present in the abdominal cavities of animals which had died of this disease, in quantities of 0.005 to 0.022 c.c. injected into the peritoneal cavity of a second animal produced death from sepsis, while one-tenth of the quantity required to induce fatal sepsis proved harmless. The virulence of this form of septicæmia is, therefore, less than that of Davaine's rabbit septicæmia. The bacillus which was found in the serous fluid resembled the bacillus of malignant œdema described by Koch. The bacillus when cultivated upon a solid nutrient medium produced spores. In all of the experiments on animals with septic microörganisms, a certain interval of time elapsed between the inoculation and the first appearance of symptoms indicating the presence of septicæmia. Another constant feature of artificial septicæmia, produced by the introduction of septic blood, products of the septic process, or cultivations from either of these, is that the symptoms became more intense as the disease progressed; both of these facts are positive proofs that the active agents which caused the septicæmia are reproduced in the body, and that the beginning of the disease takes place as soon as a certain amount of virus has been formed in the body, and the intensity of the symptoms is proportionate to the quantity of infective material circulating in the blood—in other words, septicæmia caused by the introduction into the organism of living septic germs is noted for its progressive character, resembling, in this respect, perfect true septicæmia as we observe it at the bedside; differing thus entirely from another clinical form of septicæmia which is caused by the introduction into the circulation of preformed toxic substances, and which has been designated by Mathews Duncan as sapræmia. This form of septicæmia I will illustrate by reference to the experiments which have been made for the purpose of studying

Septic Intoxication in Animals by the Introduction of Putrid Substances.

That putrid substances injected directly into the circulation produce symptoms of septic intoxication has been known for a long time, and the extensive researches of Panum threw additional light on this subject. It was believed that putrid material when introduced into the organism induced a process of fermentation to which were attributed the most constant post-mortem appearances found in septicæmic subjects—fluidity of the blood and softening of the tissues. That these changes were not necessarily caused by the action of living microörganisms was determined by experiments, as the introduction of putrid blood, or meat infusion that had been boiled for a considerable length of time, produced toxic symptoms, and when a sufficient quantity was used, death and identical pathological changes in the blood and tissues as in cases of true sepsis. A step in advance in the study of the action of putrid substances was made by the discovery of the ptomaïnes in an exhumed body by Selmi in 1872. The ptomaïnes isolated by Selmi were volatile alkaloids which were separated from a body some time after death. Gautier, independently of Selmi, and about the same time, made the same observations, but believed that the toxic substances were volatile, and that in their action they resembled the narcotics morphine and atropine, and were more closely allied to the alkaloid extracted from poisonous mushrooms.

Semmer ("Putride Intoxication und septische Infection, Metastatische Abscesse und Pyæmie," Virchow's *Archiv*, B. lxxxiii.) gives an account of the action of septic substances as studied experimentally by Guttmann in the pathological department of the Veterinary School at Dorpat. The experiments were made with putrid substances, products of inflammation, septic blood, and cultivations of septic bacteria. These researches showed that a chemical, putrid poison is formed in putrefying substances, and that a certain quantity of such poison produces symptoms of sepsis and death in animals. The blood of animals killed with such putrid poison was found to possess no infective qualities, and the usual putrefactive bacteria are destroyed in the blood, and only appear again after the death of the animal. It was claimed, even at this time, that the bacteria elaborate the poison, as experiments made with cultivations grown outside the body produced the same effects. Another conclusion arrived at was that putrid substances administered subcutaneously may produce gangrene, phlegmonous inflammation, or erysipelas, according to the stage of the putrefaction, temperature, culture-soil, etc. The infective material was never found in the blood, but always in the products of the inflammation. A sharp distinction was made between contagious septicæmia and putrid intoxication. It was clearly stated that true septicæmia is always preceded by a stage of incubation, and that its contagium is destroyed by boiling, putrefaction, and germicides.

Bergmann (*Das putride Gift*, etc., Dorpat, 1866) isolated from putrid blood a crystallizable chemical substance which he called sepsin, which when injected into animals produced a complexus of symptoms resembling true sepsis, with this important difference, however, that as soon as the toxic substance reached the circulation the maximum symptoms were observed, and if the animal recovered from the immediate effects of the intoxication, it recovered showing that the disease induced resulted from the introduction of a preformed poison, and on this account was not progressive. Later, Hiller (" Die Lehre von der Fäulniss," *Centralblatt f. Chirurgie*, 1876) produced a similar affection with ferments. Bergmann and Angerer (*Das Verhältniss der Ferment-Intoxication zur Septicæmie, Festschrift zur Feier des* 300 *jährigen Bestehens der Jul. Maximilian Universität*. Würzburg, 1882) produced a condition in animals resembling septicæmia by injecting into the circulation pepsin and pancreatin. When death occurred after intravascular injections of these ferments, fibrinous deposits were found in the heart and pulmonary vessels: these experiments were, therefore, confirmatory of the observations previously made by Edelberg and Birck, who had shown that the injection of putrid substances into the circulation materially increased the free fibrin ferment in the circulating blood.

Blumberg (Virchow's *Archiv*, B. c. Heft 3) concluded from his numerous experiments on animals that the symptoms which follow an injection of putrescent material into the circulation are not always constant; that, in fact, extreme prostration, high temperature, rapid pulse and respiration, are the only constant symptoms found. The same author also confirmed the statement that the blood of patients dying from putrid intoxication contained no microörganisms.

Samuel (*Archiv f. Exp. Pathologie u. Pharmacie*, i. 817) believes that putrid fluids from the second day until the eighth month act differently, and divides their action, according to this supposition, into three stages: 1, phlogogenic, in which they produce only inflammation; 2, septogenic, in which they produce in the living organism putrefactive processes; 3, pyogenic, in which they cause only suppuration, having lost in the meantime their other qualities.

Mikulicz (" Ueber die Beziehungen des glycernis zu Cocco-bacteria septica u. zur septischen Infection," *Archiv f. klin. Chirurgie*, B. xxii.) found that putrid fluids, according as they are free from bacteria, or contain more or less of putrefactive germs, will produce a slight inflammation, a suppurative inflammation, or a progressive phlegmonous inflammation.

Bergmann (*Verh. der Deutschen Gesellschaft für Chirurgie*, 1882) advances the idea that in cases of septicæmia the microörganisms enter the colorless blood-corpuscles and by multiplication cause their dissolution, during which the fibrin-generators are liberated, the process ending in intravascular coagulation and capillary embolism. In Koch's mice-septicæmia such a chain of pathological conditions can be readily demonstrated, but in many cases of fatal sepsis in man the microörganisms in the blood are few, no destruction of leucocytes can be demonstrated, extravasations and capillary embolism are absent, hence death cannot be attributed to fibrin intoxication. In such instances we can only assume the presence of a soluble ptomaïne, which is diffused throughout the entire body and destroys life by its toxic properties.

Fräukel (" Ueber Microörganismen der chir. Infections-Krankheiten," *Wiener med. Wochenschrift*, 1885, B. xxxv.) found but few micrococci in the blood of septicæmic patients, and observed that they greatly increased after death; but after the lapse of some further time, altogether disappeared, thus also confirming a fact previously known, that putrefaction destroyed septic germs. These observations may tend to harmonize the discrepancy of opinion growing out of the different results obtained by different experimenters by injection of putrid substances, as some of the fluids may have contained an abundance of living microörganisms, while others may have been rendered sterile by age, owing to advanced putrefaction.

Brieger (" Giftige Producte der Fäulniss-bacterien," *Berliner klin. Wochenschrift*, 1884, No. 14) and Maas ("Fäulniss-alcaloide," *Fortschritte der Medicin*, 1883) have rendered valuable service in the chemical isolation of ptomaïnes from putrid substances, and the results of their inoculation experiments established more firmly the fact of putrid intoxication by ptomaïnes. The number of bacteria in rabbits killed by septic infection is so great, that death may ensue from simple mechanical causes, while, in fatal cases of sepsis in man, the number is often so small that it seems natural to suppose that the microörganisms are capable of producing some poisonous substance which destroys the patient before they have time to multiply to the extent observed in the septicæmia of rabbits and mice.

Hauser (*Ueber Fäulniss-bacterien und deren Beziehung zur Septicæmie*, Leipzig, 1885) succeeded in isolating three kinds of schizomycetes from a putrid meat solution, which he called, respectively, proteus vulgaris, mirabilis, and Zenkeri. He claimed that all of them changed their form during their growth, appearing at different times as cocci, long and short rods, vibriones, spirilli, etc. (Fig. 5.) The variety in form, he claims, was influenced by the nature of the culture substance. All these bacteria are putrefactive agents and the proteus vulgaris and mirabilis are most frequently present. The experiments were made by removing organs, or part of organs, from animals immediately after death, and placing them in sterilized vessels, where they were inoculated with pure cultivations of the proteus. A sterilized emulsion of boiled meat and eggs, inoculated with a culture of the three kinds of proteus, was transformed into a putrescent mass in a short time. That this change was not caused by a preformed putrefactive ferment was proved by the fact that when the decomposed mass was filtered through clay cells the filtrate did not produce the same effect. A small amount of the putrescent fluid injected into the tissues of animals produced intense symptoms of intoxication and sometimes death within an hour.

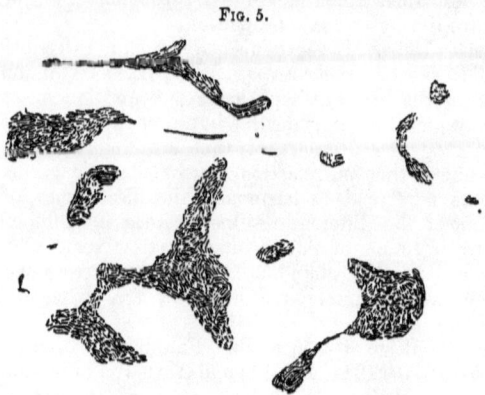

Fig. 5.

Proteus vulgaris, 285 : 1. (Hauser.)

Injections into the subcutaneous tissue of rabbits of a pure culture of the proteus vulgaris and mirabilis not infrequently caused extensive abscesses. An alcoholic extract of the putrid meat proved less toxic, but a large quantity also produced sepsis and death.

Rosenberger ("Experimentell Studien über Septicæmie," Centralblatt f. d. med. Wiss., 1882, No. 4) studied experimentally on rabbits the effect of injections of putrid blood. In one series of experiments he simply injected the putrid material, in another he boiled the fluid before injecting it. The result was the same—typical septicæmia, only that the animals infected with the boiled material required a larger dose and did not succumb so rapidly to the sepsis. Microscopical examination of the blood and culture experiments yielded the same results: the presence of bacteria and their growth upon culture media. He is of the opinion that in septicæmia bacteria are not the first or the essential etiological condition, but believes that under certain circumstances innocent bacteria which may exist in the body are transformed within twenty-four to forty-eight hours into specific septic bacteria. In his experiments he attributed the occurrence of sepsis to the introduction of the septic poison, boiled or unboiled. The most interesting proof that true progressive sepsis is not the result of the introduction into the circulation of the products of putrefaction, but of pathogenic germs, has been furnished by Tiegel (Dissertation, Bern, 1871). By injecting putrid fluids into animals he produced true progressive septicæmia. He then resorted to filtration of the fluid through clay cells, so as to separate from it any organic germs it might contain, and showed that fluid, thus treated, was rendered perfectly sterile, and that, when injected, it produced only putrid intoxication and no progressive sepsis.

Rinne ("Der Eiterungs process und seine Metastasen," *Archiv f. klin. Chirurgie*, B. xxxix. S. 21) asserts that the chemical product of pus-microbes alone, as well as sterilized putrid fluids, never produces metastases. He sterilized fluid cultures of the staphylococcus pyogenes aureus after filtration, and injected directly into the bloodvessels of rabbits as much as four grammes of this fluid, and in dogs increased the dose to fourteen grammes. Many of the animals showed slight symptoms of toxæmia, somnolence, diarrhœa, and collapse. By using still larger doses the symptoms were intensified and the animals died. Metastatic abscesses were never found.

Hoffa (*Verh. der Deutschen Gesellschaft f. Chirurgie*, 1889) has recently made some very interesting observations on the immediate cause of death in rabbits inoculated with a pure culture of Koch-Gaffky's bacillus. The animals were inoculated at the base of the ear, and immediately after death the ptomaïnes were isolated by Brieger's method. In every instance he obtained a substance called methylguanidin, which on chemical analysis was shown to consist of the formula $C_2H_7N_3$. When this substance was injected into rabbits it produced symptoms of intoxication which resembled in every respect those produced by the injection of the pure cultures obtained from septicæmic rabbits. As methylguanidin could not be produced from the cadavers of healthy animals by the same method, Hoffa naturally came to the conclusion that it was a product of the bacteria, and that death was to be attributed to the production of this substance in the tissues of the infected animals. The source of methylguanidin in the body is kreatin, and the bacteria must possess the property of oxidation, as kreatin is transformed into methylguanidin only by oxidation.

Septicæmia in man corresponds with the two forms produced, experimentally, in animals: 1. True progressive septicæmia, caused by the introduction of microbes into the tissues, where they multiply and later reach the blood; where mural implantation and capillary embolism and thrombosis take place, which directly interfere with the proper nutrition and function of important organs, and where the septic intoxication is caused by the formation of ptomaïnes in the organism. For this form of sepsis Neelsen (*Verhandl. der Deutschen Gesellschaft f. Chirurgie*, 1884), has suggested the name acute mycosis of the blood, to distinguish it from the second form—2. Sapræmia, putrid intoxication, or, as Neelsen terms it, *toxic mycosis of the blood*, in which few or no microbes are found in the blood, and in which death is due entirely to the presence of ptomaïnes. In the first variety it seems that the infection is generally due to the presence of pus-microbes which either reach the circulation directly by permeating the vessel wall, or enter indirectly through the lymphatic channels. The latter mode of infection gives rise to the most acute and fatal form of septicæmia. Sapræmia, or putrid intoxication, represents that form of septicæmia in animals in which a preformed toxic agent, as boiled putrid substances or a toxic alkaloid, is injected into the circulation, and in which the maximum symptoms are reached as soon as the poison has become mixed with the blood. This form of sepsis may be caused by any microbes, otherwise harmless, or only with slight pathogenic properties, as the bacilli of putrefaction, which cause putrefaction in any dead tissue, as, for instance, a blood-clot or contused tissue; and the symptoms arise as the ptomaïnes are absorbed, and are proportionate to the amount absorbed, and subside with the cessation of absorption and their elimination through some of the excretory organs.

Septico-pyæmia.

Septico-pyæmia is a condition in which the symptoms indicate the presence of both septicæmia and pyæmia, and in which the post-mortem appearances point to septic and purulent infection. Leube ("Zur Diagnose der spontanen Septico-pyæmie," *Deutsches Archiv f. klin. Medicin*, B. xxii. S. 335) first described this affection and called it spontaneous, because he was unable to trace the source of infection from without in the cases which came under his observation. Litten ("Ueber septische Erkrankungen interner Art, namentlich hämorrhagische Sepsis interna," *Zeitschrift f. klin. Medicin*, B. ii.), on the other hand, was always able, in his cases, to locate the infection-atrium, but the primary infection at the time acute symptoms appeared had either disappeared or its location could only be ascertained by a most careful examination. Jürgensen (*Berliner klin. Wochenschrift*, No. 18, 1888) calls it "Kryptogenetic Septico-pyræmia," as he was unable to find a tangible infection-atrium. He gave an account of one hundred cases which had come under his own personal observation. The patients were usually attacked first with an angina, and, as this stage was generally attended by a chill and a general feeling of malaise, the patients usually attributed it to a cold. In most cases the general infection was announced by a severe chill. Rapid loss of strength was one of the most prominent symptoms, so that in a very few days the patients became utterly prostrated. The symptoms which pointed to local processes during life were referred most frequently to the lungs, liver, spleen, pleura, heart, and the long bones. Whether the primary infection occurred through the pharynx, where the first symptoms were manifested, could not be definitely ascertained. In the acute cases, the symptoms were grave from the beginning and increased as the infection progressed, while in chronic cases, infection is maintained from some suppurating focus, and the disease may become prolonged for several years. Subcutaneous and retinal hemorrhagic extravasations were frequently observed. The post-mortem examinations revealed suppuration in some of the internal organs and vascular conditions which are found in cases of sepsis. These cases may be compared with acute suppurative osteomyelitis, where often the most careful inquiry and the most scrutinizing examination fail in furnishing reliable evidences for locating the primary source of infection. It is possible that the pus-microbes have gained entrance through an intact mucous surface, or through the skin, and that they have remained in a latent condition until a *locus minoris resistentiæ* is created somewhere in the body, where they localize in a soil prepared for their growth and multiplication, or, what is more likely the case, they entered through an abrasion or slight lesion, which may have been so insignificant that the patient himself failed to notice it, and produced no symptoms until, by accident or disease, a proper soil was prepared for the initiation of an acute attack in one or more of the internal organs.

CHAPTER XI

PYÆMIA.

The presence of pyæmia in over-crowded and badly-ventilated hospitals during the time before the antiseptic treatment of wounds came into use, gave rise to the general belief that the disease was due to a specific cause, and ever since bacteriology became a science diligent search has been made to demonstrate its specific microbic origin. Since the discovery of the microbes of suppuration, new light has been shed upon the etiology and pathology of this disease. Bacteriological studies of pyæmic products have shown that one or more kinds of pus-microbes are always present, thus establishing the direct relationship between a suppurating process in some part of the body, and the development of metastatic or pyæmic abscesses. Clinical experience has only corroborated the scientific investigations of this subject, inasmuch as it has shown that the frequency of its occurrence has been diminished in proportion to the lessening of suppurative inflammation in wounds under the antiseptic management of traumatic injuries and internal suppurating lesions. We are justified upon the basis of well-established facts in claiming that pyæmia is not a disease *per se*, but that its occurrence depends upon an extension of a suppurative process from the primary seat of infection, and suppuration in distant organs by the transportation of emboli infected with pus-microbes through the systemic circulation. The distant, or metastatic abscesses contain the same microbes which are found in the wound secretions or the abscess from which the general purulent infection took place. Experiments have shown that a culture of pus-microbes from a furuncle may produce pyæmia in animals, and that the microbes cultivated from a pyæmic abscess when injected under the skin of an animal cause only a localized suppurative inflammation without any general symptoms.

Artificial Production of Pyæmia in Animals.

Koch (*Untersuchungen über die Aetiologie der Wundinfections Krankheiten*, Leipzig, 1878) produced typical pyæmia in rabbits by injecting a putrid fluid, obtained by maceration of the ear of a mouse, into the subcutaneous connective-tissue in the inguinal region. A large abscess was found at the point of injection and numerous metastatic abscesses in the internal organs. In the pus of these abscesses he found a micrococcus which he considered as characteristic of this affection.

Klein (*Microörganisms and Disease*, 1885) described a micrococcus of pyæmia in mice. Certain cocci which were present in pork-broth proved fatal to mice in about a week, producing purulent inflammation at the point of injection and metastatic abscesses in the lungs. Fresh inoculations in mice again produced a fatal result with pyæmic symptoms.

Pawlowsky found that, by the simultaneous injection of sterilized cinnabar, and of cultivations of staphylococcus pyogenes aureus into the circulation, he produced abscesses in various organs—in fact, the typical picture of pyæmia. The presence of particles of foreign bodies rendered material aid in the development of metastatic abscesses, as the mere arrest of pus-microbes in the circulation without them, as a rule, is not sufficient of itself to lead to the production of true pyæmia. In rabbits, even the introduction of a large quantity of a culture of pus-microbes into the circulation does not produce pyæmia. Twenty-four hours after the injection the microbes may be found in large numbers in the pulmonary and other capillaries, but after forty-eight hours they have all disappeared from the circulation. If the cocci are suspended in an embolus, this latter, by producing alterations in the endothelia of the bloodvessels in which it has become impacted produces a *locus minoris resistentiæ* favorable to the growth of germs. In the experiments of Pawlowsky, the particles of cinnabar acted upon the endothelia lining the capillary vessels in the same manner as the fragments of a thrombus by impairing the local nutrition of the tissues at the point of impaction. If pyæmia is produced in guinea-pigs, or mice, with infectious pus, or with a pure cultivation of the same, the same local conditions are produced which invariably precede the development of pyæmia in man. Some of the veins at the seat of primary infection are invaded by pus-microbes and become blocked by a thrombus, this thrombus undergoes puriform softening, small fragments containing pus-microbes become detached, and are washed away and enter the general circulation as emboli, which, when they become arrested, establish independent centres of suppuration. In such cases the pus-microbes are present in the blood, in the tissues around the abscess, and in all purulent collections.

The Relations of Pus-microbes to Pyæmia in Man.

Rosenbach (*Microörganismen bei den Wundinfections Krankheiten des Menschen*, Wiesbaden, 1884) examined six cases of typical pyæmia with a view to determine the nature of the microörganisms present in pyæmic patients. He found the streptococcus pyogenes present in the blood and metastatic deposits in five of them; in two of these cases, staphylococci were also present, although fewer in number. In only one of them he found staphylococci alone, and this case recovered.

Pawlowsky ("Beiträge zur Aetiologie der Pyæmie," *Centralblatt f. d. med. Wissen.*, 1887, Nos. 24, 25) made a bacteriological examination of the pus of metastatic abscesses in five cases of pyæmia. In four cases he found staphylococcus pyogenes aureus, and in the fifth case, which was remarkable for the extent of the joint complications, he found the streptococcus pyogenes. He believes that the staphylococcus aureus is the usual cause of pyæmia, and especially of that form characterized by multiple abscesses in the internal organs. Large cultures of this coccus suspended in water and injected subcutaneously in rabbits caused death, and at the necropsy multiple abscesses were found. He believes that pyæmia in man occurs when disturbances in the circulation are present, so that floating cocci find places for localization within the bloodvessels. He produced these disturbances artificially by making intravenous injections of cinnabar, and ascertained that the presence of the granular material determined the localization of the microbes.

Besser, of St. Petersburg, writes in *Wratch* (*St. Louis Medical Journal*, May 2, 1888, Nos. 19 and 20) that he has examined, bacteriologically, blood, pus, and parenchymatous fluid from organs in 23 cases of pyæmia. In 8 cases, the staphylococci albi and aurei were found; in 14, the streptococci; and in 1, the streptococci and staphylococci simultaneously. The microbes were discovered, (a) during the patient's life in pus in every one out of 20 cases examined; in blood, in 11 of 12: and in parenchymatous serum, in 1 of 1; (b) after death, in pus, in 17 of 17; in blood, 4 of 9; and in organs, 9 of 14. Besser's predecessors described 23 additional cases of pyæmia, in 14 of which staphylococci were found, in 7 streptococci. Total, 46 cases: in 22 staphylococci, in 21 streptococci, in 3 both. Besser has also observed that the staphylococcus aureus could transform, under certain conditions, into staphylococcus albus, and *vice versa*. He was unable to discover the slightest difference between the microbes of suppuration and those of pyæmia.

Schüller ("Ueber Bacterien bei metastatischen Gelenkentzündungen," *Verh. der Deutschen Gesellschaft f. Chirurgie*, 1884) examined the contents of metastatic joint affections in twelve cases of puerperal pyæmia and invariably found streptococci, single and diplococci, but never bacilli.

Okinschitz (Dissertation, St. Petersburg, 1889) made the relationship which exists between the pus-microbes and pyæmia the subject of bacteriological investigation. He found that pyæmic blood invariably contained either the streptococcus pyogenes or staphylococcus pyogenes aureus, demonstrable by cultivation and ordinary microscopical examination. As the hæmic microbes seldom show any signs of fission, as compared with the bacteria at the primary focus, it is reasonable to infer that proliferation takes place mainly in the pus and not in the blood, hence the great importance of thorough disinfection and destruction of primary foci. The number of microbes in the circulating blood bears a direct relation to the gravity of the disease; if they are abundant even in the absence of metastases in internal organs the prognosis is grave, and if scanty even when metastatic foci exist the prospects of a favorable termination are better.

The occurrence of pyæmia from suppurating wounds or abscesses does not depend so much upon the kind of pus-microbes which have caused the suppuration as upon surrounding circumstances. The location and anatomical structure of the tissues in which the primary infection has taken place exert an important influence in the production of the disease. It is well known that suppurative inflammation of the medullary tissue in bone is exceedingly prone to give rise to pyæmia. Osteomyelitis, without direct infection through a wound, is always due to an intravascular infection—localization of pus-microbes in the capillary vessels of the medullary tissue. The microbes come first in contact with the endothelial cells after mural implantation has taken place, and the resulting coagulation-necrosis in the tissues of the wall of the bloodvessels leads to thrombosis. The products of the intravascular coagulation-necrosis furnish a most favorable nutrient substance for the growth and multiplication of germs, consequently the area of intravascular infection is rapidly increased. The growth of the thrombus in a proximal direction soon leads to extensive thrombo-phlebitis, and, as softening of the thrombus takes place, to embolism and metastatic suppuration. Pyæmia following a suppurative inflammation in a wound, or in the course of a phlegmonous inflammation in the connective-tissue, is the result of an extravascular infection. The pus-microbes coming first in contact with the outer coats of the veins, give rise to a phlebitis, which progresses from without inward, and which is followed by thrombosis as soon as the intima is reached. The intravascular dissemination of the pus-microbes then takes place in the same manner as in cases of intravascular infection after thrombo-phlebitis. Ordinary pyogenic microbes may and do cause pyæmia if they enter the blood attached to portions of blood-clot, or other solid materials, which after they have become impacted in bloodvessels by embolism, prepare the soil in distant organs for their localization and multiplication. The importance of thrombosis and embolism as factors in the causation of pyæmia has been clearly established by clinical observation and experimental research. Emboli may originate in the lymphatic vessels when these are the seat of invasion by pyogenic microbes, which, however, is very seldom the case. In chronic pyæmia, in which multiple metastatic abscesses are formed, embolism takes no essential part in the process, the microbes enter the circulation and are brought in direct contact by mural implantation with the tissues weakened by injury, or other debilitating causes. Experimental research has shown conclusively that the introduction of pus-microbes into the circulation is not necessarily or even usually followed by pyæmia, and their accidental entrance in the course of a suppurative inflammation is not always followed by serious consequences. There can be no doubt that some pus-microbes reach the circulation in nearly every case of suppuration, but their pathogenic action is prevented, or neutralized, by an adequate resistance on the part of the tissues with which they are brought in contact and their rapid elimination through healthy excretory organs. A limited number of pus-microbes injected into the circulation of a healthy animal, or accidentally introduced into the blood of an otherwise healthy person, are effectively disposed of by the white blood-corpuscles. If, however, the same number of microbes are present in combination with fragments of a blood-clot, the latter produce such alterations in the tissues surrounding them as to prepare the parts for their pyogenic action. The same happens if free pus-microbes localize in a part the vitality of which has been previously diminished by a trauma, or antecedent pathological changes, which constitute a *locus minoris resistentiæ* for the growth and reproduction of pathogenic microörganisms. Pyæmia, therefore, must be looked upon rather as a serious and fatal complication of suppurative lesions rather than an independent specific disease.

CHAPTER XII.

ERYSIPELAS.

HISTORY.—The contagiousness of erysipelas has been recognized for centuries, and on this account early attempts were made to include it among microbic diseases.

Nepveau (Virchow u. Hirsch's *Jahresbericht*, 1872, 1, p. 254) found micrococci in the blood of erysipelatous patients, and these were present in greatest number in blood taken from the diseased part.

Wilde (*Med. Jahrb.*, B. clv., Heft 1, S. 104) from his own investigations was able to corroborate these observations, but he also ascertained that the pus of wounds from which erysipelatous inflammation starts contains the same micrococci.

Orth (*Archiv f. Exp. Pathol. u. Pharmakol.*, B. i. S. 81) found micrococci in the contents of the bullæ of erysipelas. Recklinghausen and Lukomsky (Virchow's *Archiv*, B. lx. S. 418) found micrococci in the lymphatic vessels and the connective-tissue spaces in the structures affected by the virus of erysipelas.

Billroth and Ehrlich (Langenbeck's *Archiv*, B. xx. S. 418) found micrococci not only in the lymphatic vessels, but also in the bloodvessels of the inflamed skin.

Tillmanns (*Deutsche. med. Wochenschrift*, 1878, No. 17) found them in the skin, and Letzerich (Virchow u. Hirsch's *Jahresb.*, 1875, p. 69) in cases of erysipelas attacking vaccination-wounds, in the wound itself, in the bloodvessels, muscles, liver, spleen, and kidneys. Koch (*Investigations into the Etiology of Traumatic Infective Diseases*, London, 1880) described the specific organism of erysipelas as a small micrococcus of globular shape, united in pairs or forming short chains, and published photographic representations of them in his work, in which he also describes erysipelas in rabbits which he produced artificially by the injection, into the subcutaneous tissue of the ear, of mouse's dung softened in distilled water.

Fehleisen (*Die Aetiologie des Erysipels*, Berlin, 1883) was the first who, in 1883, discovered the essential cause of erysipelas and succeeded in cultivating the microbes on a number of nutrient media. From the appearance of the microbe and its direct etiological bearings to erysipelas, he called it the streptococcus of erysipelas. With a pure culture of this germ he produced by inoculation, not only only erysipelas in animals to prove its specific pathogenic qualities, but inoculations were also made in man for therapeutic purposes.

DESCRIPTION OF THE STREPTOCOCCUS ERYSIPELATOSUS.—Minute cocci, three to four micro-millimetres in diameter, arranged in chains, found in erysipelatous skin and in the fluid of erysipelatous bullæ. They occupy the lymphatic channels of the skin and spread along them as the disease advances. Each coccus when it is about to divide becomes larger and oval, and soon appears made up of two hemispherical masses, the two new cocci resulting from fission of the old one. Morphologically, the streptococcus of erysipelas and the streptococcus pyogenes are nearly identical, only that the coccus of erysipelas is somewhat larger, and both are somewhat smaller than the staphylococci.

CULTIVATION EXPERIMENTS.—The streptococcus of erysipelas can be cultivated upon gelatin or agar-agar. The appearances of cultures resemble very strongly those of streptococcus pyogenes. There is less tendency, however, to the formation of terraces, the margin is thicker and more irregular in outline, and the appearance of the growth is more opaque and whiter. Rosenbach mentions as another distinguishing feature between the two, that the culture of the coccus of erysipelas represents upon solid nutrient media the shape of a fern, while the outlines of the cultures of the pus streptococcus describe the shape of an acacia leaf.

The culture appears as a very delicate grayish-white film. The growth is very slow, and the individual colonies remain small. They do not liquefy gelatin.

INOCULATION EXPERIMENTS.—The characteristic erysipelatous blush is produced by inoculating these microörganisms into the ear of a rabbit.

Krause obtained positive results by inoculating gray mice. The animals died after three or four days, even when only a minute quantity of the culture was injected under the skin of the back. Passet inoculated white mice that had been liberally fed on bread, milk, and oats, and obtained only negative results. Of seven persons inoculated by Fehleisen ("Ueber die Züchtung der Erysipelcoccen auf künstlichen Nährboden und ihre Uebertragbarkeit auf Menschen." *Sitzungsbericht der Würzburger Physic. med. Gesellschaft*, 1882) the subjects of incurable tumors, with pure cultures, six developed typical erysipelas; in the seventh case, the patient had suffered from an attack of erysipelas only a few weeks previously, and was, in all probability, still protected against a new attack. This patient was inoculated a second time with a negative result. Several times a second inoculation failed after a

PLATE IV.

Chain cocci. 962 diam.
a. From erysipelas. (Fehleisen.)
b. From Phlegmon. (After Rosenbach.)

successful inoculation. The period of incubation was fixed at from fifteen to sixty-one hours. The microbe was only found in the lymphatic vessels and connective-tissue spaces, and when the culture was pure never produced suppuration.

Whitney ("Notes on Blood-changes in Erysipelas," *Philadelphia Medical Times*, 1883) claims that he found the streptococcus erysipelatosus in the blood in five out of six cases of erysipelas. Most all authorities who have studied the subject with the greatest care assert, however, that it is only found in the lymphatic vessels and never in the bloodvessels.

INOCULATION FOR THERAPEUTIC PURPOSES.—Fehleisen has seen by this treatment a cancer of the breast become smaller, a lupus disappear almost completely, while a case of fibro-sarcoma and another of sarcoma were not materially affected by this method of treatment.

Kleeblatt (*Münch. med. Wochenschrift*, March, 1890) reports the case of a lympho-sarcoma followed by infection of the cervical glands, in which the tumors diminished markedly in size under the influence of an intercurrent attack of erysipelas, but continued to develop after this had passed off. The patient was afterward intentionally inoculated with a pure culture of the streptococcus of erysipelas, but the effect was, as before, only a temporary one, as the tumors steadily increased in size, the patient dying of exhaustion. In another case of lympho-sarcoma of the neck, erysipelas was inoculated with good results, as the tumor was found to have disappeared on recovery from the disease. In a third case of lymphadenoma of the lower eyelid, the size of a pigeon's egg, this suppurated during an intercurrent attack of erysipelas, and afterward disappeared completely.

Janicke and Neisser ("Exitus letalis nach Erysipelimpfung bei inoperablem Mamma-carcinom und microscopischen Befund des geimpften Carcinoms," *Centralblatt f. Chirurgie*, 1884) have recorded a death from the erysipelas thus intentionally produced, in a case of cancer of the breast beyond the reach of an operation. A pure culture was used. At the post-mortem it was proved that the neoplasm had almost completely disappeared, and the microscopical examination of portions that had remained appeared to show that the tumor cells had been destroyed through the direct action of the microbes. Biedert (Vorläufige Heilung einer ausgebildeten Sarcomwucherung in einem Kinderkopf durch Erysipel," *Deutsche med. Zeitung*, 1886, No. 5) saw in a child suffering from a sarcoma involving the posterior part of the cavity of the mouth and pharynx, the left half of the tongue, the naso-pharyngeal space and the right orbit, the tumor disappear almost completely during an attack of erysipelas. Cases, on the other hand, have been reported in which after an accidental or intentional attack of erysipelas the malignant tumor commenced to grow rapidly, Neelsen ("Rapide Wucherung und Ausbreitung eines Mammacarcinoms nach zwei schweren Erysipel-Anfällen von 15 resp. 10 tägiger Dauer," *Centralblatt f. Chirurgie*, 1884, p. 729) describes a case of carcinoma of the breast, in which after two severe attacks of erysipelas the tumor not only commenced to grow faster, but at the same time regional infection progressed also more rapidly.

Babtchinsky (*Bulletin Médical*, 1890) made the accidental discovery that the microbe of erysipelas is a direct antagonist to the virus of diphtheria. His son, while suffering from a most severe case of diphtheria, was suddenly attacked by erysipelas. This complication, grave of itself, seemed to hasten the fatal termination of the case, and during the first few hours of the eruption the patient was much worse. But the next day the symptoms had much improved, and the patient made a rapid recovery. Following this indication, Babtchinsky inoculated a second case of diphtheria with a culture of the microbe of erysipelas grown on agar-agar, and with an equally happy result. Since this time, of fourteen cases of diphtheria treated with these inoculations, twelve resulted in recovery, and as in the two cases resulting fatally the inoculation produced no effect, these negative results only tend to confirm the efficacy of the curative inoculations. It is remarkable that in all the cases where erysipelas was produced artificially this disease pursued a mild course, and the patients recovered rapidly from both diseases.

Schwimmer ("Ueber dem Heilwerth des Erysipels bei verschiedenen Krankheitsformen." *Wiener med. Presse*, No. 15, 16, 1888) gives an account of 11 cases of lupus, in all of which no improvement was observed after an intercurrent attack of erysipelas. In a case of keloid an attack of erysipelas was followed by marked improvement, and in a case of lipoma a similarly favorable effect was observed. Syphilitic lesions he saw temporarily improved, while the erysipelas had no effect in permanently influencing the course of the disease.

Bruns (*Monatschrift f. prakt. Derm.*, B. viii., No. 4) gives an account of the effect of erysipelas on tumors in 22 patients. Amongst these, three cases of sarcoma were permanently cured. Two cases of multiple keloid after burns were also cured. In four cases of lymphoma of the neck some of the glands disappeared and some became smaller. In five cases the erysipelas was artificially produced by inoculation with a pure culture. In three cases of carcinoma of the mamma one was not changed, one became one-half smaller, and one was reduced to a small induration in the scar the size of a pea. A multiple fibro-sarcoma was greatly benefited, while an orbital sarcoma was not improved.

In view of the uncertainty of the result, and the danger which attends the intentional form of erysipelas, the danger of the disease being as great as in the accidental form, it is safe to predict that no further inoculations will be made in man until we shall have found a certain antagonistic action of the streptococcus of erysipelas against some pathogenic microbes which are the cause of some grave disease not amenable to less heroic measures.

MANNER OF INFECTION.—As the streptococcus of erysipelas produces its pathogenic effects in the lymphatic vessels and diffuses itself through these channels in the tissues, it becomes obvious that infection takes place as soon as localization is effected in the lymphatic structures, or in the spaces contributory to them. Before antiseptic surgery was practised, infection frequently occurred through accidental or intentional wounds. Even before the microbic cause of erysipelas was known, one of the closest of clinical observers (Trousseau) claimed that infection with the virus of erysipelas is only possible through some wound or abrasion of the skin; the latter may be so insignificant as to be unnoticeable, and entirely overlooked by the patient and physician. Inoculation experiments have shown that the time of incubation is from fifteen to sixty-one hours, so that we can estimate the time quite accurately in a case of beginning erysipelas when the infection occurred. In most instances infection takes place through some wound, a slight abrasion of the skin, which may, perhaps never have attracted the patient's attention, and which has become invisible at the time the disease is first noticed. Infection, however, may also take place through a mucous surface, through which the microbes enter the tissue in the same manner, and under the same conditions as when infection takes place through the skin. One of the severest cases of erysipelas that ever came under my observation commenced in the pharynx, or tonsils, and as the symptoms subsided here, a typical and severe facial erysipelas developed. The patient was suffering at the time from secondary syphilis.

Relation of Erysipelas to Puerperal Fever.

Obstetricians recognized the danger of exposing puerperal women to the infection which might emanate from erysipelatous patients, long before the tangible contagion of erysipelas was known. Since the discovery of Fehleisen, this subject has attracted renewed attention, and positive knowledge has accumulated both from accurate clinical observation, and from the fertile, and more positive field of experimentation.

Gusserow ("Erysipel u. Puerperalfieber,"*Archiv. f. Gynäkologie*, 1887, p. 169) asserted upon the basis of an extensive experience, that no direct etiological relations exist between the contagion of erysipelas and puerperal fever. He had under his care puerperal women suffering from erysipelas of the skin without any serious disturbances following in the genital tract. In ten other cases, one of them occuring during an epidemic of puerperal fever, the erysipelas was observed as a complication of septic affections of the genital organs. Gusserow asserts that, in this case, it cannot be claimed that the erysipelas could have caused the puerperal affection, as the latter preceded the former. But another point could be raised, as it might be claimed that the septic processes should be made answerable for the occurrence of erysipelas. The author has studied this subject also by way of experiment. A pure culture of the streptococcus erysipelatosus, which had been tested and found reliable in producing erysipelas by the usual methods of inoculation, was injected into the peritoneal cavity of two rabbits; in two others it was applied to an open wound of the abdomen, and in the last two animals it was injected into the subserous connective tissue of the peritoneum. In all of these animals no effect was produced, and no pathological changes were detected at the point of injection when the animals were killed some time after the inoculation. Gusserow looks upon the results of these experiments, if not as positive proof, nevertheless as strong evidence against the claim that erysipelas can cause puerperal sepsis.

Winckel ("Zur Lehre von dem internen puerperalen Erysipel," *Verh. der Deutschen Gesellschaft f. Gynäkologie*, 1 Congress, p. 78), an equally reliable and able observer, has come to entirely opposite conclusions. He cultivated from a parametric abscess which had developed after childbed, Fehleisen's streptococcus. Injections of this culture in rabbits produced typical erysipelas. The same author also observed erysipelas following, in a puerperal woman suffering from suppurative peritonitis, pleuritis, and metro-lymphangitis. The patient died on the thirteenth day. The starting-point of the erysipelas could be traced to an ulcer of the vulva.

Blood taken from the right side of the heart soon after death was inoculated upon a solid nutrient medium and produced a culture of the streptococcus of erysipelas. The same culture was obtained by inoculations with fluids taken from the peritoneal and pleural cavities, the uterus, kidneys, and liver. In three cases a culture thus obtained was injected into the peritoneal cavity of rabbits and no peritonitis followed. In one experiment it produced suppurative peritonitis. Guinea-pigs proved less susceptible to infection than rabbits. In white mice the inoculations were invariably followed by a fatal disease. From the results of these experiments, the author claims that the virus of erysipelas is one of the most virulent puerperal poisons, and believes that they prove the causal relations of erysipelas to puerperal sepsis.

Doyen (*British Medical Journal*, 1888, ix. 93, has also found, both in mild and severe cases of puerperal fever, a streptococcus similar to the one described by Rosenbach and Fehleisen. He made some inoculations to determine their relationships. The streptococcus found in the lesions of puerperal fever caused erysipelas, and the streptococcus found in erysipelas developed puerperal fever. The author believes that the microbe of puerperal sepsis is the same as that of erysipelas.

Puerperal sepsis from the virus of erysipelas can only be feared when the virus is brought in contact with an absorbing surface in the genital tract, but when this takes place and the streptococci reach the enlarged lymphatic vessels of the puerperal uterus, the most violent and fatal form of puerperal sepsis is almost certain to follow.

Relation of Erysipelas to Phlegmonous Inflammation and Suppuration.

Some difference of opinion still exists among pathologists with regard to the question whether the streptococcus of erysipelas possesses pyogenic properties. The majority of those who have studied this subject experimentally deny this, and assert that when suppuration takes place in cases of erysipelas it is the result of a secondary infection with pus-microbes, and, on this account, look upon phlegmonous inflammation as a complication, and not as a condition belonging to the erysipelatous process.

Hajeck ("Das Verhältniss des Erysipels zur Phlegmone," *Deutsche med. Wochenschrift*, 1886, No. 47) has made careful investigations to show that the streptococcus of erysipelas is neither in form nor culture materially different from the streptococcus pyogenes, but he showed, also, that in fifty-one cutaneous or subcutaneous inoculations with a pure culture of the streptococcus of erysipelas in rabbits, the result was always a superficial migrating dermatitis which resembled to perfection erysipelas in man, while similar injections with the streptococcus pyogenes produced a more intense and deeply seated inflammation, which in almost every instance terminated in suppuration. The difference in the action of the two microbes on the tissues plainly demonstrated their non-identity. Microscopical examination of the inflamed tissues showed a still more important difference as far as the localization and local diffusion of the microbes were concerned. The coccus of erysipelas was always found with the products of inflammation *within the lymphatic vessels*, and only exceptionally in the connective-tissue spaces, which anatomically are only a part of the lymphatic system. The streptococcus pyogenes penetrates the tissues more deeply; it is not only found in the lymphatic vessels, and connective-tissue spaces, but it migrates beyond the lymphatic system and infects different kinds of tissue, thus giving rise to a more deeply seated and more intense inflammation. The cocci of erysipelas are found only exceptionally in the immediatete vicinity of bloodvessels, while the streptococcus of suppuration can always be seen arranged in radiate lines around vessels entering the adventitia, the muscular coat, and often even in the lumen of the vessel. In man, the same histological differences can be seen in erysipelas and phlegmonous inflammation as in the artificial conditions in animals subjected to experiment, and the same pathological differences are also constantly found. The author asserts that Fehleisen was in error when he claimed that the formation of abscesses occurred independently of the erysipelatous infection. He affirms that, in rabbits inoculated with the virus of erysipelas after the acute inflammation has subsided, circumscribed small nodules which remain may suppurate, but the suppurative process remains circumscribed, while after injection with cultures of the streptococcus pyogenes the inflammation assumes a phlegmonous type, and the suppuration is always more diffuse. Under certain circumstances, a circumscribed subcutaneous suppuration can also take place in erysipelatous inflammation in man. When suppuration in a joint takes place, however, it is not caused by the erysipelatous infection, but is due to the presence of pus-microbes. Death following erysipelas is caused by the introduction into the blood of ptomaïnes in sufficient quantity to produce fatal intoxication, or by the entrance of the cocci into the circulation, which seldom takes place, or it results from complications incident to the disease occurring independently of it. In the discussion on this paper Eiselsberg said, from the knowledge he derived from his own personal experimental work, he would agree with Passet in that the streptococcus erysipelatosus and pyogenes do not differ in their pathogenic effects. They are not different species of microbes, but, at the most, only varieties of the same species. Passet found that the streptococcus which he cultivated from a phlegmonous abscess was different from the one described by Rosenbach, inasmuch as in culture it resembled the coccus of erysipelas.

Von Noorden ("Ueber das Vorkommen von Streptococcen im Blut bei Erysipel," *Münchener med. Wochenschrift*, No. 3, 1887) records an observation which tends to prove that the coccus of erysipelas occasionally enters the circulation, and that, when it localizes in distant parts of the body, it can produce suppuration. In the course of a severe attack of erysipelas which proved fatal, suppuration of the sheaths of the tendons of the hand occurred. Soon after death blood was taken from the heart, and with it a solid nutrient medium was inoculated, with the result of producing a culture which in every respect resembled the streptococcus of erysipelas.

Simone observed a case of pyæmia which developed in a patient suffering from erysipelas, and the bacteriological study of this case led him to assert that the streptococcus of suppuration and of erysipelas were the same. His experiments on animals with both organisms yielded the same results.

Rheiner ("Beiträge zur pathologischen Anatomie des Erysipels bei Gelegenheit der Typhus-epidemie in Zürich," 1884, Virchow's *Archiv*, B. c. S. 185) found Fehleisen's streptococcus in all cases of traumatic erysipelas which he examined, but was unable to find it in two cases of gangrenous erysipelas following typhus. In these cases he found bacilli which he believed were identical with Klebs-Eberth's bacillus of typhus.

Max Wolff ("Bacterienlehre bei accidentellen Wund-Krankheiten," Virchow's *Archiv*, B. lxxxi. S. 408), from a review of this subject, and a number of original observations, came to the conclusion that certain micrococci produced some chemical poison which occasioned erysipelas.

The distinction, moreover, between erysipelas and phlegmonous processes was formerly not accurately made, and Tillmanns even believed that the germs of erysipelas could produce septic disease.

To complicate this subject still more, Bonome and Bordini (*Centralblatt f. Chirurgie.*, No. 7, 1887) claim that they have found the staphylococcus in two cases of erysipelas. The authors assert that in the fluid removed from the bullæ of a case of facial erysipelas they found the staphylococcus pyogenes aureus and no streptococci. Culture experiments were made, and the product was a luxuriant growth of the yellow coccus. Inoculations in rabbits yielded positive results with recovery.

The second case was one of phlegmonous erysipelas of the face, from which they cultivated the staphylococcus pyogenes citreus. Inoculations with this culture were again followed by positive results. From these observations the authors conclude that other microörganisms than the streptococcus of Fehleisen can produce erysipelas. At a recent meeting of the Academy of Medicine in Paris, Doyen read a paper on the relations existing between erysipelas and puerperal fever. By means of clinical observations and experimental inoculations the author claimed to have demonstrated that the puerperal streptococcus, which is the microörganism characteristic of that affection, almost always produces erysipelas and a small abscess in the rabbit. In women, it often produces erysipelas, phlegmonous inflammation, or purulent pleuritis. The streptococcus of erysipelas produces the disease in rabbits almost invariably, and sometimes phlegmons or peritonitis in man. The streptococcus of pus sometimes produces erysipelas in the rabbit. These three streptococci are similar in cultures and appear to be one and the same whose manifestations may vary.

Smirnoff found in one case of erysipelas the specific microbe in the metacarpo-phalangeal joints of the left hand which was the seat of the disease. In the case of a man who had died of erysipelas enormous colonies of the streptococcus were found in the right shoulder and knee-joints. The synovial fluid injected into rabbits occasioned erysipelas migrans.

Verneuil and Clado ("De l'Identité de l'Erysipèle et de la Lymphangite-aigue," *Compt. rend.*, T. 108, No. 14) found in the pus of four cases of typical suppurative lymphangitis only the cocci of erysipelas, and by inoculations with them produced erysipelas artificially in rabbits. The authors consequently came to the conclusion that erysipelas and lymphangitis are only two forms of one and the same acute, infectious, parasitic disease.

Kahlden (*Centralblatt f. Bacteriologie und Parasiten-kunde*, B. i. S. 22), after a careful study of the recent literature on erysipelas and the difference in opinion on the pathogenic properties of the streptococcus erysipelatosus, remarks that the subtility in the differences between the morphology and the cultures of the microbe of erysipelas and the streptococcus of suppuration are undoubtedly the reason why no uniformity of opinion exists in regard to their specific pathogenic effects, especially as to the possibility of Fehleisen's streptococcus producing suppuration. To this I might add that not every superficial diffuse inflammation of the skin is erysipelas, and not every abscess occurring during, or soon after, an attack of erysipelas should be considered as a product of this disease. The surgeon will do well to hold to the teachings of Fehleisen, until more convincing proof shall have been furnished of the pathogenic identity of the streptococcus of erysipelas and the streptococcus of suppuration.

CHAPTER XIII.

ERYSIPELOID.

A NEW form of infective dermatitis, which in many respects resembles erysipelas, has been recently described by Rosenbach (" Ueber das Erysipeloid," *Archiv f. klin. Chirurgie*, B. xxxvi. Heft 2) under the name of erysipeloid. It attacks usually the fingers and exposed portion of the hand, and is most frequently met with in persons who handle game or dead animals, as cooks, butchers, fish-dealers, and tanners. The affection starts from some minute abrasion of the skin, as a bluish-red infiltration which slowly advances in a proximal direction. The inflamed parts are the seat of a burning, smarting sensation. While the skin at the point of infection returns to its natural condition and color, the zone of infiltration becomes larger as it continues to spread until the disease appears to exhaust itself in the course of from one to three weeks. The infectious material is contained in decomposing animal substances. The infection may take place in any abraded part of the body which comes in contact with material containing the virus. The general health is not affected and the temperature remains normal. The disease travels very slowly, so that if infection takes place in the tip of a finger, it reaches the metacarpus in about eight days, and during the next eight days it spreads over the back of the hand, from where an adjacent finger may become infected, the extension then taking a direction opposite to the lymph current. Repeated experiments to obtain a culture failed, until in November, 1886, the author succeeded in cultivating it upon gelatin from a case in which the disease could be traced to infection from old cheese. The author injected a pure culture under the skin of his own arm at three different points. After forty-eight hours he experienced a smarting, burning sensation at the points of injection, at the same time a circumscribed redness appeared around each puncture, which soon became confluent. On the fifth day each puncture was surrounded by a zone of inflammation the size of a silver dollar, somewhat elevated above the niveau of the surrounding skin. While the centre of this red patch became pale, the zone of inflammation continued to enlarge. In the inflamed area the capillary vessels could be seen enlarged, presenting in the zone an arterial hue with a slight tinge of brown, while inside of the zone the color was a livid brown. In the skin which returned to its normal pale color, slight suggillations appeared as though some of the red blood-corpuscles in the tissues had been destroyed during the progress of the disease. The disease appeared to have completely subsided on the eighth day, when the same smarting sensations returned, and a new zone appeared around the old one, On the tenth day the area measured in its transverse diameter twenty-four centimetres, and in the parallel direction of the arm eighteen centimetres. After this the affection disappeared permanently. During all this time the general health remained unimpaired, and the temperature varied from 36.8° to 37.2° C. (98.2° to 99° F.). A microscopical examination of the pure culture showed that it was composed of swarms and heaps of irregular, round and elongated bodies larger than the staphylococci. The author first believed that these bodies were cocci, but later he saw a network of intertwining threads and decided that they were thread-forming microbes. In old cultures, the threads were very abundant and arranged in every possible way and direction. These threads looked as though branches were given off, but on closer examination it could be seen that no organic connection existed between them. Terminal spores at the tips of the threads were numerous and could not be stained. Neither the microbe nor the threads manifested motile power in the culture, or when suspended in water. A gelatin culture became visible on the fourth day as a delicate cloud which increased in size very slowly at a temperature of 20° C. (68° F.). The older cultures change into a brownish-gray color, and then resemble the culture of the bacilli of mice septicæmia. In cultures four months old the growth was not entirely suspended.

The author, as yet, has not given a name to this microbe, but believes that it belongs to the "cladothrix" variety of microörganisms. He wished to ascertain the action of this microbe on lupus, but in several cases in which it was tried the inoculations failed.

CHAPTER XIV.

NOMA.

THE most violent of the local effects of bacteria are seen in various affections which terminate in gangrene, such as traumatic gangrene and noma. Here, it is not a case of death of the tissues as the result of violent inflammation, so much as a direct killing of them by the ptomaïnes of the bacteria. In acute progressive gangrene, bacilli have been found which are apparently the cause of the disease. In noma, long bacilli are present, which Lingard has demonstrated to be the cause of the disease. In gangrenous stomatitis in the calf, which affects this animal at particular seasons, he has found bacilli which are very similar in appearance to those present in noma in man. On cultivation they present characters which render them easily distinguishable from other bacteria, and on inoculation of these organisms into the calf a gangrenous stomatitis is again produced.

Ranke's ("Etiology and Pathological Anatomy of Gangrene resulting from Noma," *Archives of Pediatrics*, April, 1888) investigations on noma led to the following conclusions: Different forms of gangrene resulting from noma can unquestionably occur spontaneously in children who have a tendency to disease of this character —that is, without contact with other cases of noma. The frequent occurrence of cases of noma in public institutions, and the apparent preference of the disease for localization upon the mucous membrane of the different openings of the body, suggest that the origin of it may be referred to the penetration from without of microörganisms. In the zone of tissue contiguous to that which has undergone necrosis from noma may be found cocci which have almost the characteristics of a pure culture. At the periphery of the necrobiotic zone which has been invaded by cocci the connective-tissue is found to be in an active state of nuclear proliferation. The entire condition is suggestive of the tissue necrosis in field mice, which is caused by a chain coccus, and has been described by Koch. Up to the present time the specific nature of the cocci which are found in noma has not been shown. In the tissues which limit the necrotic areas are found peculiar degenerative processes in the nuclei which in some cases suggest karyokinesis. These changes in the nuclei appear to belong to necrosis in general.

CHAPTER XV.

TETANUS.

HISTORY.—The infectious nature of tetanus was well known and established before the discovery of the bacillus tetani. In 1859 Betoli related the case of a bull that died of tetanus after castration. Several slaves ate some of the flesh of the dead animal and of these, three were in a few days seized with tetanus, two of them dying. He adds further that in Brazil, where this occurred, the flesh of animals dead of tetanus is generally regarded as capable of transmitting the disease. In 1870 Anger reported a case in which a horse had spontaneous tetanus, after which three puppies which had been in the same stable were also affected. Kelly in 1873 had three cases in the same week, all arising in a civil hospital, and a few days later there was a fourth case in a neighboring hospital.

Larger in 1853 saw a woman who had a fall while cleaning a farm-yard, causing a slight wound of the elbow. Four weeks later, she was seized with tetanus, and on investigation it was found that a horse affected with that disease had been in a stable opening into the yard where she fell. He also mentions that in a small village where tetanus was previously unknown, five cases appeared in eighteen months under quite different climatic conditions. Of these, one had been taken to a hospital, after which two others in the same ward became affected.

Verhoogen and Baert have recently published an article upon the nature and etiology of tetanus, in which these authors cite the well-known endemic character of the disease in our Southern States, Cuba, Ceylon, a number of the Pacific Islands, and other localities, and quote a large number of circumstances that suggest the occasional epidemic type of the affection as met in man and some of the lower animals. Among a number of clinical and experimental occurrences suggesting the probability of the transmissibility of the malady, and the likelihood of the agent of transmission existing in unclean instruments, Thiriar's experience is narrated. This operator was unfortunate enough to lose ten cases of major operations by tetanus before he determined the seat of the infection to exist in his hæmostatic forceps, the thorough sterilization of which

CULTIVATION EXPERIMENTS. 143

by heat was happily followed by a complete cessation of the undesirable sequences.

Although the infectious nature of tetanus was suspected for a long time, it is only quite recently that the real microbic cause was discovered almost simultaneously by Nicolaier and Rosenbach. Nicolaier showed the exogenous origin of the disease by finding a bacillus in earth which produced tetanus in animals by inoculation. Rosenbach found a similar bacillus in the pus of a patient suffering from traumatic tetanus. The identity of the bacillus of tetanus with Nicolaier's bacillus of earth tetanus was demonstrated in Koch's laboratory April 10, 1887.

DESCRIPTION OF THE BACILLUS TETANI.—Rosenbach describes the bacillus as an anaërobic microörganism which presents a bristly appearance, with a spore at one of its extremities which gives it the resemblance to a pin, or drumstick. According to Kitasato, the

FIG. 6.

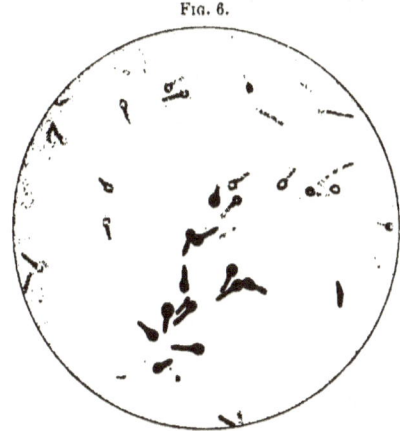

Tetanus bacilli bearing spores from an agar culture. × 1000.
(FRÄNKEL and PFEIFFER.)

bacilli produce spores in 30 hours in cultures kept at a temperature of the body. They manifest great resistance to heat, as they have been found active after an exposure of one hour to 80° C. (176° F.) moist heat, and are only destroyed by placing them in a steam apparatus heated to 100° C. (212° F.) for five minutes. The bacillus has been found in different kinds of soil and in street dust. In man, it has been found in tetanic patients in the wound secretions, in the nerves leading from the seat of infection, and in the spinal cord.

CULTIVATION EXPERIMENTS.—Rosenbach found it impossible to obtain a pure cultivation; although he resorted to fractional cultivation, it was found that the last culture was still contaminated by one or more additional microbes. Flügge claimed to have obtained a pure culture by heating for five minutes the mixed culture to 100° C. (212° F.), but after this procedure the bacillus was incapable of further propagation. After many trials it was found that sterilized solid blood-serum was the best soil for the propagation of the bacillus outside of the body. Both Nicolaier and Rosenbach observed the anaërobic nature of the bacillus, as it was found impossible to obtain a culture on the surface of the nutrient media, or anywhere else where oxygen could not be excluded. The culture appeared slowly as a delicate whitish-gray film in the track of the needle puncture below the surface of the culture substance. By a long series of cultivations, Rosenbach finally succeeded in eliminating all other microbes with the exception of a bacillus of putrefaction. The growth of the bacillus takes place most readily at an equable temperature of 37° C. (98.6° F.), and becomes first visible about the third day in the depth of the culture media.

Kitasato ("Ueber den Tetanuserreger," *Verh. d. Deutschen Gesellschaft f. Chirurgie*, 1889) has finally succeeded in obtaining a pure culture of the bacillus of tetanus from pus taken from a patient suffering from tetanus. As the bacillus will only grow where atmospheric air is excluded, he exposed his cultures to an atmosphere of hydrogen gas. Mixed cultures which had been kept for several days in the incubator were then exposed for half an hour to an hour to a temperature of 80° C. (176° F.) in a water-bath, and further growth was secured upon plate cultures in closed glass vessels filled with hydrogen gas. He succeeded in destroying all other anaërobic bacilli found in the pus by heating the mixed culture to 80° C. (176° F.), with the exception of the bacillus of tetanus, which upon gelatin plates in the hydrogen atmosphere at a temperature of 18° to 20° C. (64.4° F. to 68° F.) after a week produced a visible culture. Growth is more rapid at a temperature of 18° to 20° C. (64.4° F. to 68° F.), when a culture appears in from four to five days, and if the temperature is kept at 36° to 38° C. (95.8° F. to 100.4° F.) the development of spores and bacilli takes place most rapidly.

INOCULATION EXPERIMENTS.—Nicolaier ("Ueber infectiösen Tetanus," *Deutsche med. Wochenschrift*, 1884, No. 52) produced tetanus in rabbits and mice experimentally by inoculations with different kinds of earth. Out of 140 experiments, in 69 a disease was produced which very closely resembled tetanus in man. In the pus at the point of inoculation bacilli and micrococci were constantly found. Among the bacilli one form was invariably present; this bacillus resembled in appearance and culture the bacillus of septicæmia in mice, but was more slender. This bacillus was found in isolated places in the connective-tissue, but could not be found in the muscles, nerves, and blood. If the earth was sterilized by exposure to high temperature for an hour the inoculations, without exception, proved harmless, showing conclusively that the contagium of tetanus had been rendered inert. Inoculations with pus taken from tetanic animals were most successful. Inoculations with mixed cultures grown in solidified blood-serum yielded positive results

Rosenbach ("Zur Aetiologie des Wundstarrkrampfs beim Menschen," Langenbeck's *Archiv*, B. xxxiv. S. 306) made his experiments with a mixed cultivation grown from the pus taken from the line of demarcation of a case of frost gangrene in a patient who had died of tetanus. The inoculations proved successful. Carle and Rattone (*Giornale della R. Accademia di Med. di Torino*, 1884, No. 3) succeeded in producing tetanus in rabbits by inoculation with pus from a suppurating acne in a tetanic patient in whom the infection was traced to this source.

Bonome ("Ueber die Aetiologie des Tetanus," *Deutsche med. Wochenschrift*, 1887, No. 15) reports the case of a man suffering from paraplegia, the result of disease of the spine in the dorsal region, complicated by an extensive sacral decubitus, the seat of phlegmonous purulent inflammation, who was suddenly attacked by tetanus, which proved fatal in two days. One hour after death a small portion of the infiltrated tissue around the gangrenous part was removed, and after reducing it to a fine pulp by trituration, he injected it under the skin of a rabbit. Twenty-two hours after the inoculation the animal died with well-marked symptoms of tetanus. The products of inflammation from the point of injection thrown into the subcutaneous tissue of other animals produced the disease, while intravenous injections proved harmless. The gravity of symptoms following subcutaneous injections was commensurate with the quantity of fluid injected. Guinea-pigs proved less susceptible to infection than rabbits. In the pus taken from the dead tissue he found, besides the usual pus-microbes, a bacillus which resembled in every respect the one described by Nicolaier and Rosenbach. Hochsinger (*Centralblatt f. Bacteriologie u. Parasitenkunde*, B. ii. Nos. 6, 7) made his observations on a case of tetanus which proved fatal on the fifth day. The day before the patient died blood was abstracted from a vein under strict antiseptic precautions for microscopical and bacteriological study. No microörganisms could be found in it. With the greatest care sterilized solid blood-serum was inoculated with the blood, making with the needle both superficial tracks and deep punctures. The nutrient medium was kept at a temperature of $37°$ C. ($98.6°$ F.). On the third day, a white cloudy streak marked the direction of the deep punctures, while the superficial plant remained sterile. On the third day a portion of the culture was removed and stained with aniline gentian, and submitted to microscopical examination. Delicate bacilli measuring from 0.8 to 1.2 micromillimetres in length, composed the culture. The detection of the bacillus was quite difficult, and its growth very slow. On the sixth day the serum around the punctures had become more cloudy, and the microscope showed that the bacilli were present in greater abundance. From this time on, the cultivation ceased to increase, and the surface of the nutrient medium still remained sterile. From the original culture five other tubes were inoculated, but in only one of them could a slight cultivation be detected on the fourth day. A large rabbit was infected by injecting blood obtained from the patient during life. The blood was intimately mixed with sterilized water, and a syringeful of this mixture was injected under the skin in the iliac region, and half this quantity under the skin of the left thigh. The next day the animal appeared quite ill, and was unable to use the left hind leg, which was dragged along in walking. At this time great nervous excitability was observed, the exaggerated reflex symptoms being especially well marked in the posterior extremities which, on the slightest touch, were thrown into clonic spasms. On the following day the animal was found dead. A few hours before death well-marked symptoms of tetanus developed. No positive results were revealed at the post-mortem examination. Injections of blood from this animal produced no results in other rabbits, and cultivation experiments were equally fruitless. A syringeful of inspissated blood of the patient, kept for three weeks, thrown under the skin of a white mouse, was followed by a fatal attack of tetanus, while a second animal inoculated in a similar manner with one-half of this quantity remained perfectly well. Flügge had before observed that by injecting blood from animals rendered tetanic by inoculation, it was necessary to use a large quantity in order to reproduce the disease in other animals, and even by doing so the result was not always satisfactory. It appears from the experience of these authors that the blood of tetanic patients possesses greater toxic properties than the blood of animals suffering from the same disease. Hochsinger also made experiments with the cultivation. Eleven days after establishing the primary cultivation he injected a syringeful of the liquefied turbid nutrient material into the subcutaneous tissue of the thigh of a medium-sized rabbit. The next day the reflexes were increased, respiration more rapid, and the animal appeared otherwise quite sick. On the third day the posterior extremities were stiff, the animal dragging them in walking. Reflex irritability enormously exaggerated. On the fifth day the animal died. In another experiment he injected the primary culture, seventeen days old, into the left thigh of a rabbit. On the fourth day the left hind leg was stiff, at the same time the reflexes were intensified. On the following day both hind legs were stiff, and the animal dragged itself along with difficulty on the front legs. On the sixth day the animal appeared more sick, and two days later died with well-marked symptoms of tetanus. The liquefied nutrient serum in the glass tube, containing the only secondary culture, hypodermically injected in another rabbit, produced tetanus.

These experiments appear to prove conclusively that the patient's blood contained the essential microbe of tetanus. The bacillus found in the patient and in the affected animals corresponded in every respect with the bacillus described by Rosenbach. In rabbits, Flügge estimated the stage of incubation at from three to five days, and the duration of the illness from the time that the first symptoms were noticed to the fatal termination, from five to seven days.

Beumer ("Ueber die Aetiologische Bedeutung der Tetanusbacillen," *Berl. klin. Wochenschrift*, 1887, No. 31) gives an accurate and able description of his studies in two cases of tetanus. The first case occurred in a mechanic, who injured himself under the nail of the right middle finger with a splinter of wood. Eight days after the injury, the patient having had but slight pain in the finger, pains appeared in the neck and muscles of the back. The next morning spasms of the muscles of the chest, abdomen, and jaw developed. These attacks occurred at intervals of an hour and a half. Four days later the lower extremities were affected, also the upper, but in a less degree. The right middle finger was slightly swollen. An incision was made, and the foreign body removed, which was followed by the escape of a drop of pus; death on the fourth day. The second case was a boy six and a half years old, who was brought into the clinic with well-marked symptoms of tetanus, and who lived only a few hours after his admission. The author obtained some of the dust and splinters of wood from the place where the mechanic had injured himself, and inserted small particles under the skin of mice and rabbits. In all experiments the animals were attacked with tetanus in from two to three days after inoculation, and died during the third or fourth. The spasmodic contractions were always noticed first in the muscles nearest the point of inoculation. A portion of the sole of the foot was taken from the boy, and small fragments of it inserted into the subcutaneous tissue of six mice. In all of these, symptoms of tetanus appeared after two days, developing gradually into general convulsions and death. The same results were obtained in mice and rabbits by inoculations of particles of dust taken from the spot where the boy sustained the injury. The bacillus of tetanus was found in the wound of the second patient. Beumer is firmly convinced that a direct relationship exists between the bacillus described by Nicolaier and Rosenbach and the cause of tetanus.

At a meeting of the Imperial Royal Society of Physicians of Vienna (*British Medical Journal*, July 25, 1888) Eiselsberg gave an account of a case of tetanus in Billroth's clinic: A woman, aged forty, drove a splinter of wood into the palm of her hand while scrubbing the floor. A fragment of the splinter was extracted by her husband. During the course of the next week an abscess formed in the hand; this was opened by the attending physician. On the twelfth day after the injury the woman was admitted into Billroth's clinic with typical symptoms of a severe attack of tetanus, which lasted four weeks. She afterward recovered to a great extent, and was discharged at her own request. At that time she still presented slight contractions of the affected limb. Two months later a suppurating fistula formed, and a small splinter of wood came away in the discharge. The wound then completely healed, and the patient made a perfect recovery. Eiselsberg used the extracted piece of wood for making cultures. Two rabbits were inoculated with the culture thus obtained, one of which succumbed to tetanus on the sixth day after inoculation, while the second one, which was inoculated at a later date, showed marked symptoms of tetanus, such as increased irritability, trismus, pleurosthotonus, etc.

Giordano ("Contributo all' Eziologia del Tetano," *Giorn. della Acad. di Med. di Torino*, 1887, Nos. 3, 4) performed his experimental work on the following cases of tetanus in the laboratory of Perroneito. The patient was a man forty years old, who fell from a hayloft upon the frozen ground, and was brought to the hospital twenty hours later with a complicated fracture of the forearm. The wound, which was covered with dirt, was enlarged, drained, and partly closed; on the fourth day trismus, and on the seventh day death from well-marked tetanus. Immediately after death, blood was taken from the wound of the median nerve, and fragments of a thrombus from a vein of the affected limb were also removed and preserved in sterilized beef-tea. A piece of necrotic tissue from the wound contained microbes, but not the bacillus described by Nicolaier. Inoculations with blood and fragments of internal organs failed to produce tetanus. Inoculations with pus from the wound and fragments of thrombus caused tetanus in rabbits and guinea-pigs. Small fragments of straw taken from the place where the patient fell, inserted under the skin of a rabbit, produced tetanus in three days. The pus, among other microbes, contained few of the characteristic bacilli. Injections of pus taken from the tetanic animal produced tetanus in other rabbits. Inoculations with tissue from the medulla oblongata did not cause the disease. Successful inoculations were made from the second and third rabbits. Injections of dust taken from the place where the patient was injured, suspended in water caused no symptoms, but a culture from it, six days old, contained the bacilli, and when injected subcutaneously produced the disease. From the absence of the bacilli in the internal organs, he concluded that this microbe does not permeate the whole body, and that the disease owes its origin to absorption of toxic agents from the wound.

Ohlmüller and Goldschmidt ("Ueber einen Bakterienbefund bei Menschlichem Tetanus," *Centralblatt f. die ges. Medicin*, 1887, No. 31) made a thorough bacteriological examination of a case of tetanus following complicated fracture of the right thumb. The disease appeared the day following the injury and resulted in death after not more than seventeen hours. Soon after death inoculation experiments were made with blood taken from the heart and spleen and pus from the seat of fracture, according to directions given by Bumm. The cultivations were made in solid blood-serum and kept at a temperature of 38° C. (100.7° F.). The tubes containing blood from the heart and spleen remained sterile, but the nutrient media infected with pus showed signs of growth by liquefaction of the solid serum. The bacilli which were detected resembled those of mouse septicæmia, only somewhat larger in size. In addition to these microbes streptococci and a thicker bacillus were found. Twenty-four hours later liquefaction had increased, but the streptococci had diminished in number. The characteristic bacilli were pin-rods with globular ends, and club-shaped rods with colorless terminal spores. On the third day the serum had undergone more advanced liquefaction and at the same time a fetid odor was noticed. A slide compared with one prepared by Nicolaier showed the identity of the two microbes. In order to prove still further their identity two mice were inoculated with the mixed cultivation. Twelve hours after infection tetanus had made its appearance, followed by death in seventeen hours. It should be remarked that the spasms commenced in the tail, extended to the posterior extremities, and then gradually forward. From these animals blood-serum was taken with which other mice were infected. Again tetanus was produced and successful cultivations were made. Successive cultivations appeared to diminish the intensity of the virus. Of two mice of equal size and age, one, which received one portion, died of tetanus on the ninth day, while the other, which received a dose three times as large, died on the third day. Cultivations on agar-agar always remained sterile. Cultivations in sterilized coagulated albumen from chicken's and gooseeggs showed that the bacilli retained their properties for about a week, but later they were displaced by other organisms. An attempt was made to destroy the other microörganisms by heating the mixed cultivation to 100° C. (212° F.) for five minutes. The result was satisfactory, inasmuch as inoculations produced positive results in mice and blood-serum. Inoculations of these cultures into a ten per cent. peptone gelatin medium caused rapid liquefaction, and the microscope showed thicker rods with long processes. In some cultures, bacilli of tetanus were found as late as the ninth day. Inoculation with the last cultures had no effect. He ascertained also that inoculations with earth had so often failed because not enough material was used. He made additional experiments using a much larger quantity. The first experiment, in which a portion of earth half the size of a pea was inserted under the skin on the back, was successful. After twelve hours the mouse sickened, in twenty hours presented typical evidences of tetanus, and died soon afterward. In the pus at the seat of inoculation cocci and bacilli of tetanus were found in abundance and inoculation with the product of inflammation produced tetanus as surely as pus taken from wounds of tetanic patients. The same earth exposed for half an hour to an hour to the action of steam was rendered sterile and inoculations with it proved harmless.

Of the greatest scientific and practical interest are the observations made by Bonome ("Ueber die Aetiologie des Tetanus," *Fortschritte der Medicin*, 1887, No. 21) in reference to the causation of tetanus by infection with earth containing the bacillus discovered by Nicolaier. He had an opportunity to observe a number of cases of tetanus after the recent earthquake at Bajardo. Of the seventy persons injured in the ruins of the church, seven were attacked by tetanus. From bacteriological investigations in connection with these cases, he came to the same conclusions in regard to the cause of the disease as Nicolaier, Rosenbach, Flügge, and Beumer before him. He likewise was unable to obtain a pure cultivation by successive generations, as even the last growth was always contaminated by a bacillus of putrefaction. Of particular importance is the observation made by him, that the secretions from the wounds and the exudation from the part the seat of tetanic convulsions, when dried and preserved between two sterilized watch-glasses retained their virulent properties for at least four months. All animals inoculated with dust from the débris in the interior of the church were attacked with tetanus. Control experiments with dust from the ruins at Diano-Marina never proved successful. Of the many persons injured during the same earthquake at this place, not one was attacked by tetanus.

Beumer ("Zur Aetiologie des Trismus sive Tetanus neonatorum," *Zeitschrift für Hygiene*, B. iii. S. 242) found Nicolaier's bacillus in a case of tetanus neonatorum. He made numerous efforts to obtain a pure culture by successive cultivations, but failed, as others had before him. He found the growth contaminated by cocci and a smaller bacillus. He also made numerous inoculation experiments with different kinds of earth. Of ten experiments with soil taken from the ocean beach, tetanus followed in only two. On the other hand, of ten inoculations with garden earth and street dust, all proved successful but one. Of three cases of tetanus which recently came under the observation of Lumniczer ("Beiträge zur Aetiologie des Tetanus," *Wiener med. Presse*, B. xxx. Nos. 10–12) he was able to demonstrate the microörganism in one. In this case the attack followed a gunshot injury. After the disease had developed fragments of hemp were removed from the canal made by the bullet, and in them the characteristic bacillus was found Cultures were made to the tenth generation, and with them animals were inoculated and tetanus was invariably produced. Pus taken from abscesses produced at that point of inoculation contained the bacillus, and injection experiments made with it yielded positive results. Cultures made from the blood or organs of the tetanic animals remained sterile. Inoculation with blood from these animals proved harmless. Kitasato (*op. cit.*) experimented with a pure culture of the bacillus of tetanus on mice, rats, guinea-pigs, and rabbits, and never failed in producing the disease, provided a sufficiently large dose of the culture was administered. In mice the disease appeared, without exception, 24 hours after the inoculation, and proved fatal in 2 or 3 days. The tetanic convulsions were first always local, appearing at the point of inoculation, becoming gradually more diffuse. He was unable to find bacilli at the seat of inoculation, in the blood or any of the organs of the body. He believes that if the tetanus is produced by inoculation with a pure culture the bacilli do not remain in the body for any length of time, but are rapidly eliminated. The same question has been raised in connection with the pathogenic action of the bacillus of tetanus as with the pus-microbes, Is the disease of which it is the specific cause due to the presence of the microbe, or its products in the organism (the ptomaïnes)? Brieger, by his indefatigable labors, has demonstrated beyond all doubt that

The Ptomaïnes of the Bacillus Tetani cause Tetanic Convulsions.

Brieger ("Zur Kenntniss der Aetiologie des Wundstarrkrampfes nebst Bemerkungen über das Cholera-roth," *Deutsche med. Wochenschrift*, 1887, p. 303) has succeeded in isolating four toxic substances from mixed cultivations of tetanus bacilli in sterilized meat emulsions. The first, tetanin, in doses of a few milligrammes administered subcutaneously in mice produced the characteristic symptoms of tetanus. The second, tetanotoxin, causes first, tremors, later, paralysis and convulsions. The third, muriate of toxin, has not been designated by a special name by Brieger, it produces also well-marked symptoms of tetanus, but besides excites the salivary and lachrymal glands to increased functional activity. The last, spasmotoxin, produces severe clonic and tonic spasms which prostrate the animal at once. Besides meat emulsion, the contused brain substance from horses and cattle was used, also cow's milk mixed with carbonate of lime. It seemed that the culture substance determined to a certain extent the kind of toxin which was produced; thus in cultures grown in brain substance besides tetanin, tetano-toxin was found in greatest quantity; old cultures, in which the tetanus bacilli were dead, produced none of these toxic agents.

The same author ("Ueber des Vorkommen von Tetanin bei einem an Wundstarrkrampf erkrankten Individuum," *Berl. klin. Wochenschrift*, April 23, 1888) has very recently been successful in isolating tetanin from the amputated arm of a patient the subject of tetanus. Tetanus had developed a few days after a severe crushing injury of the hand and forearm. The first symptoms manifested themselves in the morning, and at twelve o'clock (noon) the operation was performed; at five o'clock, on the same day, the patient expired suddenly during one of the tetanic convulsions. The bacilli of tetanus were found in the serum taken from the œdematous portion of the forearm in connection with other bacilli of different length, staphylococci and streptococci. Serum containing these microbes when injected under the skin of mice, guinea-pigs, and rabbits invariably produced tetanus; on the other hand, a dog treated in the same manner, as well as after injections of tetanin, remained well. A horse inoculated with a culture of bacilli in meat emulsion showed no symptoms of tetanus, but an abscess formed at the point of inoculation. The infiltrated tissues of the amputated forearm planted on sterilized meat emulsion, solid blood-serum, and emulsion made of the flesh of fish, yielded, besides ammonia, only tetanin; no trace of tetano-toxin, spasmo-toxin, nor the unnamed toxin which could be obtained from Rosenbach's bacillus. A moderate dose of tetanin injected into the subcutaneous tissue of a horse produced muscular contractions which lasted for a considerable length of time, but the characteristic symptoms of tetanus, as it is seen in horses, did not appear.

The clinical and experimental researches quoted above demonstrate that the same bacillus is found in the wound secretions, the tissues, and, in some instances, in the blood of tetanic patients, and that tetanus in animals can be produced by injection of wound secretions of tetanic patients, or by using cultivations—facts which have sufficiently established the microbic nature of the disease. The stage of incubation, both in man and animals, appears to be extremely variable; in some instances lasting only twenty-four hours, while in others weeks may lapse between the time of inoculation and the first manifestations of the disease. This may depend on one of three things: 1. The number of bacilli introduced may be so small that a much longer time is necessary before active symptoms are produced than if a larger quantity had been introduced, as Watson Cheyne has shown that in animals the injection of a limited number of the bacilli of tetanus produced no symptoms. 2. The location of the infection-atrium and anatomical characteristics of the tissues surrounding it may influence the time which is necessary to develop the disease. 3. Brieger's investigations have shown that tetanic convulsions in animals are produced by injections of tetanin, one of the toxic ptomaïnes derived from cultivations of the bacillus of tetanus, and it is more than probable that the active symptoms of tetanus in man are due not to the presence in the tissues of the bacillus, but to the toxic action of the ptomaïnes on the spinal cord, so that the length of the stage of incubation is further modified by the capacity of the infected tissues to yield the different ptomaïnes. The degree of virulence of the bacillus of tetanus must certainly play an important part not only in determining the length of the incubation stage, but also the intensity of symptoms. There can be no doubt that both the acute and chronic forms of tetanus are caused by the same microbe, and that the clinical difference depends upon the degree of virulence of the primary cause. Whether cultivations from chronic cases of tetanus can produce an acute and rapidly fatal attack in animals remains to be determined. In this direction I have recently made an observation which, if not convincing, is, at least, very suggestive. A boy, fifteen years of age, previously in good health, was attacked with acute osteomyelitis in the lower extremity of the femur. The surgeon in attendance trephined the bone over its outer and lower aspect during the first few days and before an abscess had formed in the soft parts. A few days after the operation trismus set in, followed by typical chronic tetanus. Six weeks later, the patient entered the Milwaukee Hospital, and was placed under my charge. At this time the patient had become emaciated to a skeleton. Trismus and opisthotonus were well marked, and the lower extremities were rigid and fixed in the extended position. The slightest touch, or a draught of air in the room, would bring on intense convulsive attacks lasting for several minutes, attended by excruciating pain. Profuse fetid discharge at the site of operation; pulse 140, temperature from 99° to 101° F. (37.3° to 38.3° C.). Believing that the primary infection had taken place through the operation wound, and that osteomyelitic products served the purpose of a nutrient medium for the bacillus tetani, I determined to operate in spite of the grave symptoms. As the spinal cord at this stage of the disease was necessarily the seat of intense congestion, I resorted to chloroform as an anæsthetic instead of ether. The usual operation for necrosis of the lower end of the femur was performed, and a large triangular sequestrum removed from the lower and posterior aspect of the bone. The involucrum was imperfect and its inner side lined with a thick layer of flabby granulations. Gelatin tubes were inoculated with blood, pus, and granulation tissue. The blood cultivations remained sterile, while the two remaining tubes showed a copious growth of staphylococcus pyogenes albus which rapidly liquefied the gelatin. A portion of the granulation tissue was disinfected with a weak solution of carbolic acid, dried between layers of antiseptic gauze, and inserted under the skin of a rabbit. No suppuration followed, and the animal remained perfectly well for six weeks, when both posterior extremities became rigid so that it could only move from place to place by dragging the hind legs. The next day tetanic convulsions affecting the muscles of the back and all the limbs appeared, and on the fourth day death supervened. The interesting features in this case are that the patient recovered from the tetanus after a long illness, extending over three months; that marked improvement followed the operation, which had for its object thorough disinfection of the infection-atrium; and that the inoculation with granulation tissue in the rabbit was followed by an acute attack of tetanus six weeks after infection. In the experiments related above the animals were inoculated with cultivations, or with wound secretions from tetanic patients; the stage of incubation rarely extended over two to three days, and often only eighteen to twenty-four hours, and the disease produced death in from twelve hours to three days.

Prophylactic and Curative Treatment of Tetanus by Antiseptic Agents.

More than a year ago Sormani (*La Riforma Medica di Napoli*, January 11–13, 1890), of Naples, found that iodoform was one of the most energetic disinfectants of the virus of tetanus, and that iodol and an acid (2 per cent.) solution of corrosive sublimate were similar in their action. A second series of experiments has shown that also chloral and chloroform had a similar power. Since then Mazzuschelli has used iodoform (locally) in two cases, which in and toward the end of May, 1889, came under treatment. In one case a girl while working in a garden with a spade, inflicted upon herself a large, torn wound in the calf of the right leg. Eight days after tetanus set in and she was taken into the hospital at Pavia. After removal of the dead tissue the wound was cleansed with a 2 per cent. solution of sublimate, dusted over with iodoform, and chloral hydrate given internally. The patient died twelve hours later. In the other case, the patient had run a splinter into her foot between the great and second toe while following a path barefooted over a field. Six days later tetanus made its appearance, the splinter was removed, and the wound treated as in the preceding case; death four days later, ten days after the injury.

In the first case Sormani inoculated two rats and one rabbit with the tissue which was removed from the wound before dusting with iodoform. All these animals died from tetanus from forty-eight to ninety-six hours after inoculation. Two rabbits inoculated with a fragment of tissue from the wound after the death of the patient remained alive.

In the second case a piece of the iodoformized tissue and one from the tissues lying more deeply were used for inoculation of two rats, which, however, did not contract the disease. A culture-glass filled with agar inoculated with a fragment of the wound-tissues after the disinfection remained sterile; another tube inoculated with tissue removed more deeply developed staphylococci.

The author concludes, from these and further experiments, that where tetanus is already developed iodoform is not able to prevent its further course, but may neutralize the virus on the surface of the wound.

Baccelli (*Riforma Medica*, January 25, 1890) used subcutaneous injections of carbolic acid in doses of 1 centigramme every hour. In 1887 he cured a grave case, and now he has another such a one under treatment, where the injections have produced such an improvement that recovery is assured.

CHAPTER XVI.

TUBERCULOSIS.

OF all the microbic diseases, tuberculosis is of the greatest interest and importance to the surgeon. Of the greatest interest because the tubercular lesions which come under his care are more clearly understood from a scientific standpoint than most of the other surgical diseases, and of the greatest importance on account of their great frequency. That large class of diseases which were grouped under that indefinite and vague term, *scrofula*, in the textbooks of but a few years ago, have been shown by recent research to be identical with tuberculosis etiologically, clinically, and anatomically. It is the object of this part of the book to give a brief description from a bacteriological and clinical standpoint of such localized tubercular lesions which by general consent are regarded as surgical affections and requiring surgical treatment.

HISTORY.—The results obtained from the crude inoculation experiments, which were made by Villemin, pointed strongly toward the infectiousness of tuberculosis, and since that time diligent search was made to discover and isolate a specific microorganism which should be characteristic of this disease. Theories were advanced, microbes were found and described which were supposed to bear a direct etiological relationship to tuberculosis, but nothing definite was known on the subject until Robert Koch ("Die Aetiologie der Tuberkulose," *Berl. klin. Wochenschrift*, 1882, No. 15), in 1882, announced to the profession his great discovery. He had found and demonstrated the true cause of tuberculosis, the bacillus of tuberculosis, and in his first publication brought such convincing proof of the correctness of his claim, that, with few exceptions, it brought conviction even to the most sceptical. He had not only found the bacillus, but showed that it was constantly present in all tubercular lesions. He had isolated and cultivated the bacillus from tubercular tissue ; and, finally, he had furnished the crucial test—had produced artificial tuberculosis in animals by inoculation which was identical with tuberculosis in man. A number of pathologists, who inoculated animals with non-tubercular material, claimed that they had produced pathological conditions analogous to those found in animals which had been infected with the virus of tuberculosis. Further experimentation soon showed that these were instances of pseudo-tuberculosis; that while the gross appearances of the lesions resembled true tuberculosis, inoculations with this material never reproduced the disease, while inoculations with tubercular material could be done through a series of animals without impairing the potency of the virus, or varying the constancy of the results. Koch's discovery did not lead to such energetic search for the bacillus of tuberculosis among surgeons as physicians, because, as König asserts, the symptoms and signs of the tubercular affections coming under the observation of surgeons are so characteristic, that for practical purposes a correct diagnosis could be made in the majority of cases without a knowledge of their microbic nature and the improved methods for making a positive diagnosis derived therefrom. Koch, himself, in the publication above referred to, demonstrated the presence of the bacillus in lupus, scrofulous glands, tubercular joints, etc. He called attention to the fact that in these affections the bacillus can be constantly found in giant cells and between the epithelioid cells, while it is more difficult to find it in cheesy products, unless caseation has taken place quite rapidly.

Weichselbaum ("Tuberkelbacillen im Blut.," etc., *Wiener med. Wochenschrift*, 1884, Nos. 12, 13), Meisels, and Lustig found tubercle bacilli in the blood in cases of acute miliary tuberculosis. both during life and after death. Schuchardt and Krause (" Ueber das Vorkommen der Tuberkelbacillen bei fungösen und scrofulösen Entzündungen," *Fortschritte der Medicin*, B. i. S. 277) examined forty cases of tuberculosis of bones, joints, tendonsheaths, and the skin in Volkmann's klinik, and never failed in finding bacilli, although in some specimens careful and prolonged search had to be made. They found the bacilli in various lesions which had formerly been regarded as scrofulous affections.

Schlegtendal (" Ueber das Vorkommen der Tuberkelbacillen im Eiter," *Fortschritte d. Medicin*, B. i. S. 537) examined 520 specimens of pus from tuberculous suppurations and found bacilli present in about 75 per cent. of the cases. Mögling (*Die Chirurgischen Tuberkulosen*, Tübingen, 1884) found the bacilli never absent in tubercular pus from 53 patients.

During the last few years, surgeons have made valuable contributions to surgical literature on the subject of tuberculosis corroborative of the statements of Koch, which have placed many heretofore obscure lesions within the range of rational and successful surgical treatment.

DESCRIPTION OF THE BACILLUS TUBERCULOSIS.—The bacillus described by Koch as the essential cause of all forms of tubercular inflammation appears in the shape of very thin rods from two to eight micromillimetres in length, and rounded at the ends. They are straight or curved, and frequently beaded, occur singly, in pairs or in bundles. In the tissues they are found in the interior of giant cells and within and between epithelioid cells. The bacillli of tuberculosis are non-motile and consequently possess no power of locomotion and cannot penetrate into the tissues without assistance. Spore-formation occurs, even within the animal body, the spores having the appearance of clear vacuoles.

METHODS OF STAINING.—For section-staining Ehrlich's method is the best:

Saturated alcoholic solution of methyl-violet or fuchsin   11 parts.
Aniline water . . . . 100 "
Absolute alcohol . . . . 10 "

Sections are left for twelve hours in this solution. Treat the specimen with 1 : 3 solution of nitric acid a few seconds. Wash in alcohol (60 per cent.) for a few minutes; after-stain with diluted solution of vesuvin or methylene-blue for a few minutes; wash again in 60 per cent. alcohol, dehydrate in absolute alcohol, clear with cedar oil, mount in Canada balsam. The examination of fluids can be done rapidly and most satisfactorily by Gibbes' method.

Gibbes' magenta solution :

Magenta . . . . 2 parts.
Aniline oil . . . . 3 "
Alcohol (specific gravity 0.830) . . . . 20 "
Distilled water . . . . . . 20 "

Stain cover-glass preparations in this solution for fifteen or twenty minutes; wash in (1 : 3) solution of nitric acid until the color is removed; rinse in distilled water. After-stain with methylene-blue, methyl-green, iodine-green, or a watery solution of crysoidin, five minutes; wash in distilled water until no more color comes away. Transfer to absolute alcohol for five minutes, dry, and preserve in Canada balsam.

CULTIVATION EXPERIMENTS.—The best culture medium is solid sterilized blood-serum of the cow or sheep, with or without the addition of gelatin at a temperature of 37° to 38° C. (98.6° to 100.4° F.). The bacillus grows very slowly and only between the temperatures of 30° and 41° C. (86° and 105.8° F.). In about a week or ten days, the culture appears as little whitish or yellowish scales and grains. The bacillus can also be cultivated in a glass capsule on blood-serum, and the appearance of the growth studied under the microscope. The scales, or pellicles, are then seen to be made up of colonies of a perfectly characteristic appearance. The growth ceases after three or four weeks. The blood-serum is not liquefied, unless putrefactive bacteria contaminate the culture.

PLATE V.

Tubercle bacilli containing spores. (R. Koch.) Zeiss $\frac{1}{18}$. O. 4.

PLATE VI.

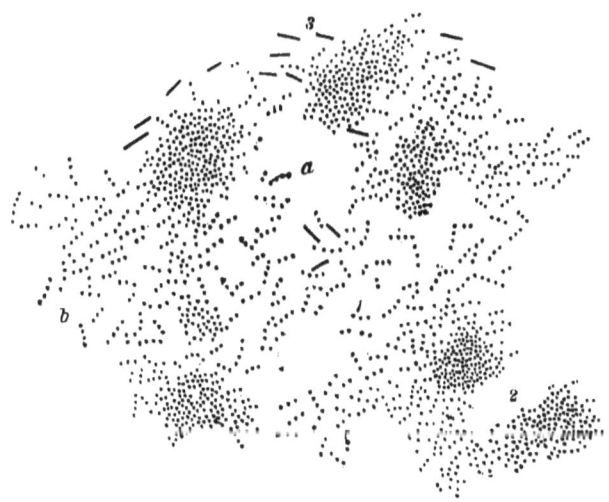

Tubercle bacilli from a tubercular cavity. Carbol-fuchsin,
nitric acid, methyl-blue. Zeiss $\frac{1}{18}$. O. 4

Nocard et Roux ("Sur la culture du bacille de la Tuberculose," *Annales de l'Institute Pasteur*, i. 1887, No. 1, pp. 19-29) have found that coagulated blood-serum is improved for the growth of the bacillus of tuberculosis by adding peptone, soda, and sugar. A further addition of 6 to 8 per cent. of glycerin favors the growth of the bacillus still more, while at the same time it prevents the formation of a crust upon the culture medium, which otherwise forms by evaporation. They also made successful cultivations upon agar-agar bouillon, to which was added 6 to 8 per cent. of glycerin, kept at a temperature of 39° C. (102.2° F.).

INOCULATION EXPERIMENTS—Even before the discovery of the bacillus of tuberculosis by Koch, genuine tuberculosis was produced in animals by inoculation with the products of what was then described as scrofula. Hueter inoculated the anterior chamber of the eye in rabbits with lupus tissue and produced tuberculosis of the iris. Schüller (*Untersuchung über die Entstehung und Ursache der scrofulösen und tuberculösen Gelenkleiden*, 1880) introduced fragments of lupus tissue into the veins of animals, and in this way produced pulmonary tuberculosis. He also claimed to have discovered the microbe of tuberculosis by fractional cultivation from lupus tissue which, when conveyed into the vessels of the lungs, produced phthisis, and when injected into joints tubercular inflammation, caseation, and, finally, miliary tuberculosis. Koch (*Mittheilungen aus dem Kaiserlichen Gesundheitsamte*, B. xi. 1883) inoculated the anterior chamber of the eyes of eighteen rabbits from five cases of lupus, and in all of them tuberculosis of the iris was produced, and, if life was prolonged for a sufficient length of time, was followed by tuberculosis of the lymphatic glands of the neck, lungs, kidneys, liver, and spleen. Similar results were obtained in five guinea-pigs.

Cornet has recently made numerous experiments in Koch's laboratory on animals to ascertain the inoculability of tuberculosis through abrasions of the skin. He found that if lupus tissue, or a pure culture of tubercle bacilli, is applied to a cutaneous abrasion, the result in most, if not in all, cases is a local tuberculosis in the adjacent lymphatic glands, and, later, a general miliary tuberculosis.

The same author ("Demonstration von tuberculösen Drüsen-Schwellungen nach Impfungen von Tuberkel bacillen bei Hunden," *Centralblatt f. d. Gesammte Medicin*, No. 29, 1889) made subsequently a long series of experiments on dogs to ascertain the different avenues through which infection is known to take place. Tuberculous sputum and pure cultures inserted into the lower conjunctival sac in healthy dogs produced tissue hyperplasia at the seat of inoculation and was followed by infection of the cervical glands on the corresponding side. Some of the glands had undergone caseation, and the presence of bacilli could be demonstrated in all of the pathological products. In other animals the tuberculous material was introduced into the nasal cavity. The cervical glands, especially those on the corresponding side, became enlarged and caseated. Infection through the mouth by depositing the tuberculous material in a depression made with a blunt instrument between the canine teeth resulted also in tuberculosis of the cervical glands. Infection of the external meatus of the ear without creating intentionally an infection-atrium was followed by infection of the lymphatic glands behind the ear and along the neck on the same side. Cutaneous tuberculosis in the form of an ulcerating lupus was produced by shaving the skin on one side of the nose and face, and scratching it with a finger-nail infected with tuberculous material. Injection of the material into the healthy vagina of bitches resulted in local tuberculosis and secondary infection of the inguinal glands. Inoculations of other parts were followed by the same train of symptoms—local tuberculosis at the seat of infection followed by extension of the tuberculous process along the nearest lymphatic channels. The lungs were found affected only in two of the animals.

Cornil and Leloir implanted lupus tissue into the peritoneal cavity of guinea-pigs, and in five cases, out of fourteen experiments, produced peritoneal and general tuberculosis. Implantations from these animals into healthy animals again yielded positive results.

Pagenstecher and Pfeiffer (*Berliner klin. Wochenschrift*, 1883) took the secretion of the conjunctiva from patients suffering from lupus of this structure and injected it into the anterior chamber of the eye in rabbits. After five to six weeks, nodules could be seen on the surface of the iris which were in every respect identical with tuberculosis of this organ. Doutrelepont (" Die Aetiologie des Lupus vulgaris," *Proceedings of International Congress*, Copenhagen) inoculated the peritoneal cavity of fifty guinea-pigs, and in eight rabbits the anterior chamber of the eye, and in all of the animals local tuberculosis was produced at the point of inoculation; and in three of the guinea-pigs, and in one rabbit, the local disease was followed by general tuberculosis. The advances in our knowledge of the etiology of tuberculosis, the discovery of the bacillus, and the production in animals of tuberculosis by implantation of lupus tissue have finally settled the identity of tuberculosis and lupus. As tuberculosis is now diagnosed wherever the respective bacillus is found, another diagnostic significance is admitted even by those who are inclined to be sceptical in regard to the etiological rôle played by it. Koch produced artificial tuberculosis in over five hundred animals with material from different tubercular lesions and examined them all with the greatest care. Of the bacillus he pro-

PLATE VII.

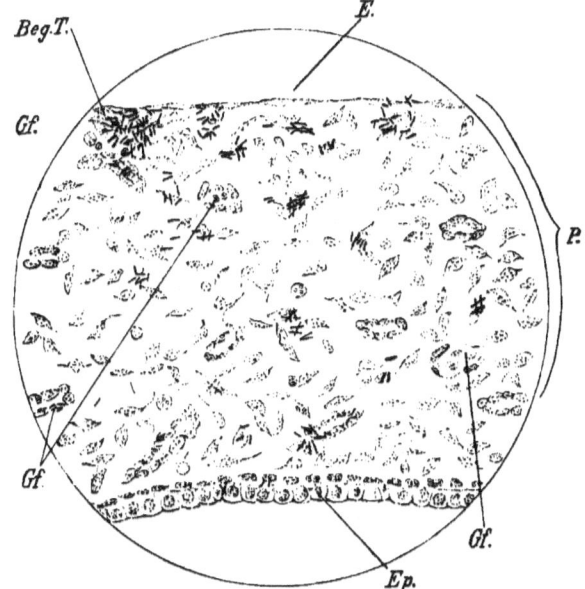

Tubercular eruption in the iris of a rabbit, fifth day after inoculation. Zeiss $\frac{1}{12}$. (Baumgarten.)

Beg. T. Formation of tubercles beginning (separation of white corpuscles.).
E. Endothelium of the anterior surface of the iris.
Ep. Epithelium of the posterior surface of the iris.
P. Iris parenchyma.
Gf. Bloodvessels.

duced forty-three pure cultures, some of which he continued through over thirty generations, occupying a period of two years. We shall see that inoculations with material from so-called scrofulous glands produce the same effect as when lupus tissue is used, and must, therefore, attribute their existence to the same cause.

Arloing (*Comptes rendus*, t. xcix. p. 661) prepared an emulsion from a simple scrofulous gland, caseous in the centre, which was taken from a boy aged fourteen. This was injected beneath the skin of ten rabbits and the same number of guinea-pigs. Visceral tuberculosis developed in all guinea-pigs, but the rabbits remained healthy, except that two showed yellow caseous granulations at the seat of inoculation. From a gland removed later, from the same boy, a similar emulsion was made and injected into the peritoneal cavity of six rabbits and six guinea-pigs. As before, the guinea-pigs all presented tubercular lesions; the rabbits, on being killed, were found to be perfectly healthy. In two instances, pus from strumous abscesses gave similar results. Some glands excised from the neck of a young woman produced tuberculosis both in rabbits and in guinea-pigs, but the patient died three weeks after the operation from miliary tuberculosis. Arloing appears to consider this case as outside the general category of strumous glands. From these experiments he inferred that either scrofula and tuberculosis were nearly allied affections, but caused by different agents, or they were derived from a single virus, of which the activity was modified in the scrofulous form.

Bollinger ("Ueber den Einfluss der Verdünnung auf die Wirksamkeit des tuberculösen Giftes," *Münch. med. Wochenschrift*, No. 43, 1889) has studied experimentally the effect of dilutions of tuberculosis material. He found that infectious milk from a tuberculous cow which produced typical local tuberculosis by intraperitoneal injections, if diluted from 1 : 40 to 1 : 100 has lost its virulence. The sputum of phthisical patients was found much more virulent and had not lost its power to produce tuberculosis on being diluted 1 : 100,000, on being injected into the abdominal cavity or the subcutaneous connective-tissue. Feeding experiments with sputum diluted 1 : 8 yielded negative results. Pure cultures remained virulent when diluted 1 : 400,000. All the experiments proved that the more concentrated the material and the greater the number of bacilli the more rapid and intense was the development of the lesion caused by the injection. In guinea-pigs it was estimated that about 820 bacilli were necessary to produce fatal tuberculosis. Intra-peritoneal injections did not always produce peritoneal tuberculosis, and when this was absent the organs affected were the lymphatic glands, spleen, lungs, liver, kidneys, and genital organs in the order of frequency named, showing that localization does not always take place at the point of primary infection.

Mr. Eve ("On the Relation of Scrofulous Gland Disease to other forms of Tuberculosis; an Experimental Inquiry," *British Medical Journal*, April 14, 1888) conducted some experiments for the same purpose, using small fragments of the glands prepared with sterilized instruments, and the possibility of infection of the animals with true tuberculosis was negatived by the precautions taken. As regards the results of the experiments on rabbits with glands from five different subjects, the material from these cases produced visceral tuberculosis; from one case, cold abscesses; and in one instance, in which a rabbit was inoculated in the anterior chamber of the eye, it escaped infection altogether. No bacilli could be discovered in the gland. Three rabbits inoculated with the infected organs from a guinea-pig, proved to be the most acute cases of tuberculosis that the author has been able to induce by strumous gland disease in rabbits, the tubercles being widely disseminated. The bacilli in the visceral tuberculosis were generally uniformly stained with even outlines, but in a strumous abscess in a rabbit which had existed for eight months they were nearly all "beaded" or uniformly stained, and often collected in groups not unlike clumps of micrococci.

Wm. Mueller ("Experimentelle Erzeugung der typischen Knochentuberculose," *Centralblatt f. Chirurgie*, 1886, p. 233) produced the characteristic form of tuberculosis in bone experimentally by the injection of tuberculous material into the nutrient artery. König for a long time had claimed that the wedge-shaped sequestrum so constantly found in tubercular foci in the articular extremities of the long bones, was due to occlusion of a small artery by a tubercular embolus. Müller's experiments were made to prove this clinical observation. He made sixteen experiments on rabbits, injecting tuberculous pus into the femoral artery, some in a peripheral, some in a central direction, without any positive results. In a second series the same material was thrown into the nutrient arteries of the femur and tibia. Of ten of these cases, two showed a tuberculous focus in the medulla of the diaphysis of the tibia, in another case miliary tuberculosis in the femur and tibia, and in the latter bone a small caseous spot in the spongy part, which contained numerous bacilli. The animals were killed eight weeks after the injection and showed no evidences of organic disease, except few tubercles in the lungs. Twenty experiments were made on young goats, five on sheep, and two on .dogs. The tuberculous material was injected directly into the nutrient artery of the tibia, the tibial artery being tied above and below this vessel. Primary union of the wound was obtained in all cases except on one dog.

In the dogs and sheep, all experiments yielded negative results. In the goats bone affections were produced which were identical with tubercular bone lesions found in man. Most frequently the disease was established in the diaphysis—cheesy masses and granulation tissue showing themselves in the medulla and cortical substance, or tuberculous osteomyelitis with, or without, sequestration. Typical lesions were also found in the ends of the bones with, and without, implication of the adjacent joints. In two of these cases the epiphysis was affected, while in three the shaft was involved. The following experiment furnishes a good illustration of the identity of the bone disease produced experimentally, and the disease as it occurs in man. Tuberculous material was injected into the tibial artery of a goat three months old. Wound healed in eight days. Some lameness four months later, gradually increasing during the next nine months. At the same time a swelling appeared at the knee-joint. Tibia painful on outer side. Animal killed thirteen months after the injection. There was found a typical fungous disease in the knee-joint most advanced at the sides, a wedge-shaped sequestrum in one of the tuberosities of the tibia, and a small granulation mass in the centre of the head of the tibia, and two similar granulation masses in the lower epiphysis of the femur. With the exception of the lymphatic glands of the knee-joint, no other organs were affected. In some cases pulmonary tuberculosis developed, twice general tuberculosis. The rest of the animals were killed when they began to show lameness—fourteen days to thirteen months after the inoculation. The tubercular lesions thus produced were examined for bacilli and these were never found absent. The starting-point in every instance must have been a tubercular embolus in one of the small arterial branches in the extremity of the bone.

The opinion that tubercle is capable of inoculation was held by many ancient writers, and Laennec himself, after a nick from a saw while making a necropsy on a phthisical subject, thought that he witnessed an example of inoculation in a small tubercle in the skin, but twenty years afterward this great physician and teacher was in good health, though finally he died of phthisis.

Inoculation-tuberculosis.

Schmidt ("Uebertragbarkeit der Tuberkulose durch cutane Impfung," *München. ärtzliches Intelligenzblatt*, 1883, Nos. 47 and 48) made a number of experiments to ascertain the effect of inoculations of superficial abrasions of the skin with the virus of tuberculosis. In guinea-pigs he made slight cutaneous abrasions of the skin to which he applied tubercular material, and covered the point of inoculation with collodium. All of his experiments failed in producing tuberculosis, while in the control animals in which the infectious material was introduced into the subcutaneous tissue, or into the peritoneal cavity, tuberculosis developed without a single exception. He believes that the results of these experiments are only corroborative of the assertion previously made by Bollinger and Koch that the susceptibility of the cutis for tubercular infection is slight. A sufficient number of well-authenticated cases, however, have been reported during the last few years to prove that in man tuberculosis is not infrequently contracted by the absorption of tubercular material through small wounds and superficial abrasions of the skin. Volkmann a number of years ago made the statement that tubercular infection never takes place through large operation wounds, or at the site of severe injuries, but that localization of the bacillus is likely to take place in parts the seat of very slight contusions, or what may appear at the time as an insignificant injury. He explained this by assuming that the active tissue-changes which take place during the process of regeneration after a severe trauma prevented the infection. In studying the cases of inoculation-tuberculosis, which will be referred to below, it will be seen that the infection-atrium was always caused by a trivial injury.

Martin du Magny (*Contribution à l'étude de l'inoculation tuberculeuse chez l'homme*, Thèse, Paris, 1886) has collected the clinical material of cases of inoculation-tuberculosis and in his comments upon them asserts that the sputum of phthisical patients and animal excretions were the usual carriers of contagion, consequently, the affection is most frequently met with among physicians, nurses, butchers, and teamsters. The external appearances manifested at the point of inoculation consist of the formation of a red nodule in the skin which increases slowly in size and forms a miliary abscess in which papillomatous proliferation takes place, and around which a new zone of infiltration forms, which in turn again suppurates and becomes papillomatous. The centre heals with the formation of a flat cicatrix, while the destructive process progresses slowly in a peripheral direction. Raymond ("Contribution à l'étude de la tuberculose cutane par inoculation directe," *France Méd.*, 1886, p. 99) reports two cases. In the first case the patient, suffering from advanced tuberculosis, inoculated himself through an abrasion of the skin of the hand by sucking a wound to arrest hemorrhage. The little wound refused to heal, and became covered with a crust under which suppuration was going on. Later, a papillomatous elevation formed, which continued to ulcerate on the surface. The margins of the ulcer were surrounded by an infiltrated zone, the skin covering it presenting a brownish color. The second case, a healthy man, injured himself by the prick of a thorn.

The wound became infected with the sputum of his phthisical wife. The puncture became the centre of a papillomatous swelling surrounded by a dark red zone. Suppuration took place at several points at the same time. The whole of the diseased portion of the skin was removed and the sections taken from it, on staining, showed the presence of numerous bacilli. In both cases the period of incubation was three weeks.

Hanot (*Archives de Physiologie*, July, 1886) has collected six observations which would tend to show that tubercular inoculation in man does take place, one case having fallen under his own observation. In the case observed by Hanot, the patient was in the third stage of phthisis, and died soon after with a tubercular ulcer on the arm of at least two years' standing; while the history of cough only dated from the last two months, which would show that the cutaneous lesion preceded the pulmonary, and was the cause of the phthisis. In the cases which he collected the sources of inoculation were necropsies on tubercular subjects, nursing phthisical patients, handling old bones, pricking the hand with a fragment of porcelain from the broken spittoon of a phthisical patient, and in four of the cases the tubercular character of the cutaneous lesion was verified by finding the bacilli.

Axel Holst mentions the case of an attendant on phthisical patients at a hospital, who had suffered for a long time from atonic ulceration of the fingers, which had been treated unsuccessfully by various external applications; no tubercle bacilli could be found with certainty in the sores. Later, the man was affected with tuberculous glandular swelling in the axilla, which contained a considerable number of bacilli, and Holst considers that it is highly probable that the patient had been infected through the ulcers.

Merklen ("Inoculation tuberculeuse localisée aux doigts," *Gazette hebd.*, 1885, No. 27) presented a case of inoculation tuberculosis to the Société des hôpitaux in which the infection of the wife of a phthisical husband could be clearly traced, and had occurred through the fissures of the fingers. At the point of inoculation hard nodules formed in which bacilli were found; this was followed by tubercular lymphangitis, which finally led to pulmonary tuberculosis. The patient had previously been in perfect health and without any hereditary taint.

Eiselsberg ("Beiträge zur Impf-tuberculose beim Menschen," *Wiener med. Wochenschrift*, 1887, No. 53), during the last few years, has observed four cases of inoculation-tuberculosis. The first case was a girl sixteen years old, in whom the disease developed in the track of a perforation of the lobe of the ear made preparatory to the wearing of an earring, and which was kept from closing by the insertion of a thread. The tubercular product appeared in the shape of a hard swelling the size of a hazelnut. The second case was a young man who injured himself with the point of a knife above the external epicondyle of the humerus. Eighteen days later a swelling the size of a pea appeared at the site of injury with an ulcerated surface covered by pale flabby granulations. In the axilla of the same side one of the lymphatic glands was found enlarged to the size of a hazelnut. The third case concerned a woman fifty years of age, who is supposed to have infected herself by washing the clothes of a person the subject of a tubercular abscess of the spine, and who with her fingers scratched an acne pustule on her face. At this point, six to eight days later, a painful swelling the size of a pea formed which subsequently became indurated and opened spontaneously in six weeks. At the end of three months the place of inoculation presented an ulcer with indurated margins. In the fourth case the inoculation followed in the track made by the needle of a hypodermic syringe in a girl twenty years of age. The swelling opened after six weeks, and a small quantity of pus was discharged. Four months, subsequently, the fistulous opening communicated with an abscess cavity the size of a silver dollar, lined by a wall of granulation tissue In all of these cases no evidences of tuberculosis could be detected in any of the internal organs, and the local disease could be traced in every instance to some antecedent lesion through which the infection had evidently taken place. The diagnosis in all cases was based on an examination of the granulation tissue for the bacillus of tuberculosis, which was always found present.

Tschering (" Inoculations-tuberculose bei Menschen," *Fortschritte der Medicin*, 1885, No. 3) reports an exceedingly interesting case of inoculation-tuberculosis occurring in a person who injured a finger with a broken spittoon used by a phthisical patient. From the little wound the tubercular infection extended to a tendon sheath and later to the axillary lymphatic glands.

Lennander (Abstract, *Annals of Surgery*, June, 1890) reports a case of tuberculosis of the skin following vaccination. The patient, a student of philosophy, thirty-five years old, presented tuberculosis of the skin of the right upper arm. The lesion had developed after the first vaccination, and later extended over a larger part of the upper arm. After repeated curetting and cauterizing the place healed over finally after grafting by Thiersch's method.

I have seen a number of cases of well-marked tuberculosis of the skin in which the diagnosis was verified by inoculation experiments and microscopical examination in every instance. I had every reason to suspect that the lesions were the result of cutaneous inoculation. I also had under treatment a well-marked case of extensive subcutaneous tuberculosis of the hand in the person of the mother of several children, who had died of pulmonary tuberculosis. The disease originated near the tip of the index finger at the site of a former abrasion in which a papillomatous swelling formed. This ulcerated and partly healed, when the disease commenced to spread along the subcutaneous connective-tissue, and had extended almost over the entire dorsum of the hand at the time she came under my care. A number of fistulous openings existed which discharged only a few drops of thin, serous pus daily. The subcutaneous tissue was transformed into a mass of granulation tissue which was removed after free incision with a Volkmann's spoon, and the wound surfaces were freely iodoformized. The process of repair was slow but satisfactory.

TUBERCULAR INFECTION OF WOUNDS.—Eiselsberg's case has already been described in which tubercular infection took place through a small punctured wound of the arm.

Middeldorpff (" Ein Fall von Infection einer penetrirenden Kniegelenks-Wunde durch tuberculöses Virus," *Fortschritte der Medicin*, 1886) reports the case of a healthy carpenter who opened his knee-joint by the cut of an axe, and dressed the wound with a soiled handkerchief. The wound healed kindly, but later the joint became swollen, tender, and painful. Resection was performed, and on examining the capsule it was found very much thickened. In the granulation tissue bacilli were found.

Czerny (*Centralblatt f. Chirurgie*, 1886) reports two cases in which tuberculosis followed in wounds treated by Reverdin's transplantation of skin. In both instances the patients were healthy, and the skin transplantation was made during the treatment of extensive burns. The skin was taken from limbs amputated for tubercular affections. In both cases tuberculosis of the adjacent joint occurred and in one of them tuberculosis of the granulating surface.

A case of tubercular infection through earrings is related (*Wiener med. Presse*, 1889) in a girl, fourteen years of age, of a perfectly healthy family, who wore earrings left to her by a friend who had died of pulmonary tuberculosis. Soon ulcers appeared on the lobes of both ears, the cervical glands became swollen, and percussion and auscultation revealed infiltration of the apex of the left lung. Tubercle bacilli were found in the ulcers and sputa. This case is only another instance of inoculation-tuberculosis where from the point of injection the disease extended along the lymphatic system and finally systemic infection occurred from the entrance of the bacilli into the general circulation.

Wahl ("Mittheilungen eines Falles von Inoculations Tuberculose nach Amputation des Unterarms," *Verh. der Deutschen Gesellschaft f. Chirurgie*, XV. Congress) amputated the arm of a boy suffering from gangrene, the result of an injury, and discharged the patient with the wound completely healed, except a small granulation surface from which the drainage-tube had been removed. At home the wound was dressed by a girl suffering from tuberculosis. The wound showed soon all the characteristic appearances of fungous disease, and the lymphatic glands became infected from this source.

König (*Die Chirurgische Klinik in Göttingen*, Leipzig, 1882) has observed sixteen cases of inoculation-tuberculosis following operations for tuberculous disease of bones, and two such cases have been described by Kraske ("Ueber tuberculöse Erkrankung der Wunden," *Centralblatt f. Chirurgie*, 1885, p. 565).

INOCULATION FOLLOWING CIRCUMCISION.—A number of cases of inoculation-tuberculosis following circumcision are on record in which the infection often occurred in the practice of orthodox Jews, who performed the operation in accordance with the directions laid down in the Mosaic laws. The loose connective-tissue of the prepuce, richly supplied with lymphatics, is an admirable surface for absorption, and when brought in contact with infectious material would furnish the most favorable conditions for the production of local lesions and the transportation of microbes along the lymphatic channels to more distant parts.

Lehmann ("Ueber einen Modus der Impftuberculose beim Menchen," *Deutsche med. Wochenschrift*, 1886) has observed ten cases of inoculation-tuberculosis in Jewish boys caused by sucking the wound after ritual circumcision by a phthisical person. Ten days after circumcision the wound became the seat of ulceration, and the inguinal glands began to enlarge. Four of the children died of tubercular meningitis, and three died after a prolonged illness caused by multiple tubercular abscesses.

Hofmokl has reported a similar case, and Weichselbaum detected the bacillus of tuberculosis in the circumcision wound.

Elsenberg (*Centralblatt f. Bacteriologie u. Parasitenkunde*, B. ii. S. 577) has described three new cases of inoculation-tuberculosis after circumcision. All the cases were infants, and the disease appeared primarily in the wound, or cicatrix, and later, in the inguinal glands. Local treatment by evidément proved successful. The diagnosis was corroborated by microscopical examination of the granulation tissue.

Meyer ("Ein Fall von Impftuberculose in Folge ritueller Circumcision," *New Yorker Med. Presse*, June, 1887) reports a case in which circumcision was performed according to the rites of the church eight days after birth by an old man, and in which four weeks after the operation an induration appeared at the frenulum and the inguinal glands about the same time began to enlarge. Syphilis was suspected, and the patient was put on a specific course of treatment. The inguinal glands suppurated and another small abscess formed in the right gluteal region. The diseased tissue about the glans penis was excised. Microscopical examination of the granulations revealed the presence of miliary tubercles and bacilli in great abundance. The above cases furnish abundant proof of the possibility of the transmission of tuberculosis by cutaneous inoculation through abraded surfaces, small wounds, and granulating surfaces, and deserve the most careful attention of surgeons in the matter of prophylaxis, diagnosis and treatment.

CHAPTER XVII.

CLINICAL FORMS OF SURGICAL TUBERCULOSIS.

It is but a few years ago since the forms of tuberculosis to which allusion will be made here were not correctly understood, and consequently a rational surgical treatment was out of the question. Most all of the localized tubercular processes were included under the general term scrofula, and were looked upon as local manifestations of a general dyscrasia, and treated in accordance with this view of their pathology. The discovery of the bacillus of tuberculosis has rendered the word scrofula obsolete, and has assigned to the tubercular processes in the various organs and tissues of the body their correct etiological and pathological significance and paved the way for their successful surgical treatment.

1. Skin.

a. Tuberculosis verrucosa cutis. Riehl u. Paltauf ("Tuberculosis verrucosa cutis," *Vierteljahrsschrift f. Dermatologie u Syphilis*, B. xiii. S. 14) have described an affection of the skin, under the name of tuberculosis verrucosa cutis, in which the bacillus of tuberculosis was constantly found, and which they attributed to local infection, because all of the patients they examined were persons handling animal products. Riehl ("Tuberkel Bacillen in einem sogenannten Leichen-tuberkel," *Centralblatt f. Chirurgie*, 1885, No. 32) has shown, by finding the tubercle bacillus in the warty affections of pathological anatomists, the probable tuberculous nature of this condition of the skin, although the infection may be a mixed one.

b. Lupus. The long debate which has been carried on for years as to the identity or non-identity of lupus and tuberculosis has terminated in favor of those who argued that lupus is a tuberculosis of the skin, hence the term lupus should be omitted from nomenclature, and the term tuberculosis, as applied to the skin and mucous surfaces, should be qualified by the anatomical name of the surface affected.

Finger ("Lupus und Tuberculose," *Centralblatt f. Bacteriologie u. Parasitenkunde*, B. xi. S. 348), in a lengthy and able article, makes a careful inquiry into the identity of lupus and tuberculosis, and after a careful review of all that pertains to this subject decides in favor of its tubercular origin and character. He alludes to the views of Hebra, Virchow, Klebs, Hueter, and others. From a clinical standpoint Hebra brought the different varieties of lupus under one common head. He separated it entirely from syphilis, but otherwise did little to fix its pathological significance. He adopted the classification of Fuchs, and the older French and English authors, who taught that it was one of the manifestations of scrofula, and that anatomically it was composed of granulation tissue. Virchow classified it with the granulomata, but denied its identity with scrofula. Rindfleisch described it as a proliferation of epithelial cells, as a sort of phthisis cutanea. Eppinger referred to it as a product of connective-tissue growth and proposed its classification with carcinoma. Klebs looked upon it as a specific affection, and histologically included it among the small-celled leucocytoses. Hueter, who in his pathological views was generally far in advance of his time, affirmed that it was a form of fungous inflammation, the specific cause of which, when introduced into the organism, produced a miliary tuberculosis. Volkmann included it among the affections composed of granulation tissue. Friedländer ("Ueber locale Tuberculose," Volkmann's *Klinische Vorträge*, 1881, No. 31) was the first to take a positive stand in asserting that lupus is a tubercular affection of the skin, and showed its histological identity with other recognized forms of local tuberculosis. He demonstrated the presence of miliary tubercles in it, and that these nodules were composed of giant and epithelial cells, the same as in tubercles of the lungs. The views entertained by Baumgarten as to the histological structure of lupus are different from those just described. He believed that in the miliary tubercle the abundance of epithelioid cells predisposed to caseation and suppuration, while he recognized in lupus, as the most characteristic distinguishing features, absence of caseation and suppuration and the presence of cicatricial tissue. He admits, at the same time, that lupus may be closely allied to tuberculosis.

Schüller ("Ueber die Stellung des Lupus zur Tuberculose," *Centralblatt f. Chirurgie*, 1881) opposed Baumgarten and emphasized the fact that caseation, characteristic as it might be for all tuberculous affections, always constitutes only a secondary condition and depends upon the soil present in and around the nodule. The absence of caseation in lupus, which in rare cases, however, has been shown to be present by Cohnheim and Thoma, could not be urged as a positive and infallible diagnostic criterion of the non-tubercular nature of lupus.

Neisser (*Die chronische Infections-Krankheiten der Haut*, Ziemssen, B. xiv., 1882) accepts fully and pleads strongly in favor of the tubercular nature of lupus. In the meantime Pantlen, Bizzozero, Baumgarten, Chiari, Hall, Janisch, Riehl, Vidal, and Finger described a new affection, a diffuse tuberculosis of the skin and mucous membranes, occurring as a sort of secondary localization in patients suffering from advanced tuberculosis. To prove that lupus and tuberculosis are identical it became necessary to furnish the necessary experimental proof, and to show the uniform presence of the bacillus of tuberculosis in the lupus tissue. The inoculation experiments with lupus tissue have already been mentioned, and from them it can be learned that with few exceptions they were followed by positive results; that is to say, implantation of lupus tissue into the subcutaneous tissue or peritoneal cavities of animals susceptible to tuberculosis gave rise to local tuberculosis at the point of implantation, and to dissemination of the process in a manner characteristic of tuberculosis in man.

Demme ("Lupus und Tuberculose," *Würzburger med. Blätt.*, 1887) reports two cases of cutaneous inoculation-tuberculosis in man. The first case, a nurse girl, contracted an ulcerating lupus from a child three years of age which died of tubercular ulceration of the tonsils, and tuberculosis of other organs. The second case was a child with eczema which slept in the same bed with its phthisical mother. Sections of the eczematous skin, stained and examined under the microscope, showed numerous bacilli. The child died later of hemorrhage of the stomach, the cause of which at the necropsy was shown to be a tuberculous ulcer of the stomach.

Demme enumerates the following reasons as showing the identity of lupus and tuberculosis:

1. Similarity of histological structure.
2. Presence of the bacillus of tuberculosis in the granulation tissue of lupus.
3. The production of typical tuberculosis in animals not immune to this disease by implantation of lupus tissue, or injection of a pure culture of the bacillus of tuberculosis obtained from lupus tissue.
4. The fact that patients suffering from lupus are frequently attacked by and die of tuberculosis of other organs.
5. The prevalence of tubercular affections among relatives suffering from lupus (hereditary predisposition).

Boeck has made the statement that, of sixteen cases of lupus, three die subsequently of pulmonary and general miliary tuberculosis. Heiberg reports death from tubercular meningitis in a lupus patient.

Rassdnitz ("Zur Aetiologie des Lupus vulgaris," *Vierteljahrsschrift f. Dermat. und Syphilis*, 1882) collected 209 cases of lupus and found that in thirty per cent. of all the cases it was associated with other evidences of tuberculosis. He placed, also, great importance on the observations that lupus is prone to develop in the scar left after the healing of a localized tuberculosis in lymphatic glands, and that lupus is often observed upon the nose or eyelids in cases of chronic nasal or conjunctival catarrh. In ten to fifteen per cent. of his cases lupus could be traced to a hereditary predisposition. Demme observed miliary tuberculosis in two of his cases after scraping lupus. Pontoppindau asserted that, in his experience, in fifty to seventy-five per cent. the patients manifested additional evidences of tuberculosis.

Quinquaud (*De le Scrofule*, Thèse, Paris, 1880) saw in three cases of lupus pulmonary tuberculosis appear as the final cause of death.

Of thirty-eight cases that came to the personal knowledge of Bessnier (*Le Lupus et son Traitement. Ann. de Dermat.*, 1883), eight of them suffered from pulmonary phthisis. Of two patients treated by Aubert, one died of acute pulmonary tuberculosis, and the other of tubercular pleuritis after scarification, Renouard was able to ascertain the existence of pulmonary phthisis in fifty per cent. of his cases of lupus.

Block (" Klinische Beiträge zur Aetiologie u. Pathogenese des Lupus," *Vierteljahrsschrift f. Dermat. u. Syphilis*, 1886) met with tuberculosis in other organs before or after the development of lupus, in 114 out of 144 cases. Bender (" Ueber die Beziehungen des Lupus vulgaris zur Tuberculose," *Berl. klin. Wochenschrift*, 1886) examined 374 cases of lupus. In 159 of these cases an accurate account could be obtained. In 99 of the latter number symptoms of other antecedent or coexisting tuberculous lesions existed. In 77 of the cases tuberculosis in an etiological or clinical aspect was present.

Leloir (" Recherches nouvelles sur les relations qui existent entre le lupus vulgaris et la tuberculose, " *Annal. de Dermat. et Syphilis*, I., viii.) observed several cases in which after years a lupus of the face gave rise to a pseudo-erysipelatous swelling of the face which disappeared after a time, to be followed by swelling of the submaxillary lymphatic glands, which remained stationary. Soon after the affection of the lymphatic glands had appeared, febrile disturbances, gastric symptoms, and evidences of pulmonary infiltration followed. In all of these cases, Leloir believes that the virus of tuberculosis had left the primary location and had migrated through the lymphatic vessels and glands into the lungs. In ten out of his seventeen cases, the tubercular nature of lupus was clinically manifest.

Sachs (Beiträge zur Statistic des Lupus," *Centralblatt f. Chirurgie*, 1887, No. 2, p. 19) ascertained that of 105 cases of lupus which he had collected, in eighty-six per cent. the patients had coexisting tuberculosis in other parts of the body, or a hereditary predisposition to tuberculosis could be shown to exist.

The above literature furnishes strong clinical evidence of the identity of lupus with tuberculosis, but the following bacteriological researches furnish the final and conclusive proof. Koch in his paper on the etiology of tuberculosis (previously quoted) states that he produced a pure culture of the bacillus of tuberculosis from a case of lupus which resembled in every respect the culture obtained from recognized tubercular lesions, and with the fifteenth generation from this source, one year after the first cultivation, he inoculated five guinea-pigs by subcutaneous injection, and produced typical tuberculosis in all of them.

Doutrelepont ("Tuberkel-bacillen im Lupus," *Monatschrift f. prakt. Dermatologie*, 1883), in seven cases of lupus, found the bacillus of tuberculosis invariably present in greater or less number, either within the cells, or dispersed in small groups between them. He never found them in giant cells, but in their immediate vicinity. In a second paper the same author ("Zur Therapie des Lupus," *Monatschrift f. prakt. Dermatologie*, 1884) reports eighteen additional cases of lupus, in each of which the presence of the bacillus could be demonstrated.

Demme ("Zur diagnostischen Bedeutung der Tuberkel-bacillen für das Kindesalter," *Berl. klin. Wochenschrift*, 1883) in six cases of lupus detected the bacillus of tuberculosis. Pfeiffer ("Tuberkelbacillen in der lupös erkrankten Conjunctiva," *Berl. klin. Wochenschrift*, 1883) demonstrated the presence of the specific microbic cause of tuberculosis in a case of lupus of the conjunctiva.

Schuchardt and Krause ("Ueber das Vorkommen der Tuberkelbacillen bei fungösen und scrofulösen Entzündungen," *Fortschritte der Medicin*, 1883) found the bacillus of tuberculosis in three cases of lupus affecting the face, ears, and legs. In examinations made of eleven cases of lupus by Cornil and Leloir, and four by Koch for the especial purpose of showing the identity of lupus and tuberculosis, the bacillus was found in every instance. In the artificial tuberculosis of animals produced by implantation of lupus tissue, the bacillus of tuberculosis was shown to exist by Pagenstecher, Pfeiffer, Koch, and Doutrelepont.

2. *Primary Tuberculosis of Iris.*

Griffith (*British Medical Journal*, Dec. 21, 1889) has related a case of primary tuberculosis of the iris which occurred in a seven months' old female child. The eye had been affected for one month; there was an enlarged gland in the neck on the same side, but there were no other physical signs of tubercle. No history of heredity.

A yellowish nodule grew from the periphery of the iris of the right eye, and numerous millet-seed bodies from its surface; the pupil was closed, but there was no acute inflammation. The local disease increased rapidly in extent. The eye was enucleated after three weeks' treatment. The disease was found to be confined to the iris and ciliary body. Under the microscope the new-growth showed the characteristic structure of tubercle. In 32 recorded cases in which microscopic and bacteriological tests left no doubt as to the tubercular nature of the disease, one eye only was affected in 29; the average age of the patients was twelve, youngest four months, oldest fifty-one years. In 10 cases bacilli were searched for, but only found in 4; in one of the remaining 6 cases, however, the inoculation test was successful. A number of patients recovered completely and permanently after enucleation.

3. *Tuberculosis of the Internal Ear.*

That an ordinary otitis media with perforation of the tympanum may occasionally be transformed into a tubercular lesion by the entrance from without of bacilli of tuberculosis there can be no doubt. Habermann (*Prager med. Wochenschrift*, March 7, 1888) has recently investigated this subject by examining, post-mortem, 18 tuberculous subjects in whom either otorrhœa or deafness without active discharge had been observed during life, and in nine of these he could demonstrate the presence of tubercular lesions in the auditory canal. In one case he found in the left auditory apparatus tuberculosis of the entire middle ear, where the tympanum was intact. In another tubercular subject, a man, thirty-eight years of age, in whom tuberculosis of the ear was observed a year and a half before death, the post-mortem revealed extensive tuberculosis of the cochlea, in the internal auditory canal, and in the superior semicircular canal, while the other semicircular canals and the vestibule were destroyed by caries.

4. *Lymphatic Glands.*

That most cases of chronic inflammation of the lymphatic glands are in their origin, course, and final termination, cases of local tuberculosis has been satisfactorily shown by clinical experience, inoculations, and cultivation experiments. The tubercular virus enters the lymphatic circulation undoubtedly most frequently through some superficial abrasion, ulceration, eczema, or some other affection of the skin, as any loss of continuity of surface may furnish the necessary *portio invasionis* for the entrance of bacilli from without. Volkmann found tubercle bacilli in the skin of an eczematous forearm. In perhaps 95 out of 100 cases of tuberculosis of the lymphatic glands the disease attacks the glands of the neck, as the scalp, face, and mouth are parts of the body frequently the seat of slight injuries and superficial lesions, and consequently often exposed to tubercular infection. The lymphatic glands are filters for the microbes which enter the body through the lymphatic vessels The pathological conditions which are produced in the interior of a lymphatic gland by the presence of pathogenic microörganisms are well calculated, for the time being, at least, to limit the affection. The lymphadenitis which is produced blocks the lymph spaces with the products of a specific inflammation which temporarily at least mechanically fixes the microbes in their location. Primary infection of a lymphatic gland by the bacillus of tuberculosis in many instances attacks different portions of the gland from the very beginning, as a number of independent centres of tissue-proliferation are established around each microbe which has become arrested on its way through the gland. These separate nodules soon become confluent and form a mass of considerable size, which soon implicates the entire parenchyma of the gland. Local diffusion of the bacillus of tuberculosis in the interior of the gland is accomplished by the assistance of the lymph stream as long as the bacillus remains free, and through the medium of cells as soon as it has become intra-cellular. Local infection is not limited to the lymphatic glands on the proximal side of the primary focus, as during the course of the disease we often observe that lymph glands become involved which are not in the direct course of the lymphatic circulation. As the bacillus of tuberculosis is non-motile, we can only explain its transportation in a direction opposite the lymph current by its conveyance in such a direction by migrating cells. The usual course of infection along the lymphatic channels is, however, in the direction of the lymph current. The course of the disease is almost characteristic. A lymphatic gland in the submaxillary or parotid region becomes enlarged, and from this centre the infection invades successively gland after gland, until the whole chain of lymphatics from the lower jaw to the clavicle has become involved. Another interesting feature is observed in reference to the regional diffusion of the tuberculous process, as the course of infection usually corresponds to the location of the gland first affected. If the infection has involved primarily one of the deep glands of the neck, the glands subsequently invaded belong to the deep lymphatics which follow the large vessels of the neck. If, on the other hand, the primary infection is located in one of the superficial glands which are being irrigated by the lymph which flows through and from this gland become the seat of successive infection, showing again that regional infection usually takes place in the direction of the lymph current. As long as the infection has not extended along the entire length of the chain of lymphatic glands the patient is protected against miliary tuberculosis, but as soon as the virus has passed all the lymphatic filters, and has direct access into the thoracic duct, miliary tuberculosis follows as an inevitable result by the entrance of the bacilli into the general circulation.

Weigert ("Die Verbreitungswege des Tuberkelgifts nach dessen Eintritt in den Organismus," *Jahrb. f. Kinderheilkunde*, B. xxi. S. 146), in his description of the process of dissemination in cases of acute miliary tuberculosis, has pointed out that in some cases the bacilli are conveyed through the lymphatic system successively until they reach the general circulation, while in others, and by far the greater number, generalization of the tuberculous process takes place more directly by the entrance of tubercular products through a vein, an occurrence which is followed at once by rapid and extensive diffusion by embolism. When the bacilli of tuberculosis have reached the systemic circulation, the intensity of symptoms and the subsequent course of the disease depend on the number of bacilli which the blood contains.

Weichselbaum ("Ueber Tuberkel-bacillen im Blute bei allgemeiner acuter Miliartuberculose," *Anz. d. ges. Wienerärzte*, 1884, No. 19), starting with the idea advanced by Weigert that every acute miliary tuberculosis is caused by the direct entrance of the tuberculous virus into a vein or indirect entrance along the lymphatic channels and through the thoracic duct into a vein, examined the blood of three such, and found that it contained numerous bacilli. In the lymphatic glands the tubercular process pursues a typical pathological course. According to the rapidity of tissue-proliferation, or in proportion to the number of bacilli present, the granulation tissue which is always the first and constant product of tubercular inflammation undergoes caseation, and at an early stage often numerous centres can be seen from where the retrograde degenerative changes take their starting-point. By confluence of such caseous foci the entire gland is formed into a cheesy mass. Liquefaction of such caseous material results in the formation of a fluid which macroscopically resembles pus, but under the microscope it presents only the product of retrograde tissue-transformation, and no well-defined pus corpuscles, and none of the microörganisms which are known to produce suppuration can be obtained by cultivation experiments, showing conclusively that the fluid possesses no pyogenic properties. The bacillus of tuberculosis finds in the granulation tissue a soil well adapted to its growth and development, but, as soon as caseation takes place, its culture soil in that portion of the gland has been destroyed, and its field of growth is limited to the surrounding zone of granulation tissue, in which it can be constantly found, while it is absent in the central mass, or, at least, only an isolated rod can be found here and there on patient, careful search. In all probability the death of the granulation tissue which precedes caseation is caused by the action of the ptomaïnes of the bacilli of tuberculosis, which by their toxic properties destroy the protoplasm and arrest further cell-growth in that portion of the gland as well as reproduction of the micro-organisms. When a true suppuration takes place in a tubercular lymphatic gland, it does so in consequence of a secondary infection with pyogenic microörganisms. A spontaneous and permanent cure is not infrequently effected by the substitution of an acute suppurative process in place of the primary specific chronic inflammation, which destroys the entire soil of the bacillus of tuberculosis, and, at the same time, effects complete elimination of the pus-microbes and the bacilli through the discharges of the abscess. The capsule of the gland furnishes an efficient protection wall to the para-glandular tissues against tubercular infection often for months and years, but when perforation has once taken place the disease attacks the surrounding tissues irrespective of their anatomical structure. When a tubercular gland becomes the seat of an acute suppuration, the suppurative inflammation appears in the form of a para-adenitis with suppuration between the gland and the surrounding tissues, which often terminates in extensive separation of the gland from the adjacent tissues. While tuberculosis of the lymphatic glands often stands in a direct causative relationship to, and precedes general, diffuse, and pulmonary tuberculosis, it is seldom observed as a secondary affection in the course of tuberculous affections of other organs. Tuberculosis of the lymphatic glands is an affection usually noted for its chronicity, often remaining stationary for many years. I have repeatedly seen patients forty to fifty years of age who during childhood had suffered from enlarged cervical glands, which eventually diminished in size, but did not disappear, which, after years, again became the seat of an active tubercular process. In exceptional cases this affection pursues an acute course. Delafield reports an exceedingly interesting case of this kind (*Medical Record*, vol. i. p. 425, 1885). The disease commenced with an enlargement of one of the cervical glands near the angle of the lower jaw with a temperature of 40° C. (104° F.) and rapid extension to the proximal glands as far as the clavicle. Symptoms of pulmonary complication were not present. Rapid emaciation and marked anæmia supervened, followed after six weeks by swelling of axillary and inguinal glands. Ophthalmic examination revealed the same conditions of retina and papilla as in leukæmia or Bright's disease. A few days after the beginning of the disease profuse diarrhœa and reduction to nearly normal temperature occurred. The diagnosis was between malignant lymphoma and tuberculous adenitis. During the further course of the disease bronchial breathing in both lungs appeared. Heart, liver, and spleen appeared to be normal. Urine normal, but increase of temperature and respirations took place during this time. Death occurred in less than five months. At the autopsy the lungs were found congested and œdematous, with red hepatization of the lower lobes, and a few miliary tubercles. The spleen contained many miliary tubercles the size of the head of a pin, and most of them in a state of cheesy degeneration. The mesenteric glands were much enlarged; and a few of them in a condition of cheesy degeneration and calcification. In the cheesy masses bacilli were found. All the cervical glands were affected with softening and cheesy degeneration in the centre. The calcification of mesenteric glands pointed to an earlier affection. The disease remained latent and recurred in the same glands, and, later, extended to the cervical glands. This case resembles the cases described by Hilton-Fagge and Pye-Smith. Next to the cervical glands the glands of the axilla are most frequently affected. In my experience the operative treatment of tuberculosis of the axillary glands has yielded more satisfactory results than when the same operation was made under similar conditions on the glands of the neck. Tuberculosis of the inguinal glands is an extremely rare affection, and when present is usually associated with primary tuberculosis of the genital organs. From a practical standpoint the following two papers are of great importance.

Fränkel ("Zur Histologie, Aetiologie und Therapie der Lymphomata Colli," *Prager Zeitschrift f. Heilkunde*, 1885, Hefte 2 und 3), for three years, made it his duty to study the histological structure of all lymphatic tumors which were extirpated in Billroth's clinic. As the result of his observations he classified all primary tumors of the lymphatic glands examined into two classes: 1. Lymphatic tuberculosis; 2. Lympho-sarcoma. He also described a simple hyperplasia of the glands, which present under the microscope only healthy glandular tissue. In the tubercular glands he found the histological structure of tubercle perfect, and never failed in detecting the bacillus of tuberculosis. In persons between fifteen and thirty years of age he found local tuberculosis of the lymphatic glands as a quite frequent affection. In his paper he gives the clinical histories of 148 cases of glandular tuberculosis. Only in 15 of this number were the lungs affected. In 18 the patients were weak and badly nourished; while in 72 the general health was unimpaired. The age of the patients varied from nine to fifty-one years. On an average the duration of the disease was from three to four years; the shortest time was two months, and the longest thirty years. Of special etiological interest is the case of a woman, fifty-one years of age, who had given birth to two children, their father being the subject of advanced tuberculosis, and both of whom died of tuberculosis. She had been in perfect health until her forty-ninth year, when she was attacked simultaneously with pulmonary and glandular tuberculosis. In 128 of the reported cases the glands were extirpated, some of the operations being quite serious; in 16 cases the internal jugular vein had to be tied. In 91 of the operations the wound healed by primary union, and in 25 the healing was retarded by suppuration. Erysipelas complicated the result five times. In one of these cases a large part of the tuberculous mass was left, and it was noticed that the erysipelas had no effect on the tubercular process. Only in 49 of the cases operated on could the final result be obtained. Taking three and a half years as the time when the patient could be considered exempt from a recurrence of the disease, it was ascertained that in 24 per cent. no relapse followed the operation; a local relapse was observed in 14 per cent., and reappearance of the disease distant from the seat of operation in 4 per cent.

As lymphatic tuberculosis, in most instances, signifies the entrance of bacilli through a loss of continuity of the skin or of the mucous membrane, or through the socket of a carious tooth, localization occurring in one of the nearest glands to the *portio invasionis*, it must be looked upon as a local process, and amenable to timely surgical treatment by the removal of all of the infected tissues. The capsule of the lymphatic glands furnishes an efficient barrier against infection of the para-glandular tissue for a long time, and perforation only takes place after the disease has made considerable progress, and has been followed by extensive caseation, and especially by suppuration. Early operations are as necessary in the treatment of tubercular adenitis as in the treatment of malignant tumors, and holds out more encouragement so far as a permanent cure is concerned. By a thorough removal of the primary focus of infection, successive infiltration of proximal glands and miliary tuberculosis are prevented almost to a certainty if the operation is performed before the para-glandular tissues are affected. If the operation is done at such a favorable time, it is not attended by any great difficulties, as the glands can be readily enucleated, and as suppuration has not taken place the wound usually heals by primary union. If, however, the tubercular inflammation has involved many glands, and has extended to the connective tissue surrounding the glands, the operation becomes one of the most formidable in surgery on account of the close proximity of important vessels which are often imbedded in the mass. Under such circumstances a complete removal is often impossible and early local recidivation is inevitable, owing to imperfect removal of the infected area.

PLATE VIII.

From encysted bronchial gland in miliary tuberculosis. Giant cell with radiating arrangement of bacilli. (Koch.) 700 diam.

Traumatic dissemination is very likely to follow all imperfect operations, in which portions of glands or infected capsules are left behind by inoculation of the surface of the wound with the bacilli. I have seen in a number of such cases as early as a week after the operation the entire wound surface covered by a thick layer of granulation tissue, which showed all the histological evidences and possessed all the bacteriological properties of tubercular tissue. As an additional testimony in favor of the operative treatment of tubercular glands, I will quote from the paper of Schnell (*Ueber Erfolge von Extirpation tuberculöser Lymphome*, Dissertation, Bonn, 1885), who collected 56 cases of tuberculosis of the cervical glands, which were treated by extirpation in the clinic at Bonn. In 37 of these cases he was able to learn the ultimate result. In 57 per cent. the operation was followed by complete recovery; in 27 per cent. the disease returned at site of operation; and in four cases death resulted from pulmonary tuberculosis. The largest number of cases were patients between ten and twenty years of age.

5. *Bones.*

Tuberculosis of the bones is an exceedingly frequent affection in children and young adults. Its favorite location is in the epiphyseal extremities of the long bones, although it is also quite frequently met with in the short bones of the carpus and tarsus, and some of the flat and irregular bones, as the ribs, scapula, ilium, and vertebræ. Direct infection is never observed, and when the disease has made its appearance it is only an evidence of the existence of tubercular infection at an earlier day, or the presence of a tubercular process in some other organ. We observe clinically what Müller has demonstrated experimentally, that when the bacilli of tuberculosis are present in the blood current, very often localization takes place near the epiphyseal cartilage in young persons by the microbes becoming arrested in one of the terminal branches of an artery, the lumen of which becomes obliterated by the presence of an embolus of granulation tissue containing bacilli, or the lumen of the vessel is gradually diminished by the formation of a mural thrombus, which forms around bacilli implanted upon the vessel wall, and the lumen of the vessel is finally completely obstructed by the growth of the thrombus. The new vessels in the vicinity of the centres of growth in the bones of young persons, on account of their imperfect structure and irregular contour, furnish the most favorable conditions for the arrest of floating granular matter and the localization of pathogenic microbes. This predisposing anatomical element goes far to explain the frequency with which we meet with tubercular foci in the epiphyseal extremities of the long bones. Before the age of puberty it is safe to state that the primary lesion in tubercular affections of joints is located in one or both of the epiphyses of the bones which enter into the formation of the joint, while in the adult primary tuberculosis of the synovial membrane is more frequently met with. As age advances and the process of ossification is completed, the predisposing localizing causes in bone seem to disappear, while the synovial membrane becomes more susceptible to primary localization.

Of 204 specimens of tuberculous joints examined by Müller and quoted by König (*Die Tuberculose der Knochen und Gelenke*, Berlin, 1884, p. 66), 158 were primary osteal and 46 primary synovial tuberculosis.

As soon as embolic infection in bone has taken place, a process of decalcification occurs around the tubercular embolus or thrombus, and the preëxisting connective tissue is transformed into embryonal or granulation tissue which imparts to the product of the specific inflammation its characteristic fungous appearance. It is not often that only a single focus of tubercular infection in bone is present; more frequently, two or three foci appear at the same time or in slow or rapid succession, and it is not unusual to find that two neighboring epiphyses are infected at the same time, or during the course of the disease. The granulation tissue in bone undergoes the same secondary degenerative tissue-changes as in the lymphatic glands, hence, in advanced cases we expect to meet with caseation, liquefaction of the caseous material, and suppuration in cases of secondary infection with pyogenic microbes. The obstruction of a small artery by an embolus or thrombus which contains the bacilli of tuberculosis usually leads to sequestration of a triangular piece of bone which maps out the area of tissue which received its blood-supply from the obstructed vessel; thus the triangular sequestra are formed that are so frequently met with in osteal tuberculosis of the epiphyseal extremities. It is seldom that tuberculosis of bone develops in the course of pulmonary tuberculosis, but pulmonary and miliary tuberculosis can often be traced to a tuberculous focus in a bone. The intimate relations which exist between the tubercular nodule in bone and the bloodvessels furnish a satisfactory explanation of the frequency with which systemic infection takes place. As soon as the granulation process reaches an adjacent vein, the tissues of the vein-wall undergo the same process, and the bacilli reach the lumen of the vessel and reënter the systemic circulation, and give rise to miliary tuberculosis in organs which are anatomically predisposed to localization. As long as the decalcification of the surrounding bone goes on, the infection is progressive; but as soon as sclerosis takes place the process becomes limited, as the microörganisms are shut in, as it were, by an impermeable wall of sclerosed bone. The most unfavorable conditions are created in cases in which the tubercular focus becomes the seat of a secondary infection with pyogenic microbes, as the suppurative process opens up to the bacillus of tuberculosis new areas for infection in which the resistance of the tissues to tubercular infection has already been greatly diminished. It is also during the suppurative stage that joint complications are most likely to arise. The clinical history of cases of tuberculosis of bone, as well as the macroscopical and microscopical appearances of the lesion, are typical of tuberculosis as found in other organs. The crucial test which proves the tubercular character of most of the chronic inflammatory affections of bone in children has been furnished by bacteriological investigations. Most of the investigators who have studied the subject agree that in tubercular bone affections it is sometimes very difficult to find the bacillus, that it is not found in great abundance, and that sometimes it has evaded even the most careful search.

Schuchardt and Krause ("Ueber das Vorkommen der Tuberkelbacillen bei fungösen und scrophulösen Entzündungen," *Fortschritte der Medicin*, B. i. No. 2, p. 277) examined a great variety of tubercular lesions and came to the conclusion that tubercle bacilli are present without exception, but, as a rule, few in number, and only to be found after long and patient search. They found them invariably present in cases of secondary and primary tuberculosis of synovial membranes, tuberculosis of bone, in tubercular abscesses, and in the latter cases not in the pus, but in the granulations of the abscess-wall.

Renken (*Jahrbuch f. Kinderheilkunde*, B. xxv.) found the bacillus of tuberculosis in all cases of spina ventosa which he examined.

Müller ("Ueber den Befund von Tuberkel-bacillen bei fungösen Knochen und Gelenkaffectionen," *Centralblatt f. Chirurgie*, 1884, No. 3) studied carefully numerous specimens of synovial and bone tuberculosis with special reference to the existence of the bacillus of tuberculosis, and although the results in a number of cases were negative, he believes that the most intimate and direct etiological relations exist between the bacillus and all tubercular lesions in bones and joints. Among others who have shown the never-failing presence of the bacillus of tuberculosis in different forms of surgical tuberculosis, including bones and joints, may be mentioned Kanzler ("Ueber das Vorkommen des Tuberkel-bacillus in scrophulösen Localerkrankungen," *Berl. klin. Wochenschrift*, 1884, pp. 23-41), Mögling (*Ueber chirurgische Tuberculosen*, Dissertation, Tübingen, 1884), Bouilly ("Note sur la présence des bacilles dans les lésions chirurgicales tuberculeuses," *Revue de Chirurgie*, 1883, No. 11), and Letulle (*Gaz. hebd. de Méd. et Chir.*, 1884, 49, 51, 52). Kanzler wished to make a distinction between scrofulous affections and tuberculosis, as he found the bacilli not as constant in the former, and observed that after implantation of tissue of scrofulous affections in animals, the process was slower than after inoculation with tubercular products. Letulle considers scrofula and tuberculosis as belonging to one and the same disease, of which the former constitutes the milder form, and appearing externally, while the latter represents the graver form, attacking in preference the internal organs. The points made by the last two authors are too unimportant in still maintaining a scientific, or even practical distinction between scrofula and tuberculosis. The surgeon must recognize every lesion as tubercular in its origin, nature, and course in which the bacillus of tuberculosis can be constantly found, and from which successful cultivations can be made, and with which the disease in animals can be produced by inoculation.

7. *Joints.*

Tuberculosis of joints is so closely related to the same disease in bone, that it often follows the latter, and when it occurs as a primary lesion it not seldom extends to the subjacent bone. As stated above, it occurs more frequently as a primary lesion in the adult than in children. Primary infection is only possible through a wound of a joint, as in the case referred to under the head of inoculation-tuberculosis. Tubercular infection of an intact joint presupposes the entrance of the bacillus of tuberculosis through some infection-atrium into the systemic circulation, or, the diffusion of bacilli through the same channel from some preëxisting tubercular focus, and the localization of floating bacilli in the synovial membrane by capillary embolism, or mural implantation. A single tubercular nodule on the surface of the synovial membrane may lead in a comparatively short time to diffuse tuberculosis over the entire surface of the membrane by local diffusion of the microbe, in which the movements of the joint play an important part. When the synovial surface has become the seat of diffuse tuberculosis, the tissues undergo the same pathological changes as during the first stage of tuberculosis in other organs, and it is the characteristic granulation tissue which has given to this form of arthritis the name of fungous synovitis and synovitis hyperplastica granulosa. During the early stages of the disease the surgeon meets with two distinct classes; in one the tubercular infection produces a pulpous condition of the entire synovial sac, with little or no effusion into the joint, the swelling being due entirely to the presence of a thick layer of granulation tissue, the true *tumor albus* of the older writers. This form of tuberculosis gives rise at an early stage to extensive deformity of the joint, flexion, rotation, and, in the case of the knee-joint, partial dislocation of the tibia backward. In the other variety, the fungous granulations are less marked, but a copious effusion into the knee-joint takes place, which simulates a catarrhal synovitis until time and the effect of treatment enable the surgeon to make a correct differential diagnosis. In this form König assures us that he has never observed a tendency to flexion, or any other form of displacement of the joint surfaces. If suppuration takes place, which is not very often the case, it begins in the granulations which cover the synovial membrane, and the pus accumulates in the cavity of the joint until perforation of the capsule takes place. During the suppurative process, the superficial granulations are destroyed, and the tubercular infection penetrates deeper, and as, during the destructive process, bloodvessels are destroyed the patient is exposed to the additional risks of general infection. If such a joint opens spontaneously, or is not incised under the strictest antiseptic precautions, the additional infection from without leads to the most serious consequences, as under these circumstances the pus-microbes are brought in contact with a surface which has been admirably prepared for suppurative and septic processes by the antecedent pathological changes.

Tuberculosis of a joint may terminate in a spontaneous cure in cases in which the intensity of the infection is slight, or the resistance on the part of the patient is so great that the fungous granulations do not undergo degenerative changes, but are converted into connective tissue. A partial or complete synechia of the joint is often one of the unavoidable results in such cases. This endeavor on the part of the organism to limit extension of the disease is often observed in cases in which the joint affection occurs in connection with osteal tuberculosis. As soon as the joint is perforated a wall of granulation tissue is thrown out around the seat of infection, and under favorable circumstances a partition of cicatricial tissue is formed, which isolates the infected from the intact portion of the joint. In such instances we have an illustration how the tubercular process is retarded, and sometimes permanently arrested by the transformation of granulation into connective tissue. For such a favorable termination to take place it is necessary that the tubercular virus should become attenuated by age, or want of a proper nutrient medium, or that the pathogenic effect of the bacilli should be neutralized by an adequate resistance on the part of the tissues before degenerative changes have occurred in the granulation tissue. An exceedingly practical observation has been recently made by a number of surgeons in reference to the influence of operative interference in tuberculous joints in the dissemination of the disease. König observed in his own practice sixteen cases of miliary tuberculosis following shortly after resection of tuberculous joints, and he believes that the generalization of the process was favored, if not directly produced, by the operation. It is not difficult to conceive the *modus operandi* of such an occurrence. The resection wound opens numerous veins in the bone, the lumina of which remain patent ready for the introduction of minute fragments of granulation tissue or bacilli, which, on entering the venous circulation, are the direct cause of metastatic tuberculosis in distant organs.

Wartmann (*Deutsche Zeitschrift f. Chirurgie*, B. xxiv. Hefte 5, 6), after giving a careful account of the results following excision of tuberculous joints in the hospital practice of Feurer, has collected from the practice of different operators 837 cases of excisions for tuberculosis. Of this number, 225 died. Of the fatal cases, in 26 death followed the operation closely, and resulted from acute tuberculosis, probably induced by the operation. In these cases we must take it for granted that a tubercular focus during the operation furnished the essential fragments of granulation tissue, or free bacilli which were aspirated, or forced into the openings of wounded vessels, and through them into the general circulation. That fungous synovitis is a genuine tuberculosis has been abundantly illustrated by clinical experience, microscopical examinations, and particularly the results obtained by implantation experiments in animals. I will only refer to the work of Tavel (Senn: *Four Months among the Surgeons of Europe*, 1887, Chicago, p. 154) in this connection, who has been studying in a systematic manner the diagnostic value of implantation of tubercular material in animals, mainly guinea-pigs. Granulation tissue from tubercular joints in his experiments on guinea-pigs invariably produced acute, diffuse tuberculosis, and death in from five to six weeks. The course of the disease in the animal is typical; at the point of inoculation a hard nodule appears first, the result of a traumatic inflammation of the tissues around the graft. Next a lymphatic gland becomes enlarged in the immediate vicinity of the primary seat of infection, which is always done in the flank, consequently the inguinal glands enlarge first. Glandular infection increases rapidly; after the whole chain of lymphatic glands in the groin are involved, the axillary glands become affected. At the post-mortem examination it was always found that of the internal organs the spleen becomes affected first, then the liver and lungs, but usually the disease is so diffuse that scarcely an organ remains exempt. When the diagnosis between syphilis and tuberculosis cannot be made, either clinically or by aid of the microscope, inoculation experiments always give positive and reliable information. When the lesion is tubercular the animal always becomes tubercular and dies. When it is syphilitic, the inoculation is harmless, and the animal remains well. So far, only one animal that was inoculated with tuberculous material has lived for five months, and in this case a large abscess formed at the point of inoculation a few weeks after the operation. Examination of the contents of the abscess showed abundant bacilli of tuberculosis; a gland in the groin remains enlarged, and the disease, if not arrested by the suppurative inflammation, may have become latent.

7. *Tendon-sheaths.*

Tuberculosis of the tendon-sheaths, or, as Hueter terms it, *tendovaginatis granulosa*, has only recently been recognized and described as an independent primary lesion. Hueter, in his most admirable text-book on surgery (*Grundriss der Chirurgie*, B. i. S. 127), says that this affection is seldom met with as a primary lesion, but that it appears usually as a complication of joint tuberculosis. Volkmann, in his classical article on diseases of joints ("Die Krankheiten der Bewegungsorgane," *Pitha u. Billroth*, B. xi. S. 866), devotes only a few sentences to this part of his subject. The first scientific treatise on this affection came from the clinic at Göttingen by Riedel. Another important paper on the same subject was published by Beger ("Die Tuberculose der Sehnenscheiden," *Deutsche Zeitschrift f. Chirurgie*, B. xxi. S. 385), who reports four cases that occurred in the clinic at Leipzig.

The chronic tendo-synovitis, or compound ganglion of the older text-books, has been shown to be, on careful clinical observation, microscopical examination, and bacteriological research, cases of local tuberculosis. The extension of tubercular processes along tendon-sheaths from a tuberculous joint after perforation has, for a long time, been known to occur, but as a primary lesion it has only been recently introduced into surgical nomenclature. When this affection occurs primarily and independently of tuberculosis of an adjacent joint, infection with the bacillus of tuberculosis takes place by localization of floating microbes in some small vessel, and subsequently the pathological processes in the tendon-sheaths resemble those of tuberculous joints. In some cases the products of the disease are massive granulations which occupy the inner surface of the tendon-sheath, in others the fungosity is less, but a copious synovial exudation is thrown out, while in a third class the granulations form hard white masses, the so-called *corpora oryzoidea*. The intrinsic tendency of the disease consists in progressive extension by continuity of structure along the tendon primarily affected, and when this tendon is part of a compound tendon the disease gradually creeps from tendon to tendon until all of the sheaths are involved. As this affection is met with most frequently in the tendon-sheaths surrounding the carpus, and as these sheaths are not infrequently in direct communication with the wrist-joint by means of small synovial sacs, it extends to the joint by continuity of surface.

König ("Die Bedeutung des Faserstoffes für die pathologisch-anatomische und die klinische Entwicklung der Gelenk und Sehnenscheiden-Tuberculose," *Centralblatt f. d. gesammte Med.*, 1886, No. 25) assigns to the bacillus of tuberculosis properties which place it among the agents which produce fibrinous inflammation. The rice bodies in the tendon-sheaths the seat of a chronic inflammation he considers as the product of a fibrinous inflammation caused by the presence of the bacillus of tuberculosis. Nicaise, Poulet, and Villard (*Nature tuberculeuse des hygromes et des synovites tendineuses à grains riziformes*, 1884) examined four cases of hygroma containing rice bodies, and found in all of them the bacillus of tuberculosis. I have observed this form of inflammation of the tendon-sheaths in the common extensors and flexors of the hand, the peroneus longus, the tendon of the patella, and the tendo-Achillis. One of these cases was remarkable for its acuity. The patient was a man sixty years of age, laborer, and addicted to intemperance. About the beginning of the year I examined him at the request of another physician, and found an oblong swelling on the dorsum of the hand corresponding to the extensor tendon of the index finger. The swelling was not painful, and but little tender on pressure. Fluctuation was well marked; on deep pressure movable bodies could be distinctly felt which were recognized as corpora oryzoidea. An operation was advised, but was declined. A few weeks ago the patient was admitted into the Milwaukee Hospital, being completely incapacitated from following his occupation. At this time the dorsum of the hand corresponding to the index and middle fingers, and the radial aspect of the forearm as far as the middle, presented a continuous swelling with well-marked fluctuation. The swelling had become painful, and exceedingly tender on pressure. Under strict antiseptic precautions the swelling was incised in its entire length, and a large quantity of synovia-like fluid and softened rice bodies escaped. The sheath of the extensor communis digitorum and extensors of the wrist were lined with a thick layer of fungous granulations, and near the annular ligament numerous loose attached rice bodies were found. The tendon-sheaths were carefully dissected out, and the whole wound, after thorough disinfection, dusted with iodoform, drained and sutured. A copious antiseptic dressing of iodoform gauze and sublimated moss was applied, and the forearm and hand fixed upon an anterior splint. Inoculations of the fluid from the sheath upon potato remained sterile. Cultivation upon solidified serum, obtained from a hydrocele, showed, after a few weeks, a scanty culture of the bacillus of tuberculosis. Implantation of one of the rice bodies into the subcutaneous connective tissue of a guinea-pig resulted in a typical tuberculosis, starting from the point of inoculation, spreading to adjacent lymphatic glands, and finally resulting in miliary tuberculosis.

8. *Peritoneum.*

Primary tuberculosis of the peritoneum has recently been described as a local tuberculosis amenable in some cases, at least, to operative treatment. Localization of the bacillus takes place in accordance with the form of tuberculosis, of which three distinct varieties are recognized by pathologists and clinicians :

a. As a part of a general diffuse tuberculosis ;

b. Extension of an adjacent tubercular process to the peritoneum ; and

c As a primary tuberculosis.

For the surgeon, only those cases have interest which are classified under the second and third varieties. The prevalence of the affection in the female sex among the cases which have been reported, points to the Fallopian tubes as the primary seat of infection with secondary invasion of the peritoneum from this source. Although the genital organs in the male are more frequently the seat of tuberculosis than in the female, so far only two cases of peritoneal tuberculosis in males have been reported, one by Kümmell and the other by Lindfors. Tuberculosis of the peritoneum by extension from a tuberculous focus in the genital organs can only mean an infection by contact, the bacillus of tuberculosis transferred from the primary seat of infection, and localization by implantation upon the peritoneal surface. Implantation experiments in animals furnish a good illustration of the manner in which the process becomes diffuse. At the point of implantation a granulation mass forms around the graft, and from here innumerable tubercle nodules take their starting-point, forming everywhere granulation-tissue, in which the bacillus may have found a new habitat. The movements of the abdominal walls and the peristaltic action of the intestines are potent factors concerned in the local dissemination of the tubercular infection. In primary tuberculosis of the peritoneum the infection takes place in the same manner as in intact joints, by floating bacilli becoming arrested in the capillary vessels of the membrane where the primary nodule forms, from which again, as from a graft, dissemination takes place. These cases are, in the true sense of the word, not cases of primary tuberculosis, as the peritoneal affection is only a local expression of an antecedent infection. In peritoneal tuberculosis we observe the same tendency to limitation of the infective process as in joints, by the formation of an impenetrable wall of connective tissue, which imparts so often to this form of peritonitis its circumscribed character.

Kümmell ("Ueber Laparotomie bei Bauchfelltuberculose," *Verhandl. der Deutschen Gesellschaft f. Chirurgie*, 1887, p. 323) looks upon this form of peritoneal tuberculosis as a purely local affection amenable to surgical treatment in the same sense as a tuberculosis of joints. That some of these cases can be permanently cured by local treatment is well shown by a case treated by Sir Spencer Wells twenty-six years ago by abdominal section, the patient having remained up to this time in perfect health.

In a recent paper on this subject Fehling ("Beiträge zur Laparotomie bei Peritonealtuberculose," *Correspondenzblatt f. Schweizer ärzte*, 1887) reports four cases of his own, and gives an account of all the operations which have been done up to that time, 21 in number. Of this number, 15 recovered, and the patients are known to have been well from 1 year to 23 years, and in a number of cases their condition was learned 4 to 5 years after the operation. 6 of the patients died; 2 of sepsis, 1 of pyæmia several months after the operation, and 3 of the continuance of the disease for which the operation was done. In 5 of the cases ascites attended the tuberculosis, and in 3 the swelling was not due to effusion but to adhesions between the intestinal loops, which were covered with miliary nodules.

Of fifty-four cases of laparotomy (*Medical News*, July 14, 1888) in tubercular patients, collected by Trzebicky, four died from the immediate consequences of the operation, while in a fifth, death occurred after the operation from acute tuberculosis, though the fluid had not re-accumulated. One case died in four months of general tuberculosis without the peritonitis disappearing; cures resulted in forty cases, though here and there evidence of pulmonary tuberculosis was reported. The majority of cases were females, which may find its explanation in the fact that most were operated upon under an error in the diagnosis of ovarian cyst. The statistics are yet too meagre, the correctness of diagnosis not entirely above doubt, and the period of observation after operation not long enough; but, in view of the results, there is no longer any justification for expectant treatment. Even though in some cases recovery was not permanent, the fluid did not re-accumulate, and the patients were relieved of their distress. Spontaneous recovery from tubercular peritonitis is exceptional, and operative interference is indicated the more, as it would seem that, in many cases, tuberculosis of the peritoneum is a primary affection, and the source of general infection.

As all other therapeutical measures are futile in such cases, and laparotomy under antiseptic principles may be considered free from danger, the operation is certainly indicated. It consists in a free incision, thorough evacuation of the fluid, and free drainage.

Lindfors, in a monograph on this subject (*Centralblatt f. Gynäkologie*, February 15, 1890), analyzes 109 recorded cases, which he divides into seven classes. The acute variety may assume the form of circumscribed, general, or suppurative peritonitis; in the chronic there may be a free or an encysted effusion, there may be simple adhesions, or the intestines may be so adherent as to cause intestinal obstruction. Lindfors thinks that the presence of acute or chronic pleurisy has an important bearing on the diagnosis of tubercular peritonitis. He is strongly in favor of laparotomy and the free use of iodoform within the cavity.

The most recent and comprehensive work on tuberculosis of the peritoneum, which has just appeared from the pen of Vierordt ("Ueber die Tuberculose der serösen Häute," *Zeitschrift f. klin. Medicin*, B. xiii. Heft 2), should be consulted by those who wish to secure for reference an exhaustive treatise on this subject.

9. *Mouth*.

We have now every reason to believe that many cases of ulceration of the tongue and cavity of the mouth which have been heretofore diagnosticated and treated as carcinoma, were not carcinoma, but tuberculosis. The recent advances made in the microscopical, bacteriological, and experimental methods of examination have succeeded in separating from syphilitic affections and malignant disease of the mouth a number of cases which belong to the long list of affections which are now classified under the head of surgical tuberculosis. The cavity of the mouth is exposed to direct infection with microörganisms which might be obtained in the air we breathe, the food we eat, and the water we drink. Remembering the frequency with which superficial abrasions and ulcerations occur in this locality, it is not strange that primary tuberculosis should occasionally develop here. The tubercle bacillus produces the same tissue-changes here as on the surface of the skin, the primary pathological product consisting of granulation tissue which undergoes molecular retrograde tissue metamorphosis followed by ulceration. The tubercular ulcer is covered by the products of coagulation-necrosis, and is surrounded by a zone of infiltration, which, however, does not present the same feeling of hardness as carcinoma. Schliferowitsch ("Ueber Tuberculose der Mundhöhle," *Deutsche Zeitschrift f. Chirurgie*, B. xxvi. Hefte 5. u. 6) gives an exhaustive *résumé* of the literature on the subject to date, and has collected all the recorded cases in which the diagnosis of tubercular disease of the cavity of the mouth could be made with some degree of certainty. The cases number 88, and include those of primary and secondary tuberculosis. From a careful study of this affection he has come to the conclusion that it occurs seldom in the very young, and that it attacks most frequently persons between forty and fifty years of age. The appearance of the tubercular ulcer is characteristic. If on the tongue, it is found on the borders near the tip of the organ. It appears as an oblong ulcer, with raised, ragged borders of firm consistence, showing the color of fresh granulations. The floor appears as if covered by a pseudo-membrane; if this covering is removed, the surface left easily bleeds. The surface of the ulcer is uneven, as if covered with papillæ. The discharge of pus is slight, and, in many cases, miliary abscesses may be found around the ulcer. Pain is not as severe as in carcinoma. Lymphatic glands may become secondarily infected, but this is not often the case. In the primary form of the disease the presence of tubercle bacilli is the safest criterion in fixing the diagnosis.

10. *Genital Organs.*

Cornet has made some experiments on tuberculosis of the genital organs in animals. In rubbing a pure culture of tubercle bacilli in abrasions of the penis, he produced a tubercular lesion of that organ. In bitches tuberculosis of the vagina and uterus could be produced by injection of a pure culture into the vagina. The local lesions were followed by general tuberculosis.

FEMALE.—Direct tubercular infection of the genital tract in women has been observed, but the cases so far reported are few. Zweigbaum (*Centralblatt f. Bacteriologie und Parasitenkunde*, B. xi. S. 558) describes a case of primary tuberculosis of the portio vaginalis uteri, which, at the time of examination, appeared in the shape of an ulcer the size of a walnut, with thick, indurated margins, and cheesy floor. Numerous tubercle bacilli were found in the secretion taken from the surface of the ulcer. Evidences of pulmonary tuberculosis were apparent at this time. After a few weeks the ulcer extended toward the left vaginal wall and left labia majora. A section of a fragment of tissue removed from these parts on staining showed numerous bacilli. This form of tuberculosis is not frequent, as the author could find only two cases of vulvo-tuberculosis in literature, although genital tuberculosis is quite a frequent affection. In the absence of tubercular lesions of the vagina and uterus, it is doubtful, if infection of the Fallopian tubes can take place by the entrance of the bacillus through the genital tract, and the relatively frequent occurrence of the disease in that part of the genital organs is only explainable by attributing it to auto-infection in the same way as we have explained the occurrence, for instance, of primary tuberculosis of joints and peritoneum. We can only safely assume that tubercular infection of the tube often, if not always, takes place upon the basis of preëxisting pathological conditions, taking it for granted that the healthy tubes do not present favorable conditions for the localization of the tubercle bacilli. A catarrhal condition of the lining membrane of the tubes, as in other organs, undoubtedly acts, in many instances, as a predisposing cause to localization of preëxisting microörganisms in the circulation.

Barbier (*Gaz. Méd.*, 1888, No. 39) believes that a woman can be infected by a tuberculous man during coitus. Bacilli have been demonstrated in the semen, as well as in the discharge attending tuberculous epididymitis. The uterus may be infected by extrusion from tuberculous growth in the vulva, without any intermediate trace of infection in the vagina. The writer even admits the possibility that tubercular infection may be transmitted by the finger of the attendant, by unclean instruments, or even through the medium of the air.

Primary Tuberculosis of the Fallopian Tube.—Kotschau (*Archiv. f. Gynäkologie*, B. xxxi. Heft 2) gives a full description of a case of primary tuberculosis of the Fallopian tube. The patient was forty-five years old, having a good family history, and had suffered for a year with pains in the abdomen, profuse metrorrhagia, and diverse nervous disturbances. She was treated for retroflexion; and subsequently had an attack of pelvio-peritonitis. Vaginal examination disclosed a firm, smooth movable tumor as large as an apple, to the right of the uterus; this was taken for a malignant ovarian growth, and laparotomy was done for its removal. On opening the abdominal cavity a quantity of turbid, purulent fluid escaped. The tumor, of oblong shape, was found lying apparently in a bed of pus; on account of its intimate adhesions it could not be removed. The patient died from shock. The autopsy showed the uterus enlarged and retroverted. The right tube was tortuous and generally thickened; near its distal end it was dilated into a swelling the size of a hen's egg, in the centre of which was a cavity containing cheesy material. Other smaller caseous deposits were found in the tubal wall in close proximity to the large swelling. The ovary on the same side was enlarged and transformed into a caseous mass; the left tube and ovary showed similar changes, though less extensive. The microscopical examination of the pathological products confirmed the diagnosis of tuberculosis.

Werth (*Centralblatt f. Gynäkologie*, July 20, 1889) in a recent paper on the subject of tuberculous disease of the Fallopian tubes, recognizes an acute and chronic form; in the former both the muscular and serous coats undergo caseous degeneration, numerous bacilli being found in the interior of the tube, while in the chronic form the wall of the tube undergoes thickening and infiltration with new cells, while its contents contain only a few bacilli. The increase in size of the tube is due to the collection of pus in its interior, as well as to the hypertrophy of the wall. When suppuration takes place in the interior of the tube the tubercular product has become the seat of a secondary infection with pus-microbes, hence the indications for operative treatment have become more urgent.

An interesting case of primary tuberculosis of the uterus and Fallopian tubes is reported by Lebedeff. The patient was the widow of a man who had died of pulmonary tuberculosis. An examination before the operation revealed a firm, nodulated, intra-abdominal tumor in the space of Douglas. An attempt was made to remove the tumor by laparotomy, but had to be abandoned as the disease had become too widely disseminated. Six weeks later the patient died with symptoms of general tuberculosis. At the post-mortem miliary tuberculosis was found in the peritoneum, lungs, colon, uterus, and Fallopian tubes. The most advanced stages of the disease were found in the uterus and Fallopian tubes, showing that the disease had commenced in these organs. Both the Fallopian tubes were dilated and filled with pus, the epithelium in parts being absent. Stained sections from the uterus and tubes showed the presence of numerous bacilli.

Jouin (*Bulletin Paris Obstet. and Gyn. Soc.*, March, 1889) believes that tuberculous endometritis from local infection is quite a common affection. Of nine cases which were observed by him, it was due to sexual contact with men suffering from genital tuberculosis. In two others the husbands were tuberculous, but had no genital tuberculosis. He calls attention to the fact that Cornil and Chantemesse have produced it in rabbits by injecting bacilli into the vagina.

Mamma.—A number of well-authenticated cases of primary tuberculosis of the mamma have recently been reported. So far as the infection is concerned, the breast must be considered as an appendage of the skin. The bacillus of tuberculosis, from without, may effect entrance into the gland through the milk-ducts, in which case the inflammatory process commences in the parenchyma of the gland, or it may enter through a fissure of the nipple, in which case the process is primarily interstitial. Where direct infection from without can be excluded, the disease is the result of auto-infection, and, on this account, the prognosis is always more unfavorable. The regional dissemination takes place along the chain of the axillary lymphatic glands. Orthmann ("Ueber Tuberculose der weiblichen Brustdrüse," etc., Virchow's *Archiv*,

B. c. Heft 3) examined the enlarged lymphatic glands in a case of tuberculosis of the mamma, and found numerous tubercle bacilli.

MALE.—In the male primary tuberculosis is most frequently observed in the epididymis, for the reason that the vessels in this structure are more tortuous and smaller than in the remaining portion of the testicle, or the vas deferens. Saltzman states (*Centralblatt für klin. Medicin*, 1888, No. 11) that these anatomical conditions are important factors in the arrest and localization of floating bacilli. That in cases of tuberculosis of the testicle we are only dealing with an external manifestation of an antecedent infection, becomes apparent by the clinical observation that not infrequently both testicles are infected, either simultaneously or some time apart, showing that the infection came from the same source. In other cases the primary localization takes place in the vesiculæ seminales, or in the structure of the prostate gland. Tuberculosis of the genital organs in the male furnishes one of the best examples of the typical course of local tuberculosis of the testicle. A small, hard nodule is first detected in the epididymis, and from this starting-point the whole structure of the epididymis is infected, when the infection slowly, but surely, extends along the vas deferens to the vesiculæ seminales, the prostate gland, and bladder, and from this viscus along the ureters to the pelvis of the kidney. As a rule, the disease remains limited to the genito-urinary organs, but, in some instances, metastatic infection takes place, either from the genito-urinary organs, or from the primary source of infection. A gentleman is now under my treatment who illustrates a number of interesting points descriptive of the clinical behavior of genital tuberculosis. He is thirty-five years of age, married for ten years; the marriage has been childless. He claims that he has never had syphilis or gonorrhœa. Tuberculosis is hereditary in the family. Nine years ago he noticed a small, hard swelling in the epididymis of both testicles. Two years ago symptoms of cystitis appeared which were not much improved by internal medication, and irrigation of the bladder. Six months ago his left knee became swollen and painful. Two months ago he commenced to suffer from severe pain in the region of the left kidney. Temperature at this time varied from 100° to 103° F. A swelling soon formed in the left lumbar region, and four weeks ago I made an incision in the lumbar region along the outer border of the erector spinæ muscles down to the kidney and evacuated a large quantity of pus. Through this incision the kidney could be distinctly felt, and by passing the finger around it, it appeared to be separated from the structures surrounding it. The left knee presents all the appearances of advanced tuberculosis. Pulmonary tuberculosis is not present. The disease in both testicles remains latent, the testicles proper are intact, the epididymis moderately swollen and indurated. the vas deferens on each side somewhat firmer than normal. The disease has extended from the epididymis to the pelvis of the kidneys, all of the intervening organs being involved in the tubercular process. The only apparent manifestation of general tuberculosis is presented by the left knee. An interesting feature in this case is the formation of a paranephric abscess around a pyelo-nephritic kidney, which must be referred to a secondary infection with pus-microbes.

Tuberculosis of Urethra and Glans Penis.—Kraske has observed a case of tuberculous ulceration of the urethra extending from the membranous portion to the neck of the bladder, in a patient thirty-three years of age. The patient was treated for chancre. The autopsy revealed advanced tuberculosis of the genito-urinary tract, and pulmonary tuberculosis. In another case, a man, forty-nine years old, a tuberculous ulceration existed on the dorsum of the glans, the size of a cent piece. This sore was also mistaken for primary lesion of syphilis. There were no signs of pulmonary tuberculosis. The glans was amputated, when it was observed that the tuberculous infiltration extended deeply into the cavernous structure. The lesion could not be traced to genital contact, and under the microscope showed the typical structure of tuberculous tissue.

Vesiculæ Seminales.—In 1829 Dalmar described a chronic inflammation of the seminal vesicles, the description of which corresponds closely to that of tuberculosis. Since then this affection has been described by Albers, Jaye, Naumann, Humphrey, and Kocher, and lately it has been studied by Rayer, Cruveilhier, and Reclus, as secondary to tuberculosis of the lungs. As a secondary affection, this trouble is not only seen in connection with tuberculosis of the lungs, but is more common after primary tuberculosis of the epididymis, either as a continuation of the cheesy degeneration in the vas deferens, or spreading by contiguity of tissue from the prostate. Primary tuberculosis of these organs is extremely rare, and still less often diagnosed, and up to the present time no surgical interference has been attempted.

Ullman (*Centralblatt f. Chirurgie*, No. 8, 1890) reports a case of primary tuberculosis of the right testicle with secondary affection of the seminal vesicles on both sides, in a lad seventeen years of age, where, after removal of the right testicle he extirpated these organs through a semilunar incision in the perineum. The general health of the patient improved, but a small urinary fistula remained, as during the operation the bladder had been opened. He is of the opinion that the seminal vesicles should be removed in primary tuberculosis of the testicle or epididymis, when no suspicious symptoms have appeared on the sound side, and when on the affected side the vesiculæ seminales are already attacked, also in cases of primary tuberculosis of the seminal vesicles. The impotence following the operation should be no contra-indication, for in all reported cases of tuberculosis of the seminal vesicles, impotence always occurs in a short time, in fact it is regarded as a cardinal symptom of the disease.

Strümpell ("Beiträge zur Diagnostik u. Aetiologie der Tuberkulose des männlichen Urogenital-apparates," *Münch. med. Wochenschrift*, B. xxxiv. No. 31, 1887), after a careful study of four cases of primary tuberculosis of the genito-urinary organs in men, came to the conclusion that infection takes place through the urethra. The tubercle bacilli, finding no place for localization and growth in the urethra and bladder, finally reach the prostate gland, or the epididymis, the whole process resembling inhalation-tuberculosis, in which the disease manifests itself not in the mucous membrane of the bronchial tubes, but in the apices of the lungs.

11. *Tubercular Abscess.*

The specific effect of the bacillus of tuberculosis on the tissues is to produce a chronic inflammation which invariably results in the production of granulation tissue. The granulation tissue must be considered in the light of a protective wall to the surrounding healthy tissue. The degenerative changes which take place in the granulation tissue are caused by local anæmia and the chemical action of the ptomaïnes of the tubercle bacilli, and consist in caseation and liquefaction of the cheesy material into a fluid which has always been regarded as pus until recent investigations have shown that it is simply the product of retrograde tissue-metamorphosis. I believe that it can now be considered as a settled fact that the bacillus of tuberculosis produces no suppuration, that its presence indicates only a specific form of inflammation, which terminates invariably in the formation of granulation tissue, and that when suppuration occurs secondary infection with pus-microbes has taken place. A tubercular abscess, without the presence of pus-microbes, does not contain pus, but the products of degenerative changes in the fungous granulations. If the bacillus meets with sufficient resistance on the part of the surrounding tissues, it finally exhausts the nutritive material in the granulations and dies, or remains in a latent condition, the granulation material is converted into cicatricial tissue, and the local lesion is cured. These are the cases which terminate most frequently in spontaneous cure. If liquefaction of the infected tissues takes place and the products of degeneration are absorbed, a similar favorable termination is possible. If the same product is evacuated by incision under antiseptic precautions, a spontaneous cure is accelerated. If, on the other hand, a secondary infection with pus-microbes takes place, the patient incurs the danger of septic infection and local and general dissemination of the tubercular process.

That the bacilli do not grow in a tubercular abscess has been definitely settled by Schlegtendal ("Ueber das Vorkommen der Tuberkel-bacillen im Eiter," *Fortschritte der Medizin*, Bd. i. S. 537). He examined five hundred and twenty specimens of pus from tubercular abscesses and found bacilli present in only 75 per cent. Garrè ("Aetiologie der kalten Abscesse; Drüseneiterung; Weichtheil- und Knochen-abscesse und der Tuberculösen Gelenkeiterungen." *Deutsche med. Wochenschrift*, 1886, No. 34) has also made an extended series of observations to ascertain the presence of the tubercle bacillus in cold abscesses. According to this author, many tubercular ulcerations and abscesses are the result of a mixed infection, as has been claimed by Hoffa for some cases of empyæma in cases of pulmonary or pleural tuberculosis. In cold abscesses, and in the pus of tubercular cavities in bone, no pus-microbes could be found, not even in cases that pursued a rapid course. Cultivations of such pus remained sterile, while inoculations produced typical tuberculosis. In such instances the pus examined under the microscope showed none of the morphological elements of pus, but was seen to consist of an emulsion composed of detritus of broken-down tissue suspended in serum. He affirmed that it is possible that in many cases of suppuration following in the course of a tubercular process pus is the result of a mixed infection, and that the pus-microbes had disappeared before the examination was made. The walls of the tubercular cavity contain the typical structure of the tubercular lesion, and the primary and essential cause of the inflammation—the bacillus of tuberculosis. The infection follows the migration of the abscess in whatever direction that may take place. If an additional infection from without takes place, following either a spontaneous discharge or after incision, the superficial granulations are destroyed by the suppurative process which is initiated, exposing the patient to the additional risks of septic infection and a more rapid local and general dissemination of the tubercular process.

CHAPTER XVIII.

ANTHRAX.

SYNONYMS: Contagious Carbuncle; Charbon; Milzbrand; Malignant Pustule.

HISTORY.—As a disease among animals anthrax has been known since the earliest records of history. The contagiousness of this disease has been recognized since the beginning of the eighteenth century. During the first part of this century it was described as a blood disease (Maladie du sang. Blutseuche der Schafe). Heusinger, in his classical work (*Die Milzbrandkrankheiten der Thiere und des Menschen*, Erlangen, 1850), declared anthrax to be a malarial neurosis. In the year 1855 Pollender ("Mikroskop. und Mikrochem. Untersuchungen des Milzbrandblutes, etc.," *Casper's Vierteljahrsschrift f. ger. u. öff. Medizin*, Bd. viii. S. 103) published his discoveries, which inaugurated a new era in the study of anthrax. As early as 1849 he had discovered in the blood of cattle suffering from anthrax a mass of innumerable fine rod-like bodies, which appeared to be of a vegetable nature and resembled vibriones. Brauell ("Versuche und Untersuchungen betreffend den Milzbrand des Menschen und der Thiere," *Virchow's Archiv*, No. xi., 1857) found the same organisms in the blood of men, horses, and sheep which had died of anthrax. He also found the same bodies during life in the blood of the diseased animals.

Delafond (*Recueil de méd. vet.*, 1860, p. 726) considered this parasite as a kind of leptothrix.

In 1863 appeared the work of Davaine (*Compt. rend. de l'Acad. des Sciences*, t. lviii. p. 22) wherein he pronounced these rod-like bodies to be bacteria, and later he called them *bacteridia*. He believed them to be the cause of anthrax, as the disease could not be propagated with blood which did not contain them.

Through the labors of Naegeli and Bollinger (*Zur Pathologie des Milzbrandes*. München, 1872) and others the microörganism of anthrax finally found a permanent place as the bacillus anthracis among the schizomycetes.

The first positive accounts of the disease in man we owe to Fournier, Montfils, Thomassin, and Chabert, who published their description between the years 1769 and 1780. Fournier first distinguished the spontaneous and the communicated carbuncle of man. The primary existence of anthrax in man was asserted by Bayle in 1800 and by Davy la Chevrie in 1807.

DESCRIPTION OF THE BACILLUS OF ANTHRAX.—Rods 5–10 micromillimetres long, and 1–1.25 micromillimetres broad, and threads made up of rods and cocci. The rods, as a rule, are straight, only when they grow to a considerable length and meeting with resistance they become curved. The rods and threads are round, and with their ends truncated at right angles appear as though they had been cut off obliquely. The interior, as long as fission does not proceed, is perfectly homogeneous and absorbs aniline dyes very readily. The development of spores in long undivided threads, as we find them in fluid culture media, takes place at regular intervals, where we find them as bright oval spots which become more and more apparent, marking the direction of the rods. Upon solid culture media the development of spores is preceded by transverse segmentation of the rods. The cell-membrane of each section finally becomes the membrane of the spore, each pole of the spore presenting a small mass of protoplasm which can be stained.

STAINING.—Cover-glass preparations of fluid specimens can be stained with a watery solution of any of the aniline dyes. They can be rapidly stained with a drop of fuchsin or gentian-violet, but more satisfactorily by floating the cover-glass for twenty-four hours. The preparations are dried and mounted in Canada balsam. The spores are not stained by the ordinary methods. Tissue sections are best stained by Gram's method, and after-stained with eosin, picrocarminate of ammonium. By double staining the rods are seen to consist of a hyaline sheath with protoplasmic contents.

CULTIVATION.—*a. Gelatin.* If a nutrient medium of 5 to 8 per cent. of gelatin is inoculated, a whitish line develops in the track of the needle puncture, and from it fine filaments spread out on the sides. In a more solid nutrient gelatin, the growth appears only as a thick, white thread. The culture liquefies the gelatin, and the growth subsides as a white flocculent mass.

b. Agar-agar. Cultures upon a sloping surface of nutrient agar-agar form a viscous snow-white plaque. Without access of air the culture does not grow, the bacilli being aërobic.

c. Potato. Inoculation of sterilized potato yields a very characteristic growth. The deep chamber, containing the potato, is placed in the incubator, and in about thirty-six or forty-eight hours a creamy, very faintly yellowish layer forms over the inoculated surface, with usually a peculiar translucent edge. On removing the cover of the damp chamber a strong, penetrating odor of sour milk is emitted.

PLATE IX.

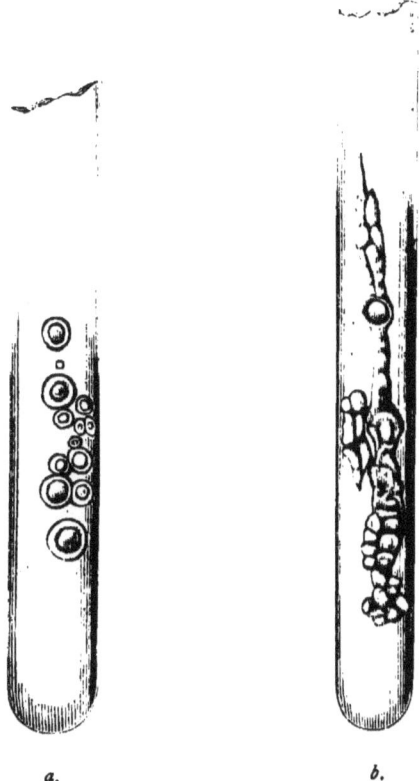

Two cultures of anthrax bacilli prepared from the same material at the same time.
a. With a pressure of 4 cm. of mercury.
b. With atmospheric pressure.

INOCULATION EXPERIMENTS.—In order to cause death of animals by inoculation of the bacillus of anthrax, the pure culture, or anthracic blood, must be injected into the subcutaneous tissue, into the circulation, or the virus may be transmitted by inhalation or by feeding. Goats, hedgehogs, sparrows, cows, horses, guinea-pigs, and sheep, can be readily infected. Rats are less susceptible. Pigs, dogs, cats, white rats, and Algerian sheep are immune. Frogs and fish have been rendered susceptible by raising the temperature of the water in which they lived.

ATTENUATION OF VIRUS.—By cultivating the bacillus in neutralized bouillon at 42°–43° C. (107.6°–109.4° F.) for about twenty days the infecting power is weakened and animals inoculated with it are protected against the disease. A still greater immunity is obtained by inoculating a second time with material which has been less weakened. The animals are then protected against the most virulent form of anthrax, but only for a time. A temperature of 55° C. (131° F.), or treatment with 5 to 1 per cent. solution of carbolic acid deprives the bacilli of their virulence. The virulence of the bacillus is also altered by passing it through different species of animals.

Woolbridge secured immunity against anthrax in animals by cultivating the bacillus in an alkaline solution at a temperature of 37° C. (98.6° F.) for two days. At this time the fluid was filtered and a small quantity of the filtrate injected into the subcutaneous tissue of rabbits, which remained well, and subsequently resisted injection of most virulent anthracic blood.

Hankin ("Immunity produced by an Albuminose isolated from Anthrax Cultures," *British Med. Journal*, Oct. 12, 1889), under the guidance of Koch at the Hygienic Institute of Berlin, isolated an albumose from anthrax cultures which, when injected into rabbits and mice in small quantities, rendered these animals immune against the most virulent cultures. The albumose was prepared from the cultures by precipitation with absolute alcohol, the precipitate was well washed in this liquid to free it from ptomaïnes, since it is known that all such substances are soluble in alcohol. After the addition of alcohol it was filtered off and dried, then redissolved, and filtered through a Chamberland's filter. Four rabbits were inoculated with virulent anthrax spores and three of them received an injection of albuminose into the ear-vein at the same time; the latter recovered, while the remaining animal not thus protected died in about forty-eight hours of anthrax. In another experiment, ten mice were each injected with the millionth part of their body-weight of anthrax albuminose and with active vaccine at the same time. Of these, three died after 108 to 116 hours; the others recovered. Three others had only the two-millionths of their bodyweights of anthrax albuminose and active culture. Two of them survived. Four control mice were inoculated, and all died of anthrax. He has come to the conclusion that when a large dose of albuminose is injected into an animal, the entry of anthrax bacilli into the system is aided, and when a small dose is administered immunity is acquired against its poisonous properties, protecting the animal against subsequent inoculations with active cultures.

INTENSIFICATION OF VIRUS.—While it is known that some chemical substances exert an attenuating influence on the virulence of the anthrax bacillus, it has also been found that an attenuated virus will again become more virulent by adding certain substances. It must, therefore, be taken for granted that the chemical composition in which the bacillus is suspended influences in one way or the other its virulence. It has been found, for instance, that the addition of a minute quantity of lactic acid to a fluid containing the bacillus in an attenuated form greatly intensifies its virulence within a very short time. Thus, Arloing, Cornevin, and Thomas found that the pathogenic power of a fluid containing these bacilli, to which $\frac{1}{500}$ part of lactic acid had been added, and the mixture allowed to stand for twenty-four hours, was increased twofold; if then a little water, containing a very easily fermentable sugar is added to the mixture, and another twenty-four hours allowed to elapse, the virulence attains its maximum, and frogs inoculated with this virus die in from twelve to fifteen hours, whereas, when inoculated with ordinary virus they live from forty to fifty hours. Kitt has repeated and confirmed these experiments.

ANTHRAX INFECTION IN MAN, AND CLINICAL VARIETIES.— The favorite location for the development and growth of the bacillus of anthrax in man and beast is in the connective tissue; it is, therefore, immaterial in what manner the microörganism reaches this tissue, as localization here marks the beginning of the disease. Buchner ("Ueber Aufnahme von Infectionserregern durch die intacte Lungenoberfläche," *Verh. des Congress f. innere Medicin*, 1888) has studied experimentally the entrance of the bacillus of anthrax through the intact surface of the lung. The bacillus and spores were administered by inhalation in the shape of dry powder and suspended in steam. On examining the bronchial mucous membrane at different stages under the microscope it was seen that the spores were transformed in a very short time into bacilli, and that the latter, by their growth, pushed themselves between the cells and into the capillary vessels. It was observed that the greater the pulmonary irritation was, the more the passage of the microbes was retarded. The entrance of the bacilli from the surface of the mucous membrane into the capillary vessels was seen to depend on an active process. Only blood bacilli, to which belong

PLATE X.

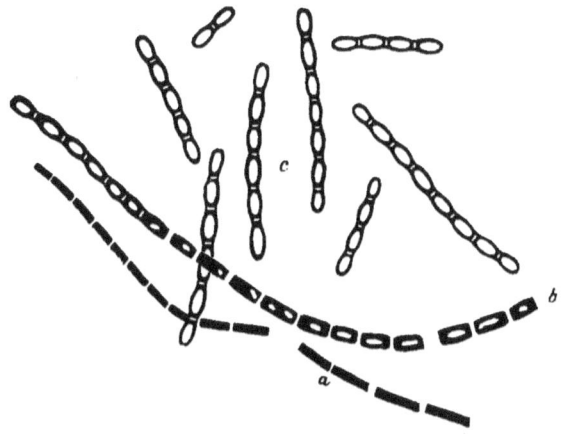

Development of anthrax spores. Culture three days old; cultivated at 40°C. (104° F.) on meat-peptone-agar. Gentian-iodine. Zeiss $\frac{1}{18}$. O. 4. Magnified 2500 diam.
   a. Threads free from spores.
   b. Threads with beginning development of spores.
   c. Threads with perfect development of spores.

the bacillus of anthrax and the spirilli of recurrent fever, possess this property. The tubercle bacillus and the bacillus of glanders are no blood bacilli, but even in infective processes with these microbes the disease can be produced by inhalation without primary localization in the lungs. Entrance of these microbes into the blood without alteration of the vessel wall has not been proved. Anthrax on the external surface of the body is the result of direct inoculation. The clinical forms vary according to the location of the disease, its extent, and the intensity of the infection. Most all authors follow Bollinger's classification, according to which all cases are brought under one of the following varieties: 1. Anthrax acutissimus, or apoplectiformis. 2. Acutus; and 3. Subacutus. The primary location of the disease is in accordance with the manner in which infection has taken place. W. Koch ("Milzbrand u. Rauschbrand," *Deutsche Chirurgie*, Lieferung 9) states that in animals and man the bacillus can enter the organism through one of the following three routes: (*a*) Through the skin. (*b*) Gastro-intestinal canal. (*c*) Respiratory passages. The microbe first multiplies at the primary point of invasion, and from here by entering the bloodvessels it is conveyed to distant parts and organs. The pathologico-anatomical conditions vary according to the primary seat of invasion and the structure of the organ the seat of the disease. The first tissue-changes are observed at the primary seat of localization. These local conditions, carbuncle and anthrax-œdema, give rise to symptoms proportionate to the importance of the organ involved. An anthrax œdema of the hand or arm is a less serious condition than when the same affection involves the face or neck. The local œdema at the point of infection is caused by vascular disturbances due to the presence of the bacilli within the bloodvessels and the interstitial inflammatory exudation caused by their presence. The local affection always becomes dangerous when the bacillus enters the bloodvessels and gives rise to general dissemination. When the microörganism enters the body through the gastro-intestinal canal with the food or drink, it gives rise to a primary anthrax of the intestinal canal, which again may become general by metastatic dissemination through the systemic circulation.

Vierhoff (*Ueber Anthrax intestinalis beim Menschen*, Dissertation, Dorpat, 1885) has collected 41 cases of anthrax intestinalis, the total number which were found reported up to 1885. The author observed himself two cases of secondary intestinal anthrax in the hospital at Liga.

Cases of secondary intestinal anthrax—that is, localization of the bacillus of anthrax in the mucous membrane of the intestines after external infection—were known to the older authors, while observations of primary localization in the digestive tract date only from the middle of the last century.

Of 63 cases of anthrax in man, collected by Slessarewskji (*Amer. Journ. Med. Sciences*, 1887), the disease showed itself 6 times on the face, 21 on the neck, and 36 in other places. Various theories have been advanced in explanation of the immediate cause of death of animals and persons infected with anthrax. In the most virulent form, the anthrax acutissimus, Bollinger believes that the rapid growth of the bacillus in the blood brings about a sudden diminution of oxygen and a surplus of carbonic acid, and that death takes place by a slow process of asphyxia. Against this theory it can be maintained that in the blood of animals which have died of the acutest form of the disease very few bacilli can be found; and further, that in the experiments made by Nencki on the blood of rabbits which had died of this form of anthrax it was found as capable of oxygenation as the blood of healthy animals. The theory that death results from purely mechanical causes due to the presence of bacilli in great abundance in the bloodvessels is likewise not tenable, because no such fatal degree of obstruction in the capillary circulation has been found at the post-mortem examinations.

As a third hypothesis, Bollinger advanced that the bacillus may generate a chemical poison which may cause death by intoxication.

In reference to the last-mentioned cause, Hoffa (*Die Natur des Milzbrandgiftes*, Wiesbaden, 1886) calls attention to the following three possibilities:

1. The bacilli of anthrax are in themselves poisonous, and with the increase in their number the quantity of the poison is increased in the same ratio. Against this supposition the results of the experiments made by Hoffa, himself, furnish the most conclusive proof. Of a pure culture of anthrax bacilli he injected a large quantity directly into the jugular vein of rabbits. The animals thus infected showed no symptoms of acute intoxication, but died in the same manner as animals infected in the usual manner.

2. The bacilli of anthrax produce a poison capable of producing fermentation in the blood, and which is soluble in the blood. The fact that filtered blood of animals which had died of anthrax did not produce toxic symptoms when injected into healthy animals speaks against this argument.

3. The bacillus of anthrax separates toxic substances from complex combinations in the organism. This last explanation appears, from analogy of the views that are now entertained of bacteria and ptomaïnes, to be the most plausible, and Hoffa went at the task to produce such substances outside of the animal body upon artificial culture media. For this purpose he cultivated the bacillus with

the greatest precautions upon sterilized meat kept for several weeks in an incubator at 37° C. (98.6° F.). The chemical product he attenuated according to the methods advised by Stass-Otto, Brieger, and after the more recent methods of Fischer. By the methods of Stass-Otto and Fischer he succeeded in finding a substance which possessed an alkaline reaction, and which produced toxic effects in animals. A strictly pure article and an accurate chemical description of it could not be obtained on account of the smallness of the quantity which could be produced. By Brieger's method, cultivations upon sterilized yolk of egg diluted with sterilized water, it was found impossible to obtain a toxic substance.

The substance produced by Stass-Otto's method was used in experimenting on frogs, mice, guinea-pigs, and rabbits, and that obtained by Brieger's method on guinea-pigs and rabbits, and both of them produced symptoms of intoxication. After a short period of intoxication with increased action of the heart and accelerated respiration the animals became somnolent, respirations deep, slow, and irregular, assisted by the action of all accessory muscles of respiration; pupils dilated; temperature below normal; diarrhœa; feces bloody; speedy death. At the necropsy the heart was found contracted, the blood was of a dark color, and ecchymoses of the pericardium and peritoneum existed; there were no microörganisms in the blood; no such toxic substance could be produced from sterilized meat alone. The same author ("Zur Lehre der Sepsis," *Verh. d. deutschen Gesellschaft f. Chirurgie*, 1889) subsequently succeeded in isolating a toxic substance from the bodies of anthracic rabbits with the formula of $C_5H_8N_2$, which he called anthracin, and to which he attributed the toxic symptoms in cases of anthrax. Injected subcutaneously in rabbits it produced first restlessness, rapid pulse and respiration, followed by somnolence, deeper and slower respiration, diarrhœa, asphyctic symptoms, convulsions, and death. These experiments leave but little doubt that the fatal termination in cases of anthrax is due to the presence of ptomaïnes, which are formed in the body in consequence of the action of the bacilli upon certain complex combinations in the organism.

Anti-microbic Treatment of Anthrax.

Lande (*Mémoires de la Société de Médecine de Bordeaux*, 1889) reports two cases of malignant anthrax saved by subcutaneous injections of carbolic acid. In the first case, a man aged twenty-seven years, the upper lip was the seat of the carbuncle; in the second, a woman aged sixty-five years, the anthrax occupied the region below the scapula. Both patients were very ill, low delirium and other symptoms of toxæmia being present. The injections were made into the subcutaneous tissue around the carbuncle. The strongest solution used consisted of 15 grammes of neutral glycerin and an equal part of distilled water, in which 3 grammes of pure carbolic acid were dissolved. The injections were made at five points around the anthrax, and represented a total dose of 50 centigrammes of the acid. The injections were painful, but rapid improvement followed. This 10 per cent. solution was stronger than any previously employed for the same purpose by Boeckel, Raimbert, and others. A 5 per cent. solution in ordinary cases is strong enough, but in grave cases the 10 per cent. solution must be used until improvement takes place, which may occur within forty-eight hours.

Kaloff, of St. Petersburg, in making experiments with anthrax on animals, accidentally infected himself, either by a needle puncture or by handling the organs of anthracic animals. The local infection appeared on the outer side of the thumb of the left hand as a small vesicle, which disappeared soon, but gave place to circumscribed infiltration on the second day. This inflammation rapidly extended and was surrounded with hemorrhagic vesicles. The indurated tissues were promptly removed by excision; nevertheless, on the next day swelling of axillary glands on the same side, fever, great prostration, also diarrhœa. The skin in the axillary region and the side of chest was much swollen, œdematous, and at different points bright red, at others bluish-red. One of the axillary glands the size of a hen's egg, and glands along the margins of the pectoralis major were moderately enlarged. All the enlarged glands were removed and the field of operation thoroughly disinfected with solution of carbolic acid, and the same solution thrown into the surrounding tissues with a hypodermic syringe. Cessation of fever and rapid healing of wound, followed by recovery. Implantations of fragments of excised glands in bouillon and gelatin yielded cultures of anthrax bacilli. Excision of infected tissue and use of carbolic acid strongly recommended.

PLATE XI.

Spleen of rabbit-anthrax.
a. Pulp.
b. Follicle.
c. Vein.

CHAPTER XIX.

GLANDERS (*Malleus Humidus*).

ALTHOUGH glanders in man is a rare affection, it presents from a bacteriological study so many points of interest that it merits more than a passing notice. It is one of the infectious diseases of which the microbic cause is now thoroughly understood.

HISTORY.—That glanders in man occurs as an infection from animals has been known for a long time. Its contagiousness among horses was asserted by Solleysel in the seventeenth century. Rindfleisch believed that he saw vibriones in the granular contents of glanderous abscesses. Klebs detected in cultures of pus taken from animals suffering from this disease small rods and granules, but further cultivations and inoculations in rabbits failed. The presence of minute organisms in cases of glanders was pointed out by MM. Christatt and Kiener, in 1868, and their observations were corroborated by MM. Bouchard, Capitan, and Charrin, who have found the organisms not only in parts exposed to the air, such as nasal ulcerations and pulmonary abscesses, but also in parts which are not so exposed, such as the spleen, liver, and lymphatic glands.

Chaveau (*Comptes rendus*, lxvii., No. 14) demonstrated by his experiments that the virus of glanders was fixed to small solid particles, as he found the sediment which formed after diluting pus with water active. This discovery marked an advance in the knowledge of the physical nature of the virus. Löffler and Schütz are the discoverers of the bacillus of glanders in horses. In 1882 they made a preliminary report of their researches (*Deutsche med. Wochenschrift*, 1882, No. 52). In 1886 Löffler published his elaborate monograph on this subject ("Die Aetiologie der Rotzkrankheit," *Arbeiten aus dem Kaiserlichen Gesundheitsamte zu Berlin*, B. i. S. 141–199). Soon after Löffler's first paper appeared, Bouchard, Capitan, and Charrin published almost simultaneously the results of their researches and observations; but it appears from Löffler's second paper that none of them had been able to produce a pure culture. Kitt and Weichselbaum were the first who, by their own labor, were able to corroborate the correctness of Löffler's discovery; the former by his observations and experiments on animals, the latter by a case of glanders in the human subject that came under his observation.

DESCRIPTION OF THE BACILLUS MALLEI.—According to Löffler, the bacillus of glanders appears as a small rod which is somewhat shorter and thicker than the tubercle bacillus; its length varies but little, and corresponds to about two-thirds of the diameter of a red blood-corpuscle; the thickness varies between one-fifth and one-eighth of its length (Fig. 7.) These bacilli are either straight or somewhat curved and rounded at their ends. Usually they are found in pairs in a parallel direction held together by a delicate unstained substance. Examined in a drop of fluid they showed active molecular movements. Spontaneous movements could not be observed by Löffler. The colorless and sometimes even somewhat dilated portions of the stained bacillus are not spores, but, as

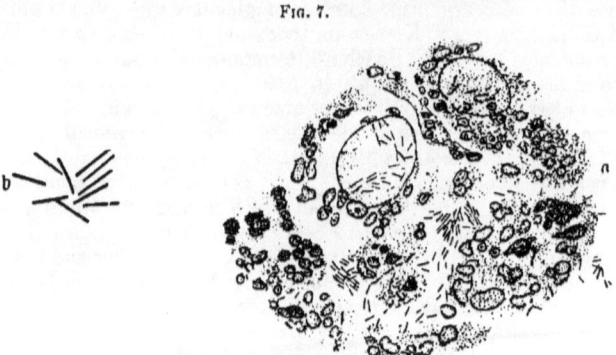

FIG. 7.

Bacilli of glanders. *a.* Section from glandrous nodule, 700 : 1. *b.* Bacilli of glanders, stained with methyl-blue. (FLÜGGE.)

Löffler affirms, are indications of commencing death of the microbe. Löffler found that dry bacilli occasionally could be made to grow after three months, but in most instances, after a few weeks, they could no longer be cultivated, which fact speaks against the existence of spores. Lundgren saw the bacilli in an agar-agar culture multiply by segmentation.

STAINING.—The manner of staining of the bacilli of glanders is characteristic, as when they are treated by basic and acid aniline dyes no effect is produced.

Method of Schütz: The sections are placed for twenty-four hours in the following mixture: Potash solution (1 in 10,000), concentrated alcohol, methylene-blue solution, equal parts. Wash the sections in a watch-glass with water acidulated with four drops of acetic acid. Transfer for five minutes to fifty per cent. alcohol, clarify in clove-oil, and mount in Canada balsam.

Löffler's method: Sections are immersed for a few minutes in a solution of potash 1 to 10,000, then for a few minutes in an alkaline solution of methyl-blue, after which they are decolorized with a solution of tropæolin in acetic acid, or, what is still better, in a fluid composed of ten cubic centimetres of distilled water, two drops of sulphuric acid, and one drop of a five per cent. solution of oxalic acid.

CULTIVATION.—When cultivated on solid sterilized blood serum at a temperature of 38° C. (100.4° F.), the growth appears in the form of minute transparent drops, consisting entirely of the characteristic bacilli.

Potato cultures, according to Löffler, form in three days a uniform amber-yellow layer which, about the sixth to the eighth day, assumes a reddish hue, resembling the color of oxide of copper, which is not easily mistaken for any other culture upon the same soil. Upon this soil the bacilli were cultivated through twelve generations, and the cultures retained their activity for a year; whether the bacillus was capable of cultivation after this time is not mentioned. The temperature at which cultures could be made to grow varied from 30° to 40° C. (86° to 104° F.).

Kranzfeld succeeded best with a nutrient medium composed of meat-peptone, glycerin, agar-agar. The bacillus is destroyed by exposure for ten minutes to a temperature of 55° C. (131° F.). A three per cent. solution of carbolic acid, a one per cent. solution of permanganate of potash, or a 1 : 5000 solution of corrosive sublimate destroys the bacilli with certainty. Lundgren succeeded in obtaining an active culture in bouillon.

INOCULATION EXPERIMENTS.—Kitt (" Der Rauschbrand, Zusammenfassende Skizze über den gegenwärtigen Stand der Literatur und Pathologie," *Centralblatt für Bacteriologie und Parasitenkunde,* Dd. i. S. 722, 1887) mentions the following animals susceptible of inoculation with the virus of glanders: catttle, sheep, goats, guinea-pigs. The horse, ass, and white rat are only susceptible to local infection, the animals recovering completely and permanently after a few days. Pigs, dogs, cats, rabbits, the common rat, ducks, and chickens possess great immunity; the inoculations at best produce only a slight local reaction. Löffler made his first experiments on guinea-pigs and the field mouse. In the guinea-pigs he observed, three to five days after the subcutaneous injection of a pure culture, an ulcer at the point of inoculation, and at the end of the first week swelling of the nearest lymphatic glands in a state of purulent softening. At this stage of the disease the process often came to a standstill and the animals recovered. In many animals the disease progressed quite rapidly to a fatal termination. Abscesses were frequently found in the testicle and the epididymis in the male, and in the breast and external genital organs of the female. The face, nasal cavity, and ankle-joint were also frequently the seat of ulcerative processes. If the disease proved fatal, death usually occurred three or four weeks after the inoculation. At the post-mortem, aside from the affections which have been enumerated, nodules were found in the spleen, lungs, and often also in the liver. Field mice proved a great deal more susceptible to the virus of glanders than guinea-pigs, as they usually died three or four days after inoculation. The post-mortem in these animals showed at the point of inoculation an infiltration from which swollen lymphatic vessels led to the nearest lymphatic glands. In the spleen and liver, which were always greatly enlarged, numerous small nodules were found, while the remaining internal organs presented a normal appearance. Glanders in guinea-pigs and field mice presents a series of pathological changes which cannot be mistaken for any other affection. The bacilli of glanders in the different organs can be detected most readily in recent specimens. In the blood bacilli were found only in very acute cases, a circumstance which explains why so many inoculations with the blood of glanderous horses proved unsuccessful.

Lundgren ("Försök till remodling af rots-mikroben," *Hygiea*, Bd. xlix., Heft 2, S. 91) took a nodule from the lungs of a horse which had died of glanders and implanted fragments of it under the skin of rabbits. The animals died about the nineteenth day after the inoculation, and the necropsy revealed induration and small abscesses at the point of infection, and small yellow nodules in the spleen, liver, lungs, testicles, and mucous membrane of the nose. These tissues stained with methyl-blue showed the bacilli of glanders. Implantations of spleen tissue into other rabbits fixed the period of incubation in this animal at from eleven to twelve days.

At a recent meeting of the Academy of Medicine, M. Cornil (*British Medical Journal*, June 7, 1890) gave an account of M. Babes's researches on the bacillus of glanders, which show that this microörganism, when obtained from pure cultivations, can penetrate the healthy tissue of animals, and thus cause glanders. Rubbing in an ointment containing the bacillus rarely succeeds, and only when the virus is very active, and obtained from a perfectly fresh cultivation. M. Nocard has repeated these experiments, and two guinea-pigs out of five were infected by this method of inoculation.

Kranzfeld (Zur Kenntniss des Rotz-bacillus," *Centralblatt für Bakteriologie und Parasitenkunde*, Bd. xi., No. 10) has recently published the results which he obtained by inoculations with the virus of glanders in an animal which had not hitherto been subjected to experimentation of this kind. He procured a pure culture from a nodule of a man who had died of glanders after a brief illness. Inoculations were made in a small rodent which is very numerous in the southern part of Russia, the spermophilus guttatus. The course of the disease in this animal was almost the same as in the field mice which were used by Löffler. Of twenty-eight animals infected with different cultures, sixteen died on the fourth day, nine on the fifth, two on the seventh, and one on the tenth. The post-mortem appearances were always characteristic; a greenish-gray infiltration at the point of inoculation, and a number of nodules in the spleen; in one animal also very small white nodules in the liver. Cultivations from these nodules yielded a pure growth of the bacillus of glanders.

If an animal is infected by the direct injection of a pure culture into a vein, no serious symptoms are produced, but if soon thereafter one or more muscles are injured subcutaneously, the microbes escape through the lacerated vessels and localize at the seat of injury and produce a grave form of the disease. It has been found by experiment that the further from the trunk prophylactic inoculations are made the less intense is the local reaction. When an animal is inoculated at a distance from the trunk and shows no general symptoms, a subcutaneous injury of any portion of the trunk will furnish conditions for a local form of infection.

GLANDERS IN MAN.—The virus of glanders can only find entrance into the organism through a wounded surface. Whether infection may not also take place through the alimentary canal has so far not been definitely ascertained. It is certain that the disease cannot be contracted by eating boiled or fried meat of animals affected with glanders. Infection through the respiratory organs is possible, as cases have been reported in which the lungs were the primary and only seat of the disease. The disease can also be transmitted from the mother to the fœtus in utero.

When man is the subject of glanders, bacilli are found more constantly in the blood than in animals. In the case described by Weichselbaum numerous bacilli could be seen in the blood. In this case a thrombus was found in one of the large meningeal veins which contained numerous bacilli, and which undoubtedly was one of the sources of the bacilli in the circulation. In man the nasal mucous membrane is not as frequently affected as in animals, although Bollinger has shown that in horses the nasal cavity is not always affected, and that it may present a normal condition even when the larynx and lungs are seriously affected. Muscular abscesses which may simulate rheumatism, are a frequent occurrence, especially in the chronic form of the disease.

A Russian medical paper of recent date (*British Medical Journal*, June 11, 1888) states that a young soldier, who had been a waggoner before his admission into the army, was received into the military hospital suffering from two foul ulcers on the hard palate, which had perforated the nasal fossa and destroyed the inferior turbinated bones. Three weeks later a swelling appeared over the eyebrow; a fortnight afterward he complained of pain on the inner side of the left knee around the internal tuberosity of the tibia. Then purulent discharge occurred from the left ear, and an abscess on the back of the right hand, which appeared as a deep purple tubercle, with a hard circumference, and sunken toward the centre; purulent discharge oozed from the surface. At first, for a short time after admission, the temperature varied, rising of an evening to 103° to 104°; later on, it fell to normal. The disease was mistaken for syphilis, and iodide of potassium was given without the least benefit. About ten weeks after admission he was in better health and left the hospital, receiving his discharge from the army. Within a few weeks he returned with extension of ulceration of the hard palate; the uvula was destroyed. The characteristic tubercles, the "farcy buds," appeared in the face, the metastatic abscess on the back of the hand remained. The patient ultimately died of exhaustion. Before death some of the tubercles were extirpated; they were found to contain microörganisms resembling the glanders bacillus of Löffler and Schütz.

Swelling of the testicles has also been frequently observed. In other cases acute or chronic pulmonary affections which simulate pneumonia or tuberculosis are the most important clinical features. If the disease attacks the nasal cavity, the mucous membrane presents hard nodules, and a copious discharge from the nose is present.

Küttner reports a number of cases in which the skin was the seat of numerous points of suppuration in the form of pustules or more diffuse purulent processes. The pus found in glanders is grayish-red, and quite tenacious in recent lesions, but when suppuration continues it assumes the characters of ordinary pus. An acute and a chronic form of glanders have been described, a classification which is accepted by Bollinger. The bacillus of glanders, after localization in the organism, produces rapid tissue-changes which at first consist in the formation of granulation tissue, which, when it becomes saturated with the ptomaïnes of the bacilli, undergoes transformation into pus, a stage of the disease which is indicated by the formation of abscesses wherever localization has taken place, either by direct infection, secondary infection by regional diffusion through the lymphatic vessels, or by metastasis through the systemic circulation. The nodules in internal organs are indications that general infection has taken place by embolism.

In cases in which no positive diagnosis can be made from a clinical aspect it becomes necessary to resort to cultivation and inoculation experiments.

CHAPTER XX.

ACTINOMYCOSIS HOMINIS.

ALTHOUGH the parasite which is the direct cause of this disease is not, properly speaking, a microörganism, as its presence in some cases can be detected by the naked eye, I shall include it in the list of surgical diseases due to the presence of microbes, as it often requires the aid of the microscope to make a positive differential diagnosis between this and other chronic infectious diseases characterized by the presence of granulation tissue.

HISTORY.—The disease as occurring in cattle was first described by Bollinger ("Ueber eine neue Pilzkrankheit beim Rinde," *Centralblatt für die medicinischen Wissenschaften*, No. 27, 1877) in 1877, as a condition in which sarcoma-like tumors were met with, associated with a peculiar growth which, from its structure, was named Strahlenpilz (ray-fungus), or actinomyces. James Israel (" Neue Beobachtungen auf dem Gebiete der Mykosen des Menschen," *Virchow's Archiv*, Bd. lxxiv., 1878) was the first to recognize the disease in man, but it was not generally understood until the appearance of the classical work of Ponfick (*Die Aktinomykose des Menschen*, Berlin, 1882) in 1882. Numerous articles on this subject have since appeared in the current medical literature, so that Partsch some two years ago brought at the end of his monograph (" Die Aktinomykose des Menschen vom klinischen Standpunkte besprochen," *Sammlung klinischer Vorträge*, Nos. 306 and 307, 1888) seventy-five references with a supplemental list containing thirty-three names furnished by Schuchardt. Since the publication of Israel's case numerous cases have been reported by different observers representing Germany, England, Belgium, Switzerland, Russia, Austria, and America, so that Partsch, in the work referred to above, estimates the whole number at not less than one hundred.

While most of the articles in medical journals contain only a description of isolated cases, it appears to have been the good fortune of some of the writers on this subject to meet with a number of cases in a comparatively short time. Thus Hochenegg (" Bui Kasuistic der Aktinomykose des Menschen." *Wiener med. Presse*, Nos. 16 and 18, 1887) reports in his paper seven cases, and Moosbrügger ("Ueber die Aktinomykose des Menschen," *Bruns' Beiträge zur klinischen Chirurgie*, Bd. ii. Heft. 2, S. 339, Tübingen, 1886) has increased the statistics by ten well-authenticated and carefully recorded cases. At the last meeting of the Versammlung Deutscher Naturforscher und Aerzte, Rotter stated that he observed thirteen cases in two years. Albert has seen not less than thirty-eight cases of actinomycosis within the past few years; of these eight have come under his observation during the last year. These cases have come mostly from Vienna and its vicinity.

DESCRIPTION OF FUNGUS.—Bollinger described as peculiar to this disease certain yellowish bodies, visible to the naked eye, which were always found in the pus of abscesses and in the middle of the tumors. Microscopically they were found to consist of threads similar to the ordinary mycelium, which terminated in bulbous ends. The threads radiate from the centre, and their clubbed extremities impart to the fungus the characteristic ray-like appearance (Fig. 8). Sometimes but one of these bulbs is connected with

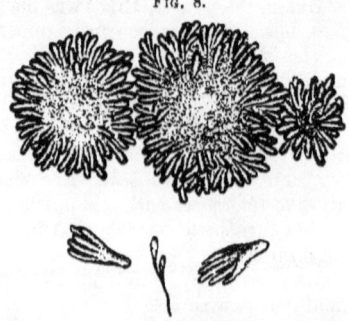

FIG. 8.

Three actinomyces from a case of pulmonary actinomycosis. Below, three finger-like buds and dichomatous branching of actinomyces threads. × 450. (BAUMGARTEN.)

a thread, at other times there may be several. In man the actinomyces occurs as small globular masses, commonly about the size of a millet-seed, usually of a pale yellow color, but at times white, brown, green, or speckled, the color being influenced by age and the consecutive pathological conditions by which it may be surrounded. In man the clubbed bodies are commonly absent, and the growth consists of the radiating filaments alone. The rays, when immersed in water or in a weak solution of chloride of sodium, become enormously swollen and lose their shape, while they effectually resist the action of acids, ether, and chloroform. For staining actinomyces Weigert uses Wedl's orseille, Marchand eosin, Dunker and Magnussen cochineal-red, Moosbrügger hæmatoxylon-alum, and Partsch, in section-staining, has had the best results with

Gram's method. Recently Babes has made beautiful dry preparations by using a two per cent. solution of safranin in aniline oil, followed by treatment with iodide of potassium. O. Israel "Ueber Doppelfärbung mit Orcein," Virchow's *Archiv*, Bd. 105, S. 169) has found that a solution of orcein in acetic acid stains the rays a Bordeaux-red, while the filaments, if decolorization is not carried too far, present a blue tinge. Baranski (*Deutsche med. Wochenschrift*, 1887) uses picrocarmine for staining fresh preparations of actinomyces bovis. A small amount of the contents of a yellow nodule, or pus from the part, is spread in a thin layer on a cover-glass, and dried in the air. The cover is then passed three times through the flame of an alcohol lamp, care being taken not to overheat the preparation. It is then floated in the picrocarmine solution, or a few drops of the staining fluid are placed on the cover. The staining is finished in two or three minutes. The cover is then carefully washed by agitating it in distilled water and alcohol, and examined in water and glycerin. The fungus takes a yellow color, while the remaining structure appears red. As regards the history of the parasite outside the body, as yet only a few facts are known. It is found in pig's meat and is peculiarly susceptible to outside influences. It cannot be cultivated in pure water or one-half per cent. solution of salt, such as is usually employed for similar forms, as in these elements the fungi swell up and assume fantastic shapes. Virchow found them as small calcareous concretions in the muscle fibres of the pig, and considered their flesh highly dangerous food, unless well cooked.

CULTIVATION EXPERIMENTS.—It has been found extremely difficult to cultivate the actinomyces, probably on account of the usual culture media not being well adapted for its growth. The first successful experiments were made by Boström (*Jahresbericht über pathologische Anatomie*, Baumgarten, 1886), of Giessen, upon plates of blood serum and agar-agar, the fungus attaining its maturity in five or six days, when it presented the typical appearances of actinomyces as found in man. O. Israel ("Ueber die Cultivirbarkheit des Actinomyces," Virchow's *Archiv*, B. 95, Heft 1) cultivated the fungus successfully upon coagulated blood serum. The culture grows very slowly, and the fungus often undergoes calcification. He made the observation that water, glycerin, blood serum, and weak saline solutions seriously impair the vitality of the fungus, and believes that the effect of these agents on the actinomyces explains the failure of previous culture and inoculation experiments. Until recently, coagulated blood serum is the only medium upon which the fungus has been successfully cultivated. If evaporation is prevented, a thin velvety layer forms on the surface of the blood serum in about eight weeks, in the vicinity of which, not before the expiration of fourteen days, the growth appears more in a downward direction than on the sides of the inoculation puncture. From the tenth to the fourteenth day numerous spores are produced, and a thick wall of club-shaped mycelia in typical centrifugal arrangement.

At a meeting of the Medical Society of Berlin, March 5, 1890 (*Berliner klinische Wochenschrift*, March 31, 1890), M. Wolf made a communication in which he described culture experiments which he and Israel made with actinomyces. He announced that they had succeeded in cultivating the actinomyces in and upon egg and agar-agar. The inoculations were made from a case of retro-maxillary actinomycosis immediately after the abscess was incised. With the yellow granules, deep and superficial, inoculations were made in agar-agar. It was found that the actinomyces is not an anaërobic fungus, as it grew upon the surface as well as in the depth of the culture soil. The agar culture appeared first as transparent little drops which, by confluence, made an opaque white mass. Under the microscope the culture was seen to be composed of short thick rods, with an admixture of other elements. The egg cultures, on the other hand, were made up of the short thick rods, besides a mass of threads, some of them twisted in the shape of a corkscrew, presenting an intricate network of threads. Three rabbits were inoculated by implantation of the pure culture into the peritoneal cavity. The post-mortem showed numerous nodules upon the parietal peritoneum, the omentum, and between the intestines, the size of a pin's head to that of a hazelnut, surrounded by a fibrous capsule. The interior of these nodules was composed of a yellow mass the consistence of tallow. In these nodules typical actinomyces were found imbedded in masses of round cells in a state of fatty degeneration.

INOCULATION EXPERIMENTS. — James Israel ("Erfolgreiche Uebertragung der Aktinomykose des Menschen auf das Kaninchen," *Centralblatt für die med. Wissenschaften*, 1883, No. 27) was successful in inoculating a rabbit from man by introducing a mass of granulation tissue into the peritoneal cavity, and Ponfick produced the disease in calves by implantation of a portion of the granulation mass into the subcutaneous tissue, the abdominal cavity, or into veins. Rotter (*Centralblatt f. Bakteriologie und Parasitenkunde*, B. lii. No. 14, 1888) experimented on calves, pigs, dogs, guinea-pigs, and rabbits, and in only one instance, a rabbit, did he succeed in reproducing the disease. In this instance a piece of granulation tissue, the size of a bean, was inserted into the peritoneal cavity, and the animal, having manifested no symptoms of disease, was killed six months after the inoculation. On opening the abdominal cavity about twenty nodules, varying in size from the head of a pin to a hazelnut were found distributed over a considerable surface, each of them showing the typical histological structure of actinomycosis. The transplanted piece of tissue was found perfectly encapsulated in one of the nodules the size of a bean. As the fungus was found in all the nodules it is only reasonable to conclude that the disease spread from the original depot by migration of some of the new fungi, which at their respective points of localization established independent centres of infection and tissue proliferation. While the actinomyces in the new nodules presented a perfect structure, and could be readily stained, the transplanted fungus in the graft had lost its structure, and could no longer be stained.

SOURCES OF INFECTION.—As the actinomyces found in man and beast resemble each other morphologically, and in their effect on the tissues, as well as in their reaction to chemical substances, it is evident that the etiology of the disease is similar in both of them. The fungus has never been found outside of the body. Israel is of the opinion that both man and animals are infected from the same source, such as vegetables or water. Jensen (*Tidskrift f. Veterinär*, B. xiii., 1883) traced an epidemic in Seeland to the eating of rye grown on land recently reclaimed from the sea; and Johne discovered a fungus closely resembling the actinomyces in grains of rye stuck in the tonsils of pigs. That the ears of barley or rye are sometimes the carriers of the contagium is well illustrated by the case reported by Soltmann (" Ueber Aetiologie und Ausbreitungsbezirk der Aktinomykose," *Jahrbuch f. Kinderheilkunde*, B. xxiv. p. 129). The patient was a boy, who had swallowed an ear of barley. The foreign body lodged in the pharynx, where it gave rise to difficulty in deglutition; afterward it perforated the pharyngeal wall, an accident which was attended by hemorrhage; and later an actinomycotic phlegmon developed, which spread rapidly, and finally opened below the scapula. Through this opening the foreign body was extracted. Piana (Virchow u. Hirsch's *Jahresbericht*, 1887, B. i. p. 293) examined the tongue of a cow suffering from a circumscribed actinomycosis, in which the disease could be traced to a similar origin—perforation of the tissues and infection by a sharp beard of the ear of barley. Actinomycosis has as yet only been found amongst herbivorous and omnivorous animals, including man, and the frequent location of the primary swelling in the mouth seems to indicate that the fungus gains entrance with food.

EFFECT OF ACTINOMYCES ON THE TISSUES.—As to the manner in which the fungus exerts its pathogenic action much yet remains to be ascertained. The most striking effect is the transformation of mature connective tissue into embryonal or granulation tissue.

The product of inflammation around each fungus consists of granulation tissue, which resembles tubercle tissue. At first the cells are round, at a later stage of the inflammation epithelioid and giant cells are formed immediately around the fungus. As the disease is almost always attended by suppuration at some time during its course, it has been customary to ascribe to the actinomyces pyogenic properties. Israel has always held that the actinomyces is a pus-producing fungus, in opposition to Ponfick and some other pathologists, who claim that when suppuration takes place it is the result of a secondary infection with pus-microbes. As cases of actinomycosis have been reported which remained stationary in the granulation stage for an indefinite time without suppuration taking place, and pus-microbes have been cultivated from the pus of actinomycotic abscesses, it appears more than probable that suppuration occurred independently of the presence of the fungus, and was produced by the specific action of pus-microbes on the granulation tissue. Firket (*Revue de Médecine*, 1884) asserts that the actinomyces does not appear to produce "coagulation-necrosis," but from a study of the earliest formed colonies he finds that the first effect of the fungus is to induce cellular hyperplasia. It is as if the tissue elements resented the intrusion of the parasite, which, however, mostly gains the upper hand, so that the result is the formation of granulation tissue, and, later, abscesses that characterize the disease. As a rule, it may be stated that the earlier suppuration takes place the more rapid the spread of the disease and the graver the prognosis, while the absence of suppuration indicates comparative benignancy, and points in the direction of a more chronic form of the affection. The localized chronic form of actinomycosis resembles in its clinical features and its anatomical locations more closely sarcoma than any other affection. In such cases it would be difficult, if not impossible, in the absence of the specific fungus, to make a differential diagnosis between it and round-celled sarcoma, even by a most careful microscopical examination, as the histological structure of both is almost identical.

CLINICAL HISTORY.—Usually the disease follows quite a chronic course and the swelling at the seat of primary localization resembles in its clinical history more a tumor than an inflammatory swelling. The extension of the morbid process takes place by diffusion of the actinomyces *in loco*, in preference along the loose connective-tissue spaces, each fungus constituting a nucleus for a nodule of granulation tissue. By confluence of many such nodules the inflammatory swelling often attains a very large size, and when suppuration occurs in the interior the further history is that of abscess. Diffusion never takes place along the course of lymphatic vessels and glands. When these structures are affected in the course of the disease, they indicate that secondary infection has taken place. In some instances the disease pursues such a rapid course that it may be mistaken for an acute phlegmonous inflammation, osteomyelitis, or, when diffused over a large surface of the body, for syphilis. A good illustration of the former class is furnished by the case reported by Kapper ("Ein Fall von acuter Aktinomykose," *Wiener med. Presse*, No. 3, 1887). A soldier, twenty-two years of age, became suddenly ill with febrile symptoms and a rapidly increasing swelling of the lower jaw. An early incision was made and liberated a large quantity of pus, which, on microscopical examination, was found to contain actinomyces. It is interesting to note that in this case the carious teeth, from where the infection had evidently taken place, contained threads of leptothrix and actinomyces. At a meeting of the Berlin Medical Society, O. Israel (*Berliner klin. Wochenschrift*, Jan. 23, 1888) gave an accurate description of the post-mortem appearances of a case of diffuse actinomycosis. The patient, a woman, forty-four years of age, had been treated for syphilis in one of the surgical clinics. The heart contained a number of minute abscesses in which the fungus could be found in large numbers. A large abscess between the diaphragm, stomach, and spleen contained thick pus of a greenish color, an unusual occurrence in cases of actinomycosis, but no actinomyces. The spleen was the seat of a large and of numerous minute abscesses, and the liver and kidneys also contained small abscesses, and in all of them actinomyces could be found. Israel claims that this case furnishes a good illustration of his views, that the actinomyces, as regards its effect on the tissues, occupies a position half-way between the bacillus of tuberculosis, which produces only granulation tissue, and the pus-microbes, which produce pus. It was impossible in this case, as in so many others in which multiple deposits have been found, to locate the primary seat of infection. The teeth were perfect and the whole digestive tract showed no evidences of disease.

SEAT OF PRIMARY INVASION.—If infection takes place by fully developed actinomyces, it can only do so by the fungus gaining entrance into the tissue through some loss of continuity in the cutaneous or mucous surface, as any other method of ingress is impossible on account of the large size of the fungus. In the cases in which no such primary infection-atrium could be found, it must be taken for granted that the local lesion had healed between the time infection took place and the first manifestations of the disease, or that infection was caused by the entrance of spores, which from their smaller size could find their way into the tissue through intact mucous surfaces. In reference to the primary localization of the disease, Moosbrügger gives the following statistics: In 29 cases the lower jaw, mouth, and throat were affected; in 9 the upper jaw and cheek; in 1 the tongue; in 2 the region of the œsophagus; in 11 the intestines; in 14 the bronchial tract and the lungs; in 7 the point of entrance could not be ascertained. Infection may take place through any abraded surface brought in contact with the specific cause, and for clinical purposes the cases can be divided into the following three groups: 1. Cutaneous surface. 2. Alimentary canal. 3. Respiratory tract.

1. *Cutaneous surface.* Partsch (*Deutsche Zeitschrift f. Chirurgie*, B. 23, p. 498) describes a case of actinomycosis in which the disease developed in the scar left after extirpation of the breast. The patient was a man, aged sixty. In June, 1884, his left breast was removed for an ulcerating carcinoma. As the wound did not heal by primary union and the process of cicatrization was very slow, a number of small skin-grafts, from a perfectly healthy young man, were transplanted. The wound was practically healed in September. Two months later, the cicatrix ulcerated and an abscess discharged itself. Actinomyces were found in the pus. The parts were excised, and the progress of the disease was apparently arrested. No explanation could be made how the infection occurred.

Hochenegg reported a case of actinomycosis of the skin of the left submaxillary region, in which he attributed the disease to an invasion of the fungus through a small atheroma.

Kaposi (*Wiener med. Wochenschrift*, Nos. 19–22, 1887) reports a very chronic case of actinomycosis which primarily started in the skin. When first noticed, it appeared as a red spot the size of a florin on the left pectoral muscle, which gradually increased to the size of a walnut, and then gradually flattened down and disappeared. Meanwhile fresh spots and lumps appeared, some as large as a pigeon's egg. Eleven years after the beginning of the disease a swelling as large as an apple appeared over the spine of the sixth vertebra, which gradually extended forward and a year later formed a large swelling behind the right axilla. A year later, this swelling had diminished in size to that of a pigeon's egg, and then again increased in size. Ulceration set in, exposing a fungous bleeding surface. At this time the entire trunk, but not the limbs, was covered with nodules, spots, and stripes. The swellings were heaped in masses. The infiltration was located in the corium. In this case it appears that secondary infection with pus-microbes only occurred at the points of ulceration.

At the meeting of the German Society of Surgeons in 1889, Leser (*Klinischer Beitrag zur Aktinomykose des Menschen*) reported three cases of primary actinomycosis of the skin which had come under his own observation in the course of a single year. In his remarks on this subject he placed special stress on the manner in which the disease extends. In the periphery of the primary lesion he found numerous nodules which later became the seat of destructive changes, resembling in this respect the clinical features of tuberculosis of the skin. The extension of the disease in the direction of the deep tissues takes place by the formation of passages the circumference of which corresponds to the size of a lead-pencil; these are filled with yellowish-gray or reddish-gray granulations which attack and destroy tissues irrespective of their anatomical structure. The lymphatic glands were always found intact.

2. *Alimentary canal.* The frequency with which the disease affects the mouth and jaws of cattle is explained by the occurrence of numerous points of injury caused by the chewing of rough food, which furnish the necessary infection-atrium through which the fungus enters the tissues.

a. Teeth. In man infection takes place frequently through carious teeth and through abrasions in the tongue and mucous membrane of the mouth.

Israel (*Klinische Beiträge zur Kenntniss der Aktinomykose des Menschen*, Berlin, 1885) found the fungus in the cavities of carious teeth, and Partsch found, in the same locality, almost pure cultures without any manifestations of disease except chronic periodontitis. The fungus occurs here often side by side with leptothrix.

b. Tongue. Hochenegg saw a case of actinomycosis of the tongue caused by an infected carious tooth. The swelling was the size of a cherry located near the apex of the organ. The affection had existed for two months. The growth was excised, and on examination was found to consist of granulation tissue with a central yellow mass the size of a millet-seed. Beside this case only three cases of actinomycosis of the tongue are on record; one primary, one secondary to disease of the jaw, and one metastatic.

c. Jaws. That carious teeth furnish a frequent infection-atrium in maxillary actinomycosis is well known, and in many instances the disease in its early stages has been mistaken for an ordinary dental affection, and patients have often sought relief at the hands of a dentist. The lower jaw is most frequently affected, the growth being connected with the bone, or situated close to it, or it has already extended to the submental or submaxillary region. As soon as the loose tissues of the neck are reached, rapid extension takes place in a downward direction along the inter-muscular septa. Israel refers to a case in which an actinomycotic swelling in the submaxillary region extended, from the month of August to December, to the level of the thyroid cartilage. When the disease is primarily located in the upper jaw, which, however, occurs only in exceptional cases, it tends to invade rapidly the adjacent soft parts, and even to implicate the base of the skull and the brain.

The prognosis is always more serious when the disease affects the upper than the lower jaw, as the tendency here to invade the deep structures is much greater.

Two cases of actinomycosis in man have come under my observation, and as both of them originated in the mouth and represent from a prognostic point two distinct classes, I will describe them briefly.

CASE I.—This patient was a man, thirty years of age, German by birth, and a soda-water manufacturer by occupation. His business required him to make frequent trips into the country by team. He had no recollection of having come in contact with cattle suffering from "swelled head" or lumpy jaw. During the winter of 1886 he suffered from what he supposed was an ordinary cold: the right side of the lower jaw was swollen and painful. As one of the molar teeth showed evidences of decay and had become loose it was extracted. The pain and swelling, however, did not improve, and the attending physician extracted all of the molar teeth of the lower jaw. At this time a fungous mass commenced to appear over the surface of the edentulous bone. The cheek on the affected side was also greatly swollen. The patient was admitted into the Milwaukee Hospital about six months after the first symptoms had showed themselves. At this time the lower jaw, in the mouth, presented a fungous mass extending from the angle of the bone to the first bicuspid; the swelling extended as far as the tonsils. The cheek was enormously swollen from the angle of the mouth to the lower margin of the parotid gland. The skin over the swollen part presented a glossy appearance, and the superficial veins were considerably dilated. Around the margin of the swelling no distinct border-line could be felt, the infiltrated parts fading gradually into the healthy surrounding tissues. Free suppuration from the surface of the fungous granulations, and a number of small abscesses had discharged themselves into the cavity of the mouth. As some doubt existed as to the character of the inflammation, careful and repeated examinations were made of the pus removed from the small abscess cavities, and on several occasions fragments of actinomyces were found. The discovery of the specific cause of the inflammation cleared up the diagnosis and furnished a strong indication for operative treatment. An incision was made along the lower border of the jaw, from just below the articulation to near the symphysis, and after arresting all hemorrhage it was carried into the cavity of the mouth. The alveolar processes of the jaw were affected and were removed; wherever the periosteum showed signs of infiltration it was carefully scraped away, and finally the whole exposed bone surface was thoroughly cauterized. The infiltrated soft tissues were dissected out with knife and scissors, the deepest portions extending as far as the tonsil. The deep portion of the wound was dusted over with iodoform and filled with iodoform gauze, while the external wound was sutured. The entire external wound healed by primary union, and the cavity in the mouth closed slowly by granulation. The patient's general health improved rapidly until six weeks after the operation, when the neck below the scar became swollen, followed in a short time by the formation of abscesses reaching from the angle of the jaw to the clavicle, and posteriorly as far as the spine of the scapula. Numerous openings were made and efficient drainage established, but suppuration continued and the patient became extremely emaciated. The suppurative process extended, and four months after the first operation the patient died, the symptoms during the last days of life pointing to a hypostatic pneumonia. Actinomyces were constantly found in the pus during the entire course of the disease. I believe that the recurrence of the disease after operation was due to an imperfect removal of infected tissues in the posterior and lower portion of the pharynx.

CASE II.—This case came under my care during the summer of 1887. Patient was a young man who was employed on a farm. About five months before he was admitted into the Milwaukee Hospital he had a number of teeth extracted from the right upper jaw under the belief that the teeth, some of which were decayed, were the cause of the pain and swelling in that region. The physician in attendance diagnosticated sarcoma of the upper jaw, and sent the case to me for operation. On my first examination I found a swelling involving the right side of the face, extending from the zygomatic arch to near the lower border of the lower jaw, involving the deep tissues and being connected with the alveolar processes of the posterior portion of the upper jaw. The swelling was firm, and without well-defined margins. No evidences of suppuration. The history of the case, and particularly the location, extent, and physical properties of the swelling, led me to the opinion that it was the result of actinomycotic infection. All infected tissue was thoroughly excised through a large external incision, the jaw-bone scraped and cauterized. The entire thickness of the cheek, with the exception of the skin and superficial fascia, appeared to be transformed into granulation tissue. In the granulations numerous yellow seed-like bodies were found, which under the microscope showed the typical structure of the ray-fungus. The mycelia were not as bulbous as we find them pictured in the books, but the distal extremity appeared to be surrounded by dust-like bodies presenting the appearance of a small brush. Some of the mycelia appeared to be more covered than others. These minute bodies I looked upon as spores. In the first case in which suppuration had taken place I never succeeded in finding the actinomyces perfect and complete; in the second case suppuration had not taken place, and the fungus always was found in a perfect state and in a condition of spore-production. These cases present a striking contrast both in regard to the local conditions and the ultimate termination. In the first case a secondary infection had already taken place, and the phlegmonous inflammation induced by the pus-microbes prepared the tissues again for the diffusion of the actinomycotic process; while in the second case the process had not passed beyond the stage of granulation, presenting a more distinct boundary-line between healthy and diseased tissues, a most important factor in the operative treatment. The first case died from a recurrence of the disease in the vicinity of the operation wound and extension to the neck and chest, while in the second case the wound healed and the patient has since remained in perfect health.

d. Intestines. In primary intestinal actinomycosis the disease must be due to mural implantation of the fungus and infiltration of the tissues by its progressive growth. Arrest and implantation of the actinomyces are determined by antecedent pathological changes.

Chiari ("Ueber primäre Darmaktinomykose des Menschen," *Prager med. Wochenschrift*, No. 10, 1884) described the post-mortem appearances in a case of primary intestinal actinomycosis. The patient was a man thirty-six years of age, the most prominent symptom being progressive marasmus. At the necropsy chronic tuberculosis in the apices of the lungs and a few tuberculous ulcerations in the lower portion of the ileum were found. The large intestine presented a very remarkable appearance, the mucous membrane of which, except the cæcum and ascending colon, was covered with whitish deposits, forming round and oblong patches some of them one cubic centimetre in diameter, and five millimetres in thickness. In some of these patches could be seen minute yellowish-brown and yellowish-green granules. The patches were firmly adherent and when removed left a loss of substance in the mucous membrane. The mucous membrane throughout was in a state of catarrhal inflammation. On microscopical examination the granules proved to be actinomyces. The mycelium had penetrated into the tubular glands and showed calcified club-shaped conidia. The calcification of the club-shaped extremities had undoubtedly prevented deeper penetration of the fungus. No other organs presented evidences of actinomycosis. Hochenegg ("Fall von Actinomykose," *Wiener med. Wochenschrift*, No. 44, 1886) presented a case of actinomycosis to the Medical Society in the person of a man forty-three years of age, who had sustained an injury of the abdomen nine months previously, and had since that time noticed a painful swelling at the seat of injury. In the region of the umbilicus a fistulous opening formed which continued to discharge a thin secretion in which actinomyces were constantly found. The patient was very much emaciated and many of the teeth carious. There was no swelling about the jaws or neck. Examination of organs of chest and the sputum contributed no additional information. The author expressed the opinion that the inflammatory swelling caused by the contusion furnished the necessary conditions for the localization of actinomyces from the intestinal canal.

Zemann ("Ueber die Aktinomykose des Bauchfells und der Baucheingeweide beim Menschen," *Wiener med. Jahresbericht*, Hefte 3, 4, 1883) reports five cases of actinomycosis of the abdomen. In four of them the disease commenced with sharp lancinating pains in the abdomen, and during their course presented the clinical picture of chronic peritonitis. Swellings could be found in one or more places in the anterior abdominal wall, and the abscesses were either incised or opened spontaneously; in three cases they communicated with the intestinal canal.

The first case was a woman, thirty years of age, who had a fistulous opening in the anterior abdominal wall, which communicated with a swelling in the left parametrium. The patient stated that this swelling appeared soon after her last childbed. A constant discharge of yellowish-red pus was maintained, in which, under the microscope, numerous actinomyces could be seen. The patient died of exhaustion, and at the post-mortem chronic parametritis and perimetritis were found, with extensive pus cavities which communicated with the rectum and bladder. The second case occurred in a person eighteen years of age, who, during life, had suffered from a large abscess in the abdominal cavity under the right lobe of the liver, which communicated with the intestinal canal and had led to numerous fistulous openings in the anterior abdominal wall. At the necropsy a loop of the ileum was found perforated and in communication with the abscess cavity. The pus contained numerous actinomyces. In the third case the diagnosis was made post-mortem by the discovery of actinomyces in the pus. The disease was located in the lower portion of the ileum and cæcum, where it had caused suppuration and numerous adhesions.

The most remarkable and interesting history is connected with the fourth case. A robust, well-nourished woman, forty years of age, was attacked quite suddenly with pain in the stomach, high temperature, diarrhœa, and vomiting, followed by cerebral symptoms and death. At the necropsy the right Fallopian tube was found transformed into a large abscess, both extremities of tube closed, walls of sac lined with granulations containing actinomyces. The fifth patient was fifty years of age, and had suffered for a long time from lancinating pain in the abdomen; a fistulous opening formed in the umbilical region and discharged a thin, yellowish-green pus. The post-mortem showed actinomycosis of the peritoneum, small intestine, left ovary and liver, large abscess among intestinal coils, perforation of small intestine and bladder. In the upper part of the small intestine small pigmented cicatrices were found. In all of the above cases the microscopical examination revealed the presence of actinomyces in the granulation tissue, as well as in the pus of the abscess cavities.

But three cases of actinomycosis had been reported from Switzerland until Langhans (*Correspondenzblatt f. Schweizerärzte*, June, 1888) recently increased this number by three more which came under his own personal observation. One affected the mastoid region, while in the two others the entrance of the fungus took place through the alimentary canal. In one of these latter cases the process evidently started from the appendix vermiformis, which was four centimetres in length, the end of which appeared as if transversely cut in an abscess cavity the size of a walnut. The abscess was on the right side of the bladder, and so deep in the pelvis that during life it could not be located. The abscess pursued a chronic course, and the walls were well defined; no sign of chronic or acute peritonitis. Furthermore, the mucous membrane of the appendix was studded with cicatrices, and presented a slate-color. The principal seat of the actinomycotic process was in the liver. The second case presented marked symptoms of perityphlitic abscess during life.

The necropsy showed perforation of the cæcum and ascending colon. No cicatrices in the mucous membrane or surrounding tissues. In all probability the perforations occurred from without inward.

Luening and Hanau (*Correspondenzblatt f. Schweizerärzte*, 1889, No. 16) report a very interesting case of primary actinomycosis of the colon, with metastatic deposit in the liver. The patient was a man twenty-eight years of age, who in 1880 suffered from an acute abdominal affection which at the time was diagnosticated as typhlitis. Four years later he suffered from a second attack, which presented the appearances of intestinal obstruction. He was very ill for eight days, when the symptoms of obstruction subsided, and he made a slow recovery. During the year 1887 he had a third attack, attended by high fever and absolute constipation for eight or ten days. During the month of December of the same year he had another but less severe attack, and at this time a hard swelling made its appearance in the right side of the abdomen. From this time until he was admitted into the hospital, April 5, 1888, he was confined to bed. The patient was at this time greatly emaciated, with a temperature from 38.4° to 39.8° C. Swelling the size of a fist in the right side of the abdomen half way between umbilicus and anterior superior spine of the ilium. Externally this swelling presented redness and œdema. Fluctuation indistinct. Deep palpation showed that swelling extended to right hypochondrium. Abdomen not tympanitic. Swelling painful and tender, pain extending to spermatic cord and testicle on same side. A few days later abscess was incised and nearly a pint of brownish pus having a feculent odor escaped. Digital exploration revealed an irregular cavity the walls of which at some points were plainly lined with intestinal coils. Disinfection and drainage. As the symptoms did not improve materially the abscess cavity was again scraped out and disinfected four weeks later. After this operation it was noticed that the pus contained yellow granules, which under the microscope were shown to be actinomyces. The abscess was incised a third time, but the patient showed no improvement, and died October 9th. The autopsy revealed primary actinomycosis of ascending colon with multiple fistulous perforations, metastatic actinomycosis of liver with perforation of one of the foci into the hepatic vein, and multiple metastases in lungs.

These observations warrant the opinion that the mucous membrane of the intestinal canal is a frequent seat of primary localization of the actinomyces, thus corroborating the statements made by Johne in reference to this disease in animals.

3. *Inhalation actinomycosis.* The case of actinomycotic abscess of the lung caused by the inhalation of an infected tooth, reported by Israel, has already been cited as an illustration in showing that decayed teeth frequently serve the purpose of an infection-atrium in actinomycosis of the mouth. Cases of primary actinomycosis of the lungs, however, have been observed in which no such carrier of the contagium could be found, and in which infection must have occurred by the direct inhalation of the fungus or its spores.

Szénásy (" Ein Fall von Lungen-Aktinomykose," *Centralblatt f. Chirurgie*, 1886, No. 41) found, in the case of the wife of a butcher, who had suffered for nine years from severe pain in the right side of the chest, latterly attended by a severe cough in the right mammary region, a fluctuating swelling the size of a hen's egg covered with normal skin. On the outer side of this swelling, in the intercostal space between the third and fourth ribs, another swelling existed double in size and elongated in shape and with indistinct margins. This latter swelling had been noticed for nine years, and was tender to the touch. Auscultation over the fourth and fifth intercostal spaces on the healthy side revealed bronchial breathing and diffuse bronchial râles. Temperature 38.4° C. (101.1° F.). Urine contained a trace of albumin. By aspiration one hundred and fifty cubic centimetres of thick yellow pus were removed, which contained colonies of actinomyces. Actinomyces were also found in the sputum. The patient had carious teeth, but no signs of actinomycosis could be detected in the mouth.

Canali (quoted by Partsch) communicates the clinical history of a girl, fifteen years of age, who had suffered for eight years from a cough, attended by a scanty fetid expectoration. Inspection and percussion yielded only negative results. Auscultatory symptoms pointed to a diffuse catarrh. Under the microscope the sputum was seen to contain pus corpuscles, epithelial cells, and numerous actinomyces. No primary source of infection could be found in the mouth, pharynx, or nose. Other cases of primary actinomycosis of the lung have recently been reported by Laker ("Beitrag zur Charakteristik der primären Lungen-Aktinomykose," *Wiener med. Presse*, 1889). Lindt ("Ein Fall von primärer Lungenspitzen-Aktinomykose," *Schweiz. Correspondenzblatt*, Nos. 9–12, 1889), and Rütimeyer ("Ein Fall von primärer Lungen-Aktinomykose," *Berl. klin. Wochenschrift*, No. 3, 1889). In all of these cases the disease proved fatal by the extension of the actinomycotic process, followed by the formation of large abscesses. In Lindt's case the disease extended from the lungs to the muscles of the neck. Moosbrügger interprets the mechanism of the ingress of actinomyces by assuming that the fungus enters the bronchial tubes during inspiration, and becomes at first deposited upon the mucous membrane, in which its presence and growth cause a destruction of the epithelial cells, when it reaches the submucous and peribronchial tissues, in which a nodule of granulation tissue is produced, which by pressure induces degenerative changes and gradual destruction of the bronchial wall for further infection. He believes that the peribronchial lymphatic vessels and glands take an active part in the local diffusion of the process, as they furnish an avenue for the distribution of the germ or its spores. He claims the existence of an actinomycotic lymphangitis, but confesses that he has never seen the fungus inside of lymphatic vessels. As soon as the fungus reaches the pulmonary tissues it gives rise to parenchymatous inflammation, the first product of which is always granulation tissue, which, at a later stage and under the influence of a secondary infection with pus-microbes, undergoes transformation into pus corpuscles and the formation of abscesses.

ACTINOMYCOSIS OF BRAIN.—Bollinger has placed on record the first case of primary actinomycosis of the brain ("Ueber primäre Aktinomykose des Gehirns beim Menschen," *Münchener med. Wochenschrift*, 1887, No. 41. The patient was twenty-six years of age. The *intra vitam* diagnosis was tumor of the brain; the most prominent symptoms were severe headache, paralysis of the left abducens, congestion of optic papilla, and momentary unconsciousness. The swelling in the brain, found on autopsy, presented the characteristic features of a cysto-myxoma in the third ventricle; all of the ventricles were found considerably dilated. The swelling contained numerous colonies of actinomyces in all possible stages of development. The tendency to suppuration of the tissues usually found in all cases of actinomycosis in man was entirely absent in this case. This case, if any, appears to be one of crypto-genetic infection, as the fungus or spores must have entered somewhere through the cutaneous or mucous surface without producing the disease at the *portio invasionis*, and, localizing in the brain by embolism, produced primary actinomycosis in this organ.

Keller (*British Medical Journal*, March 2), 1890) reports a case of actinomycosis of the brain secondary to same disease in chest-wall in which a correct diagnosis was made during life, and an operation performed followed by temporary improvement. The patient was a woman, forty years of age, who suffered from pleurisy followed in six months by an abscess over the cartilage of sixth rib and also the eleventh. Both were incised, contents removed by sharp spoon, and drained. They did not communicate with the pleural cavity, nor were the ribs affected. These abscesses healed, leaving one small fistula. Two years later she complained of increasing paresis of left arm. Diagnosis of actinomycosis in the motor area was made, but patient declined operation. Convulsions of the left arm soon set in several times taking the course of cortical epilepsy. The paresis extended to left lower extremity and left side of the face. Headache, vomiting, and complete loss of consciousness followed, which developed into deep coma, and when apparently moribund operation was consented to. Dr. Burger, without any anæsthetic, trephined the skull over the middle of the right ascending parietal convolution, incised the dura mater and discolored brain substance and removed two ounces of thin greenish pus which contained great quantities of actinomyces. Soon after opening the abscess she recovered from the deep coma, and called for water. On the following day consciousness returned. On the eighth day facial paralysis disappeared and she could move the leg. Six months after the operation she began to walk around. During the next few months, the paralytic lesions materially improved, but there still remained paresis of left arm and slight contraction of the fingers. In two months the wound had healed, and the patient felt very well. Several (eleven) months later, grave symptoms of increasing paralysis, headache, and convulsions returned. Dr. Burger reopened the brain and removed a considerable quantity of pus. This was followed by no material improvement, and the patient died a few days afterwards. At the post-mortem the middle third of the right frontal and parietal convolutions was occupied by a large mass of newly-formed tissue protruding over the surface of the brain, reaching down into the substance of the brain for one inch. Underneath it, deeply buried in the white substance, an unopened encapsulated abscess the size of a nutmeg was discovered.

CHAPTER XXI.

GONORRHŒA.

HISTORY.—In no other disease has the suspicion of a specific infective cause been so general and entertained for such a long time as in gonorrhœa. Hallier, Donné, Jouisson, Salisbury, and many others, made diligent search for its contagium, which, by them, was believed to be a living organism. Following Koch's improved method of investigation, Neisser finally discovered the specific microbe in gonorrhœal pus in the year 1879 (*Centralblatt f. d. med. Wissensch.*, 1879, No. 28). He called the new microbe gonococcus. He described the microörganism as a diplococcus which differed from the other varieties of this species of parasites in being always found in clumps of from ten to twenty, surrounded by a mucous envelope. Neisser's communication was soon followed by a number of exhaustive publications, and to-day the literature on the gonococcus has become quite extensive.

DESCRIPTION OF THE GONOCOCCUS.—The gonococcus always occurs in pairs, and is, therefore, a diplococcus. The cocci appear as hemispherical bodies with their flattened surfaces in apposition, which imparts to the microbe the characteristic biscuit-shaped appearance. The gonococci are found in clusters or clumps upon, or, what is more probable, as Bumm asserts, within the pus corpuscles of gonorrhœal pus. Their intracellular location was shown by Bumm by examining pus corpuscles in water: when, after imbibition of water, the cells became swollen, the cocci could be seen between the molecular granules of the protoplasm. The microbes within the corpuscle may become so numerous as to fill the entire space, with the exception of the nucleus. Bockhardt and Haab asserted that they found them, also, inside of the nucleus, but this has not been confirmed by some of the ablest and most careful bacteriologists.

Legrain ("Recherches sur les Rapports qu'affecte le Gonococcus avec les Eléments du Pus blenorrhagique," *Arch. de Physiol. norm. et pathol.*, 1887, No. 6) on studying the behavior of the cellular elements of gonorrhœal pus and the gonococci, has found that in the very incipiency of the disease the secretion contains an abundance of epithelia and few pus corpuscles; the gonococci are abundant on the surface of the epithelia and few in the interior of the

PLATE XII.

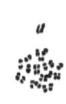 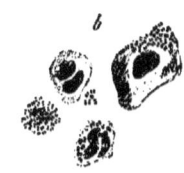 

Gonococcus.   (After Bumm.)
a. From a pure culture.
b. From a blennorrhœic conjunctival secretion; an epithelial cell covered with cocci; a pus cell with cocci in the protoplasm; a pus cell completely filled with cocci; a free mass of cocci in close proximity to a pus cell (Seibert). Oc. 2.
c. Scheme of development of gonococci.

pus corpuscles. At this stage inflammation affects only the most superficial portion of the mucous membrane. In a short time, however, the gonococcus penetrates into the lymph spaces and the inflammation extends more deeply. From the second to the third day, as the secretion becomes more profuse, the epithelial cells are less numerous and the pus corpuscles are in proportion more abundant, and from two to three of them to every hundred contain gonococci, usually two to three congregated closely together in one corpuscle. During the progress of the acute stage the epithelial cells become more and more scanty, while the pus corpuscles containing cocci increase in number so that finally one in every five contains the specific microbes.

During the subacute stage, about the fourth week, the epithelial cells again have become more numerous, about in the proportion of one to every eight or ten pus corpuscles. But few of the cells contain gonococci, while the pus corpuscles are freely charged with them. In chronic cases of gonorrhœa the epithelial cells again are more numerous, and most of them contain gonococci while the pus corpuscles almost disappear. Extra-cellular gonococci are also numerous. He believes that gonococci can enter the nucleus.

The size of the gonococcus varies according to its age, and depends, also, upon the nature of the nutrient medium. In the mucous membrane in man they attain the largest size, and measure 1.6 micromillimetre from pole to pole, while the width of each separate coccus is 0.8 and 0.6 micromillimetre.

According to Welander ("Ueber den Einfluss der Kenntniss der Gonococcen auf die Behandlung der Gonorrhœen," *Hygiea*, Bd. lxvii., 1885), the gonococcus multiplies by division. The first change that is observed is that the coccus assumes an oblong shape, in the centre of which a constriction takes place, which becomes deeper and deeper until the segmentation is completed. This process he has only observed in cocci free in the secretion, only exceptionally within epithelial cells, and never in cocci in the interior of pus corpuscles or their nuclei. He was unable to satisfy himself of the correctness of Bockhardt's observation that the gonococcus multiplies in the interior of white blood-corpuscles.

STAINING.—The gonococcus stains readily with basic aniline dyes, but by using gentiana, dahlia, and methyl-violet, the staining becomes so intense that the fissure between the cocci becomes obscured, while this is not the case by using concentrated fuchsin solution. Bumm claims that the staining can be done most successfully with aniline-violet, and that the staining is less intense by using fuchsin.

Smirnoff stained the gonococci for microscopical examination as follows: The slide is covered with a thin layer of pus, dried, and immersed for a minute in a solution of methyl-blue, composed of 1 part of methyl-blue, 33 parts of alcohol, and 66 parts of water, after which it is washed in water, and after drying mounted in Canada balsam.

Schuetz (*Münch. med. Wochenschrift*, No. 14, 1889) recommends the following method : The carefully prepared cover-glass preparations are immersed for from five to ten minutes in a cold, filtered, and saturated solution of methylene-blue in five per cent. carbolic acid water. They are then washed in distilled water, and afterward placed for a moment in a solution of five drops of acetic acid in twenty cubic cm. of distilled water, and again washed in pure water. Everything in the specimen is now decolorized, except the gonococci, which remain distinctly blue. Double staining with safranin can now be done when the gonococci and epithelial cells show a blue color, while the pus cells and their nuclei are found salmon-colored.

CULTIVATION.—The cultivation of the gonococcus is associated with many difficulties. Bumm (*Der Microörganismus der Gonorrhœischen Schleimhaut-Erkrankung, Gonococcus Neisser*, Wiesbaden, 1887), succeeded best in obtaining a pure culture upon blood serum of the calf and sheep, especially if to the coagulated serum a little serum from human blood is added, and after coagulating again the nutrient medium is kept at an even temperature of 30° to 34° C. (86° to 93.2° F.). If the temperature exceeds 38° C. (100.4° F.), the gonococci are invariably destroyed. By using the above nutrient substance Bumm observed, eighteen to twenty-four hours after inoculation, the whole surface covered with the growth. The culture grows only upon the surface and does not liquefy the soil. The addition of very mild antiseptics to the soil completely prevented the growth of the microbe. Cultivations on agar-agar and gelatin never proved successful.

Kreis ("Beiträge zur Kenntniss der Gonococcen," *Wiener med. Wochenschrift*, 1885, Nos. 30–32) succeeded in cultivating the gonococcus upon agar-agar with an admixture of Kemmerich's meat-peptone, kept at a temperature of 30° to 40° C. (86° to 104° F.). The culture soil was not liquefied and the size of the cocci was uniform; addition to the soil of 2 per cent. alkali arrested further growth of the culture.

Krause ("Die Micrococcen der Blenorrhœa neonat.," *Centralblatt f. d. prakt. Augenheilk.*, 1882, p. 134) made cultivation experiments upon meat-infusion-peptone-gelatin, but failed in many instances. After many trials with serum of animal blood he finally succeeded in obtaining a culture by placing the glass tubes in an incubator, in which the temperature was kept from 30° to 38° C. (86° to 100.4° F.). The culture appeared on the surface, starting from the puncture as a yellowish-gray film which grew very slowly. Leistikow and Löffler also succeeded best with blood serum.

INOCULATION EXPERIMENTS.—1. *In man.* Bumm found two females who were willing to submit to inoculation experiments. In one a pure culture obtained upon human blood-serum, and in the other a pure culture grown upon animal blood-serum, were applied to the urethra. In both cases a typical gonorrhœa was developed. Numerous gonococci could be found in the pus. These experiments afforded him a reliable proof of the etiological significance of the gonococcus.

Bókai (*Allg. med. Central Zeitung*, 1880, No. 74) produced gonorrhœa in two men by the injection of a pure culture into the urethra.

Bockhardt injected into the urethra of a man, forty-five years of age, suffering from a fatal disease, a pure culture of the gonococcus grown upon meat-infusion-peptone-gelatin, and produced a typical gonorrhœa. On the third day the secretions were examined for the specific microbes, which were found in abundance. On the tenth day the patient died of hypostatic pneumonia. The post-mortem revealed the existence of gonorrhœal cystitis and nephritis. In the pus of these organs, and in the sections from the fossa navicularis, numerous gonococci were found in the nuclei of the white blood-corpuscles, while the connective-tissue spaces and lymphatic channels appeared to be almost completely blocked by them. The complications which were found, he regards as the direct result of the diffusion of the specific microbes.

Welander ("Einige Versuche zur Feststellung der Vitaltität der Gonokokken ausserhalb des menschlichen Körpers," *Schmidt's Jahrbücher*, B. ccxiv. p. 39) studied the resisting power of the gonococcus outside of the organism by inoculation experiments. As early as 1884 he made inoculations with negative results with dried gonorrhœal pus removed three hours to eight days before directly from the urethra. During the year 1886 the same experiments were repeated. In all cases in which inspissated pus was used the inoculation proved harmless. The result was the same whether the pus was used in the dry form, or moistened with water. These experiments prove that the gonococci lose their virulence during the process of drying. To ascertain how long the gonococci will retain their virulence in fluid pus, he removed the pus directly from the diseased urethra and preserved it in capillary glass tubes in the same manner as vaccine virus is preserved. Four experiments were made. In three, the inoculations were made from one to several days after removal; in all the results were negative. In the fourth case, the inoculation was made with pus removed three hours before. On the third day the infected patient complained of a burning, smarting sensation in the fossa navicularis, which was followed by a typical gonorrhœa two days later. In the last case the pus was kept at the temperature of the body, which was not done in the other experiments. Two experiments he made with pus exposed to a temperature near the freezing-point, with negative results in both instances.

2. *In animals.* No uniform results have been obtained by inoculation experiments in animals, not because the gonococci are not the cause of gonorrhœa, but on account of the immunity of most animals to this form of infection. Lundström claims that he obtained a pure culture upon Koch's gelatin, and that inoculations with this produced typical gonorrhœa in dogs; in the purulent secretion numerous gonococci could be found.

Bumm states that pure gonorrhœal pus can be injected into the subcutaneous cellular tissue of animals without causing reaction, and that if, after twenty-four hours, an incision is made, and some of the pus which was injected is removed, it will be found that the cells are still in good condition, but that the cocci have disappeared.

That the gonococcus has a special predilection for the mucous membranes is well shown by the regularity with which purulent ophthalmia is produced by the infection of the conjunctiva with gonorrhœal pus.

The ease and regularity with which gonorrhœa can be produced in man by inoculation with a pure culture of the gonococcus, furnishes the *punctum saliens* which characterizes this microbe as the *materia peccans* of gonorrhœa.

ACTION OF GONOCOCCUS ON THE TISSUES.—The presence of the gonococcus so far has been demonstrated in the urethra, bladder, kidney (Bockhardt), in perimetritic abscesses following gonorrhœa, in the purulent contents of joints in gonorrhœal synovitis, the conjunctiva, rectum (Bumm), in the uterus, cervix, vagina, vulva, and in Bartholin's glands. The real seats of gonorrhœal infection are mucous membranes lined by columnar epithelium, or epithelium which closely resembles it. The grouping of the microbe in the pus corpuscles, from a diagnostic point of view, is more important than its diplococcus form. For the purpose of studying the effects of this microbe on the tissues, Bumm examined twenty-six specimens of gonorrhœal conjunctivitis. When the gonococcus is brought in contact with the conjunctiva or the urethra, the first growth takes place upon the surface of the epithelial layer, as its entrance into the deeper layers meets with difficulty, and Bumm asserts that it penetrates deeper only after the epithelial layer has become somewhat loosened by inflammatory changes—that is, after intercellular passages have formed. A dense, compact layer of epithelial cells furnishes a safe protection against gonorrhœal infection, so much so, that Bumm claims that a true gonorrhœa in the vagina of adults is impossible, while it does occur in children. The gonococcus reaches the deeper tissue layers exclusively by its growth into the intercellular passages, consequently the advance is very slow. The first effect of the infection upon the mucous membrane is an increase of the physiological secretion—a thin mucous fluid. Suppuration is not the result of the cocci, but of chemical substances which are produced by them, as pus corpuscles appear before the microbes have reached the vascular layer of the mucous membrane. As the pus corpuscles enter the epithelial layer the latter becomes still looser and some of the cells exfoliate. After this stage the pus cells are the structures in which the gonococci are developed. Cessation of the suppurative process is not always a sign that infection has also ceased, as during the latent stage only a catarrhal secretion is present, but suppuration may be lighted up again at any time under the influence of additional causes which produce an aggravation of the chronic inflammatory process.

In serous cavities gonorrhœal pus produces, as a rule, a circumscribed abscess. Sinclair, in his excellent monograph (*Gonorrhœal Infection in Women*, London, 1888, p. 79), after describing the gonorrhœal infection from the vagina, says: "The proper character and results of the pathogenous activity of the gonorrhœic microbes are therefore seen, pure and unadulterated, in the tubes. They cause purulent inflammation of the mucous membrane, but the surrounding connective tissue remains free from them. The gonorrhœic tubal pus is evacuated into the peritoneum, and whereas in other conditions the bursting of an abscess into the abdominal cavity is followed by the gravest consequences, in this case the whole process terminates with a circumscribed inflammation, encapsulating the exuded pus. The cause of this difference is the varying pathogenic value of the organisms which are contained in the pus. A puerperal pelvic cellulitic abscess bursting into the peritoneum causes general peritonitis, because it contains pyogenous streptococci, which rapidly multiply in serous cavities and are capable of exerting the most deleterious effects. Gonorrhœal tubal pus cannot do this; its microbes do not find in the peritoneum conditions for their increase; the pus, therefore, acts as an aseptic foreign body, becomes encapsulated, and is finally absorbed."

That this favorable termination does not always follow gonorrhœal infection of the peritoneal cavity is well shown by a case reported by Lovén ("Fall von Gonorrhœ bei einem fünf jährigen Mädchen; Peritonitis; Tod," Schmidt's *Jahrbücher*, B. ccxiv. p. 39) which is by no means an isolated one. The source of infection could not be learned in this case, but the diagnosis of gonorrhœic ascending infection was positive. The disease commenced as an ordinary vulvo-vaginal blennorrhœa, which consecutively extended to the uterus and Fallopian tubes, and terminated in pelvic and diffuse peritonitis. It is possible that in this case a secondary infection had taken place, as at the necropsy chain cocci were found in the peritoneal cavity.

THE GONOCOCCUS AND PURULENT OPHTHALMIA.—Haab (*Der Micrococcus der Blenorrhœa neonat.*, Wiesbaden, 1881) showed that the microörganism found in gonorrhœal pus and the secretion of purulent ophthalmia are identical. He placed great stress on the fact that according to his own observations the gonococcus is always present in the secretions of purulent ophthalmia, and that it is never found in the simple inflammatory or catarrhal form.

Widmark ("Bakteriologische Studien über purulente Conjunctivitis und gonorrhœische Urethritis," *Hygœi*, B. xlvi., 1884) examined twenty-four cases of purulent conjunctivitis in reference to the existence of the gonococcus, using Welander's method. In most of the cases he found the microbe. The gonococcus was found free in the secretions, the pus corpuscles, and the epithelial cells. He believes, with Welander, that during the period of incubation the microbes remain attached to the epithelial cells, and are reproduced there, whence they later penetrate into the deeper tissues. After a time they almost disappear in the secretions, without, at the same time, any improvement taking place in the catarrhal condition. From a practical standpoint these observations are important, as they prove the importance of early treatment before the microbes have passed beyond the reach of local applications of antimycotic remedies.

Among other prominent ophthalmologists, Sattler, Lebert, and Hirschberg recognize the gonococcus of Neisser as the specific cause of gonorrhœa and its identity with the coccus found in specific purulent ophthalmia.

THE GONOCOCCUS IN ABSCESSES.—Opinions are divided in reference to the pyogenic properties of the gonococcus. In infections of the mucous membrane its pus-producing property is well known, but at present is not attributed to its direct effect on the tissues, but to the action of ptomaïnes which it produces. A number of cases have been reported which appear to show that under certain circumstances the microbe enters the circulation and is the cause of metastatic abscesses, and on this account should be classed with the pus-microbes.

Horteloup relates the case of a man, twenty-seven years of age, who had suffered from gonorrhœa for several months, in whom an abscess formed in the clavicular region, which was incised, and in the pus numerous gonococci were found.

Schwarz (*Sammlung klin. Vorträge*, 1886, No. 279) asserts that the gonococcus is constantly found in the effusion of joints in gonorrhœal rheumatism, in abscesses caused by gonorrhœa, and in the glands of Bartholin. He further states that it is found free in the fluid, but mostly it is incorporated in the protoplasm of pus-corpuscles. Petrone detected the specific microbe in the effusion of joints, and in the blood, in two patients suffering from gonorrhœal rheumatism. He regarded the joint complications as metastatic processes caused by the gonorrhœal infection.

Kammerer (" Ueber gonorrhœische Gelenkentzündung," *Centralblatt f. Chirurgie*, 1884, No. 4) examined the fluid from two cases of gonorrhœal arthritis, and found micrococci present in one and absent in the other.

At a meeting of the Gesells. d. Aerzte, in Zürich, Prof. Haab (Correspondence, *British Medical Journal*, June 2, 1888) read a paper on a generalized or constitutional gonorrhœal infection developing consecutively to specific urethritis. He often met cases of gonorrhœal iritis and irido-cyclitis, as well as of conjunctivitis, originating without any direct transmission of the urethral discharge. In cases of the latter kind no gonococci were present in the conjunctival secretion, the disease usually running a mild and favorable course. He further cites a striking case of general infection after urethral gonorrhœa, in which death was threatening in consequence of extreme exhaustion, caused by a continuous, uncontrollable fever of many weeks' duration. The man was suffering from effusion into his knee- and elbow-joints, abscesses in the left axilla, prolonged and obstinate cystitis, and double ophthalmia of the severest kind, ending in complete destruction of one eye and a serious disorganization of the other. No microbes were found in the articular exudation, but the axillary abscess was found to contain staphylococcus. Prof. Haab seems to think that this case is an instance of a mixed general infection with gonococcus plus staphylococcus.

Bergmann (" Gonitis gonorrhœica mit Coccen," *St. Petersburger med. Wochenschrift*, 1885, No. 35) found gonococci in pus removed from a knee-joint of a patient suffering from gonorrhœa, three weeks after the beginning of the urethritis.

Haslund (" Beitrag zur Pathogenese des gonorrhœischen Rheumatismus," *Vierteljahrsschrift f. Derm. u. Syph.*, B. ix. p. 359), on the other hand, was unable to find the specific cocci of gonorrhœa in eleven cases of synovitis in gonorrhœa patients.

Hoffa (" Bacteriologische Mitth. aus der chir. Klinik des Prof. Maas, Würzburg," *Fortschritte der Medicin*, B. x. p. 75) found no gonococci in a purulent synovitis occuring in a patient suffering from gonorrhœa, but in two cases in which bubo formed he found staphylococcus pyogenes albus and aureus in the pus of the abscesses, but no gonococci. These were undoubtedly cases in which suppuration in the glands was caused by a mixed infection. Smirnoff (*Wratch*, 1886, No. 31) examined the sero-purulent contents removed from the knee-joint by aspiration, in a patient twenty-eight years of age, six weeks after gonorrhœal infection and two weeks after the commencement of the joint affection, and found numerous clusters of gonococci within the pus-corpuscles. Afanasieff made similar observations.

Sahli has recorded a very interesting case of gonorrhœal metastasis of the skin. In a patient who for two months had suffered from gonorrhœa, there appeared two abscesses, as large as a man's fist, situated in the region of the knee-joint. On incision a large quantity of sero-sanguinolent pus escaped. Microscopical examination revealed the presence of typical gonococci, imbedded, as usual, within pus-corpuscles.

Some pathologists, among them Watson Cheyne ("Lectures on Suppuration and Septic Diseases," *Brit. Med. Journ.*, February 15, March 3, 10, 1888), assert that, when in cases of gonorrhœa suppurative adenitis and para-adenitis in the inguinal glands take place, pus-microbes are present in the pus, and the suppuration must be considered as the consequence of a mixed infection.

DIAGNOSTIC VALUE OF THE GONOCOCCUS.—Neisser, Bumm, Bockhardt, Eschbaum, Newberry, Campona, Aufrecht, Schwarz, Lundström, Weiss, Ehrlich, Brieger, Hartdegen, and others have never failed in finding the gonococcus present in gonorrhœal discharges. The best ophthalmologists rely upon its presence in differentiating between specific and simple conjunctivitis. Only a few authorities have arrayed themselves against Neisser's claim.

Sänger ("Gonorrhœal Disease of the Uterine Appendages and its Operative Treatment," *Archiv f. Gynäkologie*, B. xxv. Heft 1) states that the hope aroused by the discovery of Neisser, that in the gonococcus we should find the means of diagnosing chronic gonorrhœa, had proved to be in vain, and holds it to be an established fact that gonorrhœa can exist without the demonstrable presence of gonococci. The absence of the gonococcus proved nothing against the gonorrhœal nature of the disease; whilst the presence of diplococci, in view of the occurrence of non-pathogenic forms, did not prove the gonorrhœal nature of the disease. If the cocci cannot be found, they may have been somewhere broken up, while a ferment produced by them may still be active; or they are absent from the secretion while present in the tissues; or there exists—and this would render the high degree of infectiousness of a comparatively trifling amount of secretion in latent gonorrhœa the most intelligible—a permanent form (Dauer-form) of the gonococcus not yet discovered.

Eugen Fränkel ("Bericht über eine bei Kindern beobachtete Endemie infektiöser Kolpitis," Virchow's *Archiv*, B. xcix. p. 251) claimed to have found a diplococcus in the secretions of non-gonorrhœal colpitis in children which could not be distinguished from the gonococcus of Neisser. He therefore argued that for diagnostic purposes the simple presence of the microbe could not be relied upon, and cultivation and inoculation experiments became necessary in order to differentiate between the specific and non-specific diplococcus. Later, after resorting to another method of staining, he satisfied himself that the diplococcus was identical with Neisser's gonococcus.

Sinety and Henneguy have made special observations in reference to the presence of the gonococcus in the pus of gonorrhœal urethritis in the female and found it present only in cases in which the urine was of alkaline reaction, while they failed to find it when the urine was acid.

Pott (*Archiv f. Gynäkologie*, B. xxxii. Heft 3) has examined 96 cases of purulent vulvo-vaginitis in children, more than one-half of whom were under five years of age. He believes that they were all the result of some specific infection. Small endemics occurred from the infection of several children in the same family, through the medium of soiled clothing, sponges, etc. The writer has only observed three cases of gonorrhœal infection by direct communication of the virus. Bacteriological examination of the discharge usually revealed the presence of the specific microbe of gonorrhœa. In the discussion of this paper, Prochownik stated that he had found the cocci in 17 out of 21 cases of vulvo-vaginitis in children. In all of the cases urethritis was a prominent symptom. A girl, three and a half years of age, that came under the care of Sänger developed intense peritonitis in consequence of an attack of gonorrhœa. Sänger is of the opinion that cases of pyosalpinx and old localized pelveo-peritonitis in young virgins might possibly be referred to gonorrhœa contracted in childhood through indirect infection.

Spaeth (*Münchener med. Wochenschrift*, 1889) examined the pus in 21 cases of vulvo-vaginitis occurring in girls between three and eleven years old, and found gonococci in 14. In the non-specific catarrh the inflammation never implicated the mucous membrane of the urethra. In adult females affected with gonorrhœa the greatest number of the specific microbes is always found in the pus from the urethra. In children it is not always easy to discover the source of infection. In 11 of the above cases the mother had gonorrhœa; in 2, the father; in 3 only had the child been violated. In children the disease seldom extends to the uterus and tubes, although a few cases of gonorrhœal pyosalpinx have been reported.

Steinschneider (*Centralblatt f. d. med. Wissensch.*, 1890, No. 39) regards Gram's method of staining as necessary to make the microscopic examination of gonorrhœal pus of diagnostic value, as the gonococcus is not stained by this method, while nearly all other diplococci found in the urethra are colored thereby.

These conclusions, which agree with those of Roux, are the result of the examination of 86 patients with acute and chronic gonorrhœa. The almost entire certainty of this test is rendered absolute by the observation of the further characteristic of the gonococci, namely, that they are found within the pus-corpuscles.

Sternberg (*The Medical News*, January 20, 1883) cultivated a micrococcus from gonorrhœal discharges in bouillon which in its morphology resembled the gonococcus described by Neisser. He made numerous inoculation experiments in animals and a few in man with only negative results. Subcutaneous injections of a pure culture also proved harmless. He came to the conclusion that the micrococcus which he found corresponded to the micrococcus ureæ of Cohn ; the pathogenic effect of which has been shown to be the cause of the alkaline fermentation of urine (Pasteur).

Leistikow (*Deutsche Medicinal-Zeitung*, September 7, 1882) has observed that during the first stage of gonorrhœa, when the discharge is thick and abundant, but few gonococci could be found. They were found abundant in the thin and scanty secretion of the later stages, sometimes even when the disease had existed for a year. All authorities who have studied the relations of the gonococcus to gonorrhœa with the greatest care, insist that, for diagnostic purposes, it is not only necessary to demonstrate its presence, but to ascertain its intra-cellular location and the manner in which this microbe arranges itself in groups in the protoplasm of the cell between the nucleus and the envelope of the cell, and these conditions should be studied with the greatest care in all medico-legal cases in which a positive opinion must rest on a microscopical examination of the secretion.

CHAPTER XXII.

SYPHILIS.

THE infectious microbic nature of syphilis is so evident that no one for a moment would dare to question it, and yet with all modern improvements for bacteriological research and the prevalence of this affection at all times, and all over the world, it is strange that, so far, it has not been possible to furnish positive and convincing proof of the existence of a definite, specific microörganism in all syphilitic lesions, and to demonstrate its etiological relation to this disease. It is interesting and profitable to know what has been done during the last few years in the bacteriological study of syphilitic lesions, and although the claims which have been made are in all probability unfounded, I will give a brief *résumé* of the literature on this subject.

In 1884 Lustgarten (*Wiener med. Wochenschrift*, 1884, No. 47) announced that he had found a bacillus in two cases of initial sclerosis and in a syphilitic gumma, for which he claimed specific pathogenic properties. Nearly at the same time, and without knowledge of Lustgarten's work, Doutrelepont found a bacillus in a primary hard chancre, two broad condylomata, and in one case of syphilitic papular eruption of the skin, which resembled closely the bacillus of tuberculosis. He found it difficult to stain this bacillus, but finally succeeded with gentian-violet. In the beginning of the year 1885, Lustgarten ("Die Syphilisbacillen," Mit. 4 Tafeln, *Wiener med. Jahrb.*, 1885) published his second paper, in which he gave an accurate description of the bacillus and the results of a more extended investigation of the subject. He had in the meantime examined numerous specimens of syphilitic lesions, and as he had invariably been able to demonstrate the presence of the bacillus in them and its absence in two soft chancres, he expressed his firm conviction that the bacillus was the specific cause of the disease.

DESCRIPTION OF LUSTGARTEN'S BACILLUS OF SYPHILIS.—The bacilli are rods 3 to 4 micromillimetres in length, and 0.88 micromillimetre in thickness, resembling somewhat the bacilli of leprosy and tuberculosis. The rods are not straight, but somewhat curved, or S-shaped. After staining, light oval spots were seen within their protoplasm which were thought to be spores. The bacilli were never seen free, but were always found in the interior of nucleated cells which are more than double the size of leucocytes. The bacilli have been observed in the discharge of the primary lesion, and in the hereditary affections of tertiary gummata.

STAINING.—*Lustgarten's Method.* Sections are placed for from twelve to twenty-four hours in the following solution at the ordinary temperature of the room, and finally the solution is warmed for two hours at 60° C. (140° F.):

> Concentrated alcoholic solution of gentian-violet . 11 parts.
> Aniline water . . . . 100 "

The sections are then placed for a few minutes in absolute alcohol, and from this transferred to a 1.5 per cent. solution of permanganate of potassium. After ten minutes they are immersed for a moment in a pure concentrated solution of sulphurous acid. If the section is not completely decolorized, immersion in the alcohol and in the acid must be repeated three or four times. The sections are finally dehydrated with absolute alcohol, cleared with clove-oil, and mounted in Canada balsam. Giacomi has simplified and improved this method. He immerses cover-glass preparations by staining them for a few moments in a solution of fuchsin, after which they are washed in water to which a few drops of a solution of chloride of iron have been added. Complete decolorization is effected in a concentrated solution of chloride of iron.

Doutrelepont and Schütz obtained good results by staining in a one per cent. gentian solution, decolorizing in a solution of nitric acid 1 : 15, and after-staining with safranin; after which they are dehydrated in a sixty per cent. solution of alcohol, cleared in clove-oil, and mounted in Canada balsam. After this process the bacilli are stained blue and the tissues red.

Gottstein immersed the sections for twenty-four hours in fuchsin solution, after which they were washed in water and transferred into pure or diluted tincture of chloride of iron, dehydrated in alcohol, and cleared up in oil of cloves or xylol, when they are ready to be embedded in Canada balsam. This method stains the bacilli a red or dark violet color.

CULTIVATION EXPERIMENTS.—Lustgarten's cultivation experiments did not succeed. Few attempts to reproduce the bacillus upon different nutrient media have yielded positive results. Klebs succeeded best in cultivating them upon a gelatin prepared from the bladder of a kind of sturgeon (Hansenblasen-gallerte) found in some of the rivers of Russia. Inoculations of gelatin kept at the ordinary temperature of the room produced, after thirty-seven days, around the implanted piece of tissue a grayish-yellow culture. In one instance the same culture medium was inoculated with the blood of an infected monkey, when on the fifth day a brownish

PLATE XIII.

Syphilis bacilli from a papule, after a preparation from Lustgarten.
2500 diam.

zone had formed around the streak made by the needle, which on microscopical examination was seen to be composed exclusively of small bacilli. Embedded in the culture, granules were also found, which were thought might be spores. Similar results have since been obtained by Martineau and Hammonic (*Les Bactéries*, par Cornil et Babés, p. 774) and Birch-Hirschfeld (*Centralblatt f. d. med. Wiss.*, 1882, Nos. 33, 44), who cultivated the bacillus from the primary lesion, condylomata, and gummata of internal organs.

INOCULATION EXPERIMENTS.—The transmission of syphilis to animals by implantation of syphilitic tissue, or by inoculations with cultures, has, so far, not met with uniform success, as it has been found difficult to find animals susceptible to syphilitic infection.

Kleb's experiments on monkeys yielded positive results. Martineau and Hammonic experimented on the same animal, and produced the disease by inoculation with a pure culture. They observed twenty-eight days after inoculation of the prepuce two indurations which were followed by general secondary symptoms. In one case in which Klebs used a culture for inoculation an abscess formed, and the animal remained apparently well until the seventh week, when a granulation-swelling formed at the base of the upper jaw which ulcerated, and from this point a cheesy infiltration extended to the base of the skull; the same deposits were found between the base of the skull and the dura mater. From the abscess which still contained cheesy masses lymphatic infection had taken place, and the glands also contained cheesy material. No miliary tuberculosis could be found. Bacilli were found in the cheesy material which were identical with those contained in the inoculation material. In another case a piece of tissue from a hard chancre was implanted under the skin of one of the posterior extremities. Six weeks after the implantation general and febrile symytoms supervened, attended by a papular eruption on the forehead and face. Death occurred five months after inoculation, and at the post-mortem syphilitic lesions were found in the skull and lungs.

Bacteriological and Experimental Study of Syphilis.

Doutrelepont and Schütz ("Die Bacillen bei Syphilis," *Deutsche med. Wochenschrift*, 1885, No. 9) are more reserved in their statements than Lustgarten in reference to the causation of syphilis by the bacillus just described. In regard to the uniform presence of a bacillus in syphilitic lesions they concur in the views advanced by Lustgarten. They found the bacillus between and within cells. They consider it as the probable cause of syphilis, but they thought that further proof must be furnished by culture and inoculation experiments before this question could be definitely settled.

Doutrelepont found the bacillus taken from the blood near a syphilitic lesion of the skin, and from a syphilitic primary sclerosis of the upper lip produced a feeble culture upon solidified hydrocele fluid.

Matterstock (*Sitzungsbericht d. phys.-med. Gesellschaft zu Würzburg*, 1885, und *Ueber Bacillen bei Syphilis*, Würzburg, 1885), after a careful examination of 100 sections and 150 specimens of secretions from syphilitic lesions, supported in the main the views held by Lustgarten, and places great importance upon microscopical examination of syphilitic products as a diagnostic measure. Conjointly with Bitter, of Osnabrück, he found the disintegrated bacilli described by Doutrelepont in the shape of granular masses, which outlined the size and shape of the bacilli. He often saw the bacilli between the cells and the connective-tissue fibrillæ. He states that the bacillus is not present in large numbers in the specimens, and unless carefully searched for may not be found. As it can be found in the lesions of all three stages of the disease, he asserts that its etiological importance can be no longer doubted, even although cultivation and inoculation experiments have so far not furnished the crucial test. Later the same author found similar bacilli in the secretions of the genital organs in patients in whom syphilis could be excluded with certainty.

Markuse (*Vierteljahrsschrift für Dermat. und Syph.*, 1888) on examining 109 cases found Lustgarten's bacilli in 10 of 23 cases of hard chancre, in 43 of 57 of anal condylomata, and once in 19 cases of papules about the mouth examined. In 8 gummata and 2 pustular syphilides no bacilli were found. Smegma bacilli were found in 125 cases examined. They differ from the bacilli of syphilis in being more easily decolorized by acids.

Andromico ("Ueber die parasitäre Genese der Syphilis," *Vierteljahrsschrift f. Derm. u. Syph.*, 1886, p. 475) claims to have cultivated from a flat nodule of the skin in a syphilitic patient, coccobacteria which, injected under the skin of a rabbit, produced a typical indurated ulcer, followed by glandular infiltration. He also inoculated a cat with fluid contents of a syphilitic pemphigus with the result of producing a hard, painless swelling followed by a papular eruption of the skin, and loss of hair of the skin covering the abdomen. The fluid injected and the syphilitic lesions in the animal contained the same bacteria which he had cultivated from the syphilitic nodule of the skin.

Doutrelepont ("Ueber die Bacillen bei Syphilis," *Vierteljahrsschrift f. Derm. u. Syph.*, 1887, p. 101), after more extended observations, has come to the positive conclusion that bacilli are always present in syphilitic lesions, and even in the blood of patients suffering from syphilis, and insists that in spite of the discovery of the bacillus of smegma they must be regarded as the cause of syphilis.

Zeissl ("Untersuchungen uber den Lustgarten'schen Bacillus in Syphilisproducten und Secreten derselben," *Wiener med. Presse*, 1885, No. 48) examined sections from nine cases of initial sclerosis for bacilli, and found them present only in one, and in this one, only two were to be seen. Baumgarten readily detected Lustgarten's bacillus in a specimen taken from a hard chancre by resorting to De Giacomi's method of staining after he had failed in numerous instances with Lustgarten's method.

Haberkorn claims to have found in the blood of syphilitic patients a microörganism which he describes as a round, oval, or short cylindrical-shaped spore of darkened color, 0.001 to 0.002 millimetre in diameter, which is also found in close proximity, or adherent to white blood-corpuscles. He also claims to have cultivated this microbe.

Marcus (Thèse de Paris, *Annales de Derm. et Syph.*, 1885) discovered another microbe which he believed to be the cause of syphilis. He found in syphilitic lesions cocci arranged in groups of from six to seven, which could be readily stained with gentianviolet. They resist the action of alcohol, and are easily deprived of their color in acidulated alcohol. He made successful cultivations, but his inoculation experiments yielded only negative results.

Another bacillus which was supposed to be the cause of syphilis was described by Eve and Lingard (*On a Bacillus Cultivated from the Blood and from Diseased Tissues in Syphilis*, 1886). This bacillus contained spores and was readily cultivated. It appeared in the form of rods somewhat variable in length, which on staining showed that the different segments were unequally stained. The most successful results in staining were obtained with Humbolt-red, in aniline oil, and decolorization in alcohol. Gram's method was also found efficient, while Lustgarten's yielded only negative results. The cultures grew upon solid blood serum or hydrocele fluid and agar-agar.

Disse and Taguchi ("Ueber das Contagium der Syphilis," *Deutsche med. Wochenschrift*, 1885, No. 48, and same, 1886, No. 14) examined the blood of patients suffering from secondary syphilis, and found in it almost constantly cocci of one micromillimetre in diameter, isolated, or in colonies between the blood corpuscles. The cultures upon different nutrient media appeared as grayishwhite masses. All culture media with the exception of solid blood serum were liquefied by them, and it is the first time that we have an account of the liquefaction of agar-agar by microbes. In the cultures the microbes appear in pairs which give them the appearance of short rods with a light space in the middle at the point of junction of the two cocci. They exhibited active movements, and reproduction was seen to take place by simple segmentation. Gram's method of staining proved most successful. Isolated cocci, the authors believe, exist in the blood of patients the subject of latent syphilis, and in cultures made from them the diplococcus appears. With a pure culture seven rabbits, two sheep, five dogs, and one white mouse were inoculated. Under strict antiseptic precautions the microbes were introduced into the subcutaneous tissue through a minute incision. Diplococci were found in the blood of the dogs and rabbits after ten days, in the sheep after three weeks, and their presence remained constant for several months. Among the dogs were four females; two of them were pregnant at the time of inoculation, and all of the pups died soon after they were born. Some of the inoculated animals died at different times, and the others, with the exception of two, were killed in from two to eight months after inoculation. In none of the animals could any infection of the skin or mucous membranes be found. Induration at the point of inoculation was noticed in only one, and it disappeared eighteen days after the inoculation. In all of the animals characteristic pathological changes were found in the heart, lungs, liver, and kidneys. The authors believe that the syphilitic virus in animals, like in man, acts upon the bloodvessels, only that in animals the morbid changes take place in other organs. In animals the vessels of the skin remain exempt, while the vessels in the heart, lungs, liver, and kidneys are diseased. Quite frequently the vessels of the brain were also found diseased, but very seldom the vessels of the intestines. Syphilitic gummata were also found less frequently than in man. The authors claim that the microbes live in the circulating blood, and they found them in great abundance in the affected tissue, in which they appeared to have only a limited existence. They believe that syphilis becomes latent as soon as the cocci cease to grow.

From the above extracts of the current literature on the etiology of syphilis, it can be seen that different microbes have been found in the blood and tissue lesions of patients suffering from syphilis, and that the discoverers of each of them claimed to have found the microbic cause of syphilis. So far, no uniform results have been obtained, and it must be left for future research to discover a new microbe, or to substantiate some one of the claims made in the past. This much we can say for Lustgarten, that he has at least pointed out the proper method for future investigations.

The Bacillus of Smegma.

Soon after the publication of Lustgarten's second paper, Tavel (*Archiv de physiol. et path.*, No. 7, 1885) announced that he had found a bacillus in the smegma and secretions of the mucous membranes of the external genital organs, which in shape and its reaction to staining material proved identical with the bacillus described by Lustgarten. Klemperer, from his own observations, came also to the conclusion that the smegma bacillus resembled Lustgarten's bacillus, that their identity appeared more than probable. Both Tavel and Klemperer mention as a distinguishing feature between the smegma bacillus and the bacillus of tuberculosis, that the former is completely decolorized by washing the dried and stained preparation for a minute and a half in a $33\frac{1}{3}$ per cent. solution of nitric acid, and afterward for half a minute in absolute alcohol. In specimens thus treated, the bacilli disappear, while the tubercle bacilli similarly treated remain deeply stained.

In September, 1885, Doutrelepont published his observations on the smegma bacillus, wherein he corroborates Tavel's observations (Tavel's "Zur Geschichte der Smegmabacillen," *Centralblatt f. Bact. u. Parasitenkunde*, B. i., No. 23).

Secondary Infection of Syphilitic Lesions with Pus microbes.

Kassowitz and Hochsinger ("Ueber einen Mikro-organismus in den Geweben hereditär syphil. Kinder," *Wiener med. Blätter*, B. ix., 1886) reported during the early part of the year 1886 that they had found, by means of a modification of Gram's method of staining, a chain coccus in the contents of pemphigus bullae, the bones, liver, pancreas, lungs, and thymus gland of five syphilitic children who had died soon after birth. The microbes were found within the vessels and the paravascular spaces; the finer capillaries were sometimes seen to be completely filled with them. The cocci were found in close contact with the red blood-corpuscles, but never in their interior, or within other cells. They were only found in organs in which syphilitic lesions could be detected by naked-eye appearances, or by microscopical examination, and could never be found in the same places in non-syphilitic children. From the streptococcus of erysipelas these microbes differed as they were found either within or in the immediate vicinity of bloodvessels, and from the streptococcus pyogenes in that they did not cause suppuration. The authors do not claim that this microbe is the cause of syphilis, but look upon this discovery as an important episode in the study of syphilis.

Kolisko has examined the bodies of numerous syphilitic children who died one or two days after birth, and was unable to find this streptococcus or any other microörganism. In one case in which the child lived for fourteen days, and had suffered from furunculosis, he found a streptococcus which he took for the streptococcus pyogenes. He believed that this microbe had entered the body through an open pemphigus blister, the umbilicus, or had been derived from the furuncles. In another case in which he found the same microörganism in the liver, he obtained a pure culture of the streptococcus pyogenes from blood taken from the liver.

Chotzen (*Vierteljahrsschrift f. Derm. u. Syphilis*, 1887, S. 109) states that in searching for the same microbe, in Neisser's clinique, it was found once in bone, five times in the skin, four times in the liver, once in the mucous membrane of the intestine; in the last-mentioned organ the microbe was found in great abundance, not in the vessels, but in the paravascular and lymph spaces. In bone it was found where syphilitic lesions are usually met with. As the result of his observations he looks upon the streptococcus only as an accidental, and not as an essential, condition of hereditary syphilis. In cases of acquired syphilis it was never found. As this streptococcus must be different from the microbe of erysipelas and of suppuration, it still remains an open question as to its relation to syphilitic infection. From a hypothetical standpoint he regards the presence of the streptococcus as an evidence of a mixed infection, and believes that it produces death by giving rise to sepsis. The presence of the coccus in the medullary tissue of the epiphyseal extremities of the long bones would furnish an explanation regarding the frequency with which lesions are found here in children suffering from hereditary syphilis. As points of entrance of the streptococcus he enumerates coryza and inflammation of the naso-pharyngeal space.

Doutrelepont ("Streptokokken und Bacillen bei hereditärer Syphilis," *Centralblatt f. Bact. u. Parasitenkunde*, B. ii., No. 13) agrees with Chotzen in regard to the character and etiological significance of the streptococcus found in children suffering from congenital syphilis. He has found it in the papular eruptions of the skin, and believes that it enters the lymph spaces here, and through these channels reaches the general circulation and distant organs. The frequency with which suppuration in bones and other organs is met with in children soon after birth, and sometimes in utero, would suggest that the secondary infection results from the introduction, sometimes through the maternal circulation, at others through lesions after birth of pus microbes, which become localized in organs the seat of syphilitic lesions, where, in the course of time, suppuration takes place. The presence of an abundance of pus microbes may destroy life by sepsis before a sufficient length of time has elapsed for them to manifest their specific pathogenic properties on the tissues.

CHAPTER XXIII.

ON THE ALLEGED MICROBIC ORIGIN OF TUMORS.

SMALL, round-celled sarcoma resembles in its histological structure granulomata so closely that under the microscope it would be impossible to make a positive differential diagnosis between the two; at the same time all malignant tumors in their clinical behavior have so many things in common with infective swellings that it does not appear strange that the microbic nature of sarcoma and carcinoma has been suspected for a long time, and that during the recent strides which the modern science of bacteriology has made, this subject has been studied by the most improved methods of investigation.

The veteran surgeon and pathologist, Sir James Paget (*Lancet*, 1887, No. 19), not long ago called attention to the resemblance existing between cancer and benign tumors on the one hand, and specific and micro-parasitic affections on the other. He believes that cancer is allied to the group of specific microbic diseases, including syphilis, tuberculosis, glanders, leprosy, and actinomycosis. He claims that cancer and these specific infective diseases constitute a group of growths which are self-sustaining, have special modes of degeneration and of ulceration, to which they all tend; are all, at some time, either infective to tissues at a distance by transportation of parts of the growth through lymphatics or bloodvessels, or to adjacent parts by invasion, or to other beings by inoculation; and finally they all occur by preference in tissues or organs the subject of local injury or irritation. He is strongly inclined to the supposition that, like in the infective inflammatory affections, the cause of cancer is owing to the presence of a specific microbe. As the essential predisposing cause for the localization of the as yet unknown cancer microbe he regards the existence of a susceptibility in some part of the organism which determines localization in the same manner as is furnished by the brain in hydrophobia and spinal cord in tetanus.

Microörganisms have been found in the tissues of superficial carcinoma and in secondary carcinoma of internal organs, but even Afanasieff, who studied this subject with the greatest care in Klebs's laboratory, at Prague many years ago, was unable to find them in the minute miliary nodules of disseminated carcinoma. Weigert found bacilli which behaved on staining like tubercle bacilli in rapidly growing lympho-sarcoma, and Klebs in leukæmic lymphomata.

In the fall of 1887, Scheuerlen (*Deutsche med. Wochenschrift*, 1887, No. 48) read a paper before the Medical Society at Berlin which at the time attracted a great deal of attention, and which was immediately noticed by every medical journal on both continents. In this paper he put forth the claim that he had discovered the microbe of carcinoma. From ten cases of carcinoma of the breast he had made inoculations upon solid sterilized ascites and hydrocele fluid. Usually on the third day a growth, in the form of a colorless film, was seen on the surface of the serum, which later changed into a yellowish-brown color. On microscopic examination it was found that this culture was composed of short bacilli, with spores which could be readily stained with the ordinary reagents. These bacilli could not be seen in stained sections, but could be demonstrated in fresh cancer juice. Six bitches were inoculated by injecting material from the pure cultures directly into the posterior breast gland, and fourteen days later a circumscribed tumor the size of a walnut had formed at the point of inoculation. In the animals killed four weeks after inoculation the tumors examined under the microscope were seen to be composed of epithelial cells, in which the spores of the bacilli could be readily identified.

In the discussion which followed the reading of this paper, Guttmann agreed with the author that the bacilli were the cause of cancer. Fränkel thought that the microbes entered the tumor and their presence only indicated that the tumor tissue had determined localization of floating microbes. In the same journal Schill, of Dresden, states that he has found in cancerous and sarcomatous tissues rod-shaped microbes which he succeeded in cultivating on gelatine.

Freire claims priority in the discovery of the microbe of carcinoma. Perrin and Barnahei and Sanarelli, both of Siena, have also found a bacillus in connection with the etiology of malignant growths.

Ballance and Shattock (*British Med. Journal*, Oct. 29, 1887) made cultivations from twenty-two cases of carcinoma, the majority of which only yielded negative results. When a growth did occur the organisms were the same as those described and figured by various observers as occurring in healthy tissues.

Francke's ("Ueber Aetiologie und Diagnose von Sarkom und Carcinom," *Münch. med. Wochenschrift*, 1887) experiments began in November, 1887, and he had already seen and demonstrated the bacillus of sarcoma when Scheuerlen's discovery was announced. Francke has examined nine cancers since then, and in all has observed the carcinoma bacillus and its spores as described by Scheuerlen. His observations on the bacillus of sarcoma were based on the examination of three cases.

The bacillus found in them was thinner and longer than the cancer-bacillus. The cancer-bacillus, on the average, is 2 micromillimetres long, and 0.4 micromillimetre broad, while the sarcoma bacillus measures 3 to 4 by 0.4 micromillimetres. The spores of the sarcoma bacillus also resemble the spores of the cancer-bacillus, except that they are a little larger and have a sharply-contoured pole. The two microbes were cultivated upon the same nutrient soil, producing a brown pigment. Inoculations of the pure culture of the sarcoma bacillus have produced no results as yet, but Francke thinks that four weeks is too short a time for sarcoma to develop, and he will make another report later. Pfeiffer ("Der Scheuerlen'sche Krebs-bacillus ein Saprophyt," *Deutsche med. Wochenschrift*, 1888, No. 11) holds that Scheuerlen's carcinoma bacillus is a saprophyte and identical with the *proteus mirabilis* of Hauser.

Baumgarten ("Ueber Scheuerlen's Carcinom-bacillus," *Centralblatt f. Bacteriologie u. Parasitenkunde*, B. iii., No. 13) found a potato bacillus which in size and shape resembled Scheuerlen's bacillus of cancer, only that it exerted a somewhat liquefying effect on gelatine. He found a similar bacillus in a sarcoma of the skull and breast and in a neuroma of the hand. That Scheuerlen's bacillus has no direct etiological bearing to cancer, he claims is proved both by its occurrence in other tumors, and that it is not always present in carcinoma. He believes that it is a potato bacillus which has a wide distribution and consequently likely to contaminate culture media.

Senger ("Studien zur Aetiologie des Carcinoms," *Berl. klin. Wochenschrift*, 1888), in comparing malignant tumors with infective processes, affirms that their resemblance is marked. Thus carcinoma, like a phlegmonous inflammation, follows the lymphatic system, and, as in pyæmia, gives rise to metastasis. Again, in rare cases, carcinoma appears as a miliary affection, which in a very short time becomes diffuse and leads to serious disturbances of the digestive and circulatory organs. Such cases can only be explained by assuming that minute infective particles are introduced into the circulation which give rise to innumerable embolic obstructions in the capillary vessels where independent centres of growth are established. As a last resemblance may be mentioned those numerous cases of tar, paraffin, and chimney-sweeper's cancer, which often attack young persons in whom no predisposition to cancer exists. Senger does not believe that the cause of cancer is a microbe. Under strict antiseptic precautions he made implantation experiments of carcinoma tissue in mice, rabbits, and dogs, and on examining the site of implantation at different times noticed that the malignant graft was in process of absorption, or had disappeared completely, leaving only a minute cicatrix at the point where it had been embedded. The experiments yielded no better results if the animals were subjected to starvation before inoculation, or if their health was impaired from any other cause.

Thoma (*Fortschritte der Medecin*, June 1, 1889) has found in the muscles of carcinoma of rectum, stomach, and mamma, small, unicellular protoplasmic cells, refracting light strongly, with nucleus and sometimes a nucleolus also, which stain with hæmatoxylin, eosin, saffron, and alum carmine, which he considers undoubtedly parasitic. In shape they are irregularly rounded, or, more commonly, oval; occur singly or in groups of from four to six in an epithelial nucleus which appears as a hollow bladder with granular portions besides the parasites. In birds epithelial tumors are produced by similar bodies, hence inference that they are the cause. This is questionable, because Steinhaus and Heidenhain have found similar organisms in the epithelial lining of the salamander's bowel without any pathological significance.

At the meeting of the German Congress of Surgeons in 1888, Hahn ("Ueber Transplantation von carcinomatöser Haut," *Berl. klin. Wochenschrift*, May 4, 1888) gave an account of a successful experiment which he had made on a woman suffering from a recurring carcinoma of the breast which had advanced beyond the reach of operative treatment. The breast had been previously extirpated and she was now suffering from an extensive local recurrence with innumerable nodules in the skin. He removed six small nodules with the entire thickness of the skin, and then planted them upon the opposite healthy breast, from which six similar pieces of skin had been removed. The healthy skin was grafted upon the places where the nodules had been excised. The dressing was not changed until the thirteenth day, when all the malignant grafts were found adherent and slightly elevated above the niveau of the surrounding skin. Four weeks after the grafting minute nodules were found around two of the grafts. All the grafts continued to increase in size, and at the time of death (ten weeks) had attained the size of cherry-stones. Microscopical examination of the nodules surrounding three of the grafts revealed the histological structure of true carcinoma. He believed that the success which attended his experiment was due to the fact that only new cells were taken by the excision and implantation of recent nodules. In referring to the literature on the subject of cancer inoculations, he states that the first successful inoculation was made by Langenbeck, who injected cancer juice into the jugular vein of a dog, and two months later it was claimed that two cancerous nodules were found in the lung. Follin and Lebert asserted that they had been similarly successful in transmitting the disease from man to animals. From sixty to seventy grammes of finely triturated cancer juice taken from a cancer of the breast were mixed with distilled water and injected into the jugular vein of a dog. The animal died two weeks after the inoculation, when a few firm elastic nodules, the size of a pea, were found in the walls of the heart, and numerous nodules, the size of the head of a pin, throughout the liver. Billroth, Maas, Doutrelepont, Alberts, Senger, and others obtained only negative results. Doutrelepont made inoculations from animal to animal with no better effect. In cases of cancer of the peritoneum it has been known for a long time that local dissemination takes place by detached minute particles leaving the primary matrix and becoming implanted upon the peritoneal surface forming new independent centres of growth. Implantation of cancer cells upon the surface of wounds made for the removal of a carcinoma has been recognized as one of the ways in which the disease returns. Traumatic dissemination of carcinoma is recognized by Billroth, Waldeyer, Bergmann, Reincke, and others.

At a meeting of the Medical Society in Zürich, Hanau (*British Med. Journal*, October 19, 1889) claimed that he had succeeded in inoculating carcinoma from rat to rat. On November 28, 1888, he excised two pieces of a carcinomatous lymphatic gland from a female rat with malignant disease of the vulva, and transplanted these into the right half of the scrotum in two male rats. One of the latter died on January 14, 1889. The necropsy revealed a generalized carcinosis of the peritoneum and omentum. In the other, which was killed on January 25th, there were found two typical cancroid nodules, the size of a pea, on the gubernaculum testis and the cauda of the epididymis. The transplanted malignant growths proved to have precisely the same structure as those which had spontaneously originated in the first animal. According to Hanau's theory, the carcinomatous infection should be attributed to the agency of young epithelial cells and not to any pathogenic microbe. This theory would explain satisfactorily the few instances where transplantation of malignant grafts yielded a positive result.

During the last eight years I have made numerous implantation experiments of carcinoma and sarcoma tissue in dogs, cats, rabbits, and guinea-pigs, and have invariably observed that the graft failed to increase in size. The apparent increase in size a few days after implantation was due to the formation of a wall of granulation tissue around the graft in which it becomes embedded. The experiments were made by taking a graft the size of a pea from the periphery of the malignant tumor and inserting it under the skin, usually in the inguinal region. After three to four weeks without exception the graft was found absorbed and its place occupied by a subcutaneous cicatrix. At least twenty-four such experiments were made with the same uniform negative results. The comparative rarity of malignant tumors in animals made it difficult for me to carry out the plan of inoculating from animal to animal. I finally came in possession of a dog which had a sarcoma of the lower jaw. Implantation experiments were made on the same animal and a number of other dogs, but in all of them the grafts were absorbed as promptly as when tumor tissue from the human subject was used. I finally decided on making an implantation experiment on man in the first case in which such an experiment would be justifiable from the hopelessness of the case. The case that I selected was a large carcinoma of the leg which had taken its origin from the skin and in which amputation was objected to. The fungous surface was scraped and cauterized and from the deepest portion of the growth a piece of tissue, the size of a peach-stone, was inserted through a small incision into the connective tissue over the posterior aspect of the leg and the wound sutured with fine catgut. Primary union took place, with circumscribed infiltration of tissue around the graft from the second day to the end of the first week. After this time gradual disappearance of the graft by absorption occurred, until, at the end of four weeks, it had entirely disappeared.

At the meeting of the German Congress of Surgeons in 1889, Wehr ("Weitere Mittheilungen über die positiven Ergebnisse der Carcinomueberimpfungen von Hund auf Hund") admitted that in the animals which he showed at the previous meeting the nodules had since become smaller and smaller and finally disappeared completely; but he made the statement that since that time he had not only been able to inoculate a dog successfully by implantation of carcinomatous tissue, but that, in one instance at least, the animal died of carcinoma. His transplantations were made from animal to animal. He infected twenty-six dogs, using tissue from five bitches affected with vaginal carcinoma and two dogs suffering from carcinoma of the penis. Under strict antiseptic precautions a small incision was made in the skin, and then a trocar was inserted to the depth of three or four cm., and through the canula the fragment of cancer tissue was deeply embedded. Six of the animals had to be killed three and four weeks after the inoculation, because they had been bitten by a dog which was supposed to have hydrophobia. In the remaining animals he obtained twenty-four nodules, five of which were examined at a time when possibly they might have continued to increase in size, while the remaining nodules were all absorbed. In a bitch which was inoculated December 12, 1887, in four places in the region of the mammary gland, the nodules increased in size until they were as large as a plum or hazelnut. In April of the following year the tissues around the tumors became the seat of inflammation, which, however, soon subsided. The tumors increased in size, and the following June the animal lost flesh rapidly and died the middle of the same month. The autopsy showed metastatic tumors in the pelvis and along the lumbar portion of the spine. The tumor in the pelvis had compresed the urethra so firmly that the bladder ruptured—an accident which was the direct cause of death. Two carcinomatous nodules were also found in the chest. Makra has been making a series of bacteriological studies on malignant tumors in the laboratory of Kovacs, but, so far, only negative evidence has been obtained.

Galippe and Landouzy ("Note sur la presence de parasites : 1, dans les tumeurs fibreuses (myomes) uterines ; 2, dans le liquide des kystes ovariens et sur leur rôle pathogénique probable," *Gaz. des Hôp.*, 1887, No. 24) have extended the bacteriological investigations of the microbic origin of tumors to the benign variety. In two cases of extirpation of myomatous tumors of the uterus they inoculated a nutrient medium with the tumor tissue, and three days later found numerous microörganisms; the most conspicuous was a diplococcus arranged in groups, or in long chains; also streptococci and bacilli. The same microbes were found in both tumors. The same authors also found microörganisms in the contents of two ovarian cysts. They believe that solid and cystic tumors develop in consequence of the entrance of microbes into the organism in the same manner as parasites cause excrescences in plants.

In 1888 Neisser (*Vierteljahrsschrift für Dermatologie und Syphilis*, 1888) published an elaborate paper on the microbic origin of tumors containing the results of his own observations, in which he places the parasite in the coccidia groups of the sporozoa.

The following year Darier (*Archiv de Méd. Expérimentale et d'Anatomie Pathologique*, March 1, 1890) made two reports of the result of his work in Malassez's laboratory. In the first of these he intimated the recognition of a coccidium in a case of *acné cornée*, and defines the condition as a *psorospermose cutanée*; in the second he announced that he had found a parasite belonging to the same class in a case of Paget's disease of the nipple. He claims that this parasite presents all the different morphological phases of development characteristic of this organism; at first a naked mass of protoplasm, afterward surrounded by an envelope, then the protoplasm dividing into numerous granules, which being surrounded by the cell-membrane, presents the appearance of a cyst. He regards the disease in which these formations are formed as parasitic, a *psorospermose*. The same year Albarran (*La Semaine Médicale*, 1889, p. 101) informed the society that he had found a similar parasite in two epithelial tumors of the jaw.

About the same time Thoma (*Fortschritte der Medicin*, June 1, 1889) published a short paper on "A Characteristic Parasitic Organism in the Cells of Carcinoma." He describes the parasite which he found in malignant growths as a unicellular organism, consisting of protoplasm and a nucleus, with sometimes a nucleolus. In shape it is round or oval, and he found it most frequently in the vacuolated nucleus, in other cases near the nucleus.

In January, 1890, Wickham (*Archives de Médecine Expérimentale et d'Anatomie Pathologique*, 1890, p. 46) published an exhaustive paper on "The Pathological Anatomy and the Nature of Paget's Disease of the Nipple," in which he describes and figures a parasite which he regards as coccidia or psorospermiæ.

This parasite consists of a double-contoured capsule either filled with protoplasm, or the protoplasm is gathered into a mass in the centre. As the sporozoa reproduce themselves by the formation of spores, the author does not appear to have observed this process in connection with the organism which he described.

During the last year Klebs (*Deutsche med. Wochenschrift*, Nos. 24, 25, 28, 1890) published a series of papers on "The Nature and Diagnosis of Cancer-formation," in which he discusses, with his usual ability and thoroughness, the microbic origin of carcinoma. He describes hyaline bodies in carcinoma tissue, which, however, he regards as products of degenerative changes. He met with these bodies, presenting either an angular or rounded form in the proliferating epithelial tubes, and also, but less numerously, in the stroma.

Sjöbring (*Fortschritte der Medicin*, No. 14, 1890) describes a "Parasitic Protozoa-like Organism in Carcinoma," which he found in six cases of mammary carcinoma.

The most recent publication on this subject is from the pen of William Russel, of Edinburgh ("An Address on a Characteristic Organism of Cancer," *British Medical Journal*, December 13, 1890). He has found in carcinoma tissue and a few other pathological products certain bodies, which for want of reliable knowledge concerning their nature and on account of their specific reaction to certain staining materials, he calls "fuchsine bodies." The following are his directions for staining : " 1. Saturated solution of fuchsine in 2 per cent. carbolic acid in water. 2. 1 per cent. solution of iodine-green (Grüber's), in 2 per cent. carbolic acid in water. Place section in water; then stain in fuchsine ten minutes or longer; wash for a few minutes in water; then wash for *half a minute* in absolute alcohol. From this put the section into the solution of iodine-green, and allow it to remain well spread out for *five minutes*. From this rapidly dehydrate in absolute alcohol, pass through oil of cloves and mount in balsam. Tissues stained by this process containing the "fuchsine bodies" presented a characteristic and striking ap-

ALLEGED MICROBIC ORIGIN OF TUMORS. 257

Fig. 9.

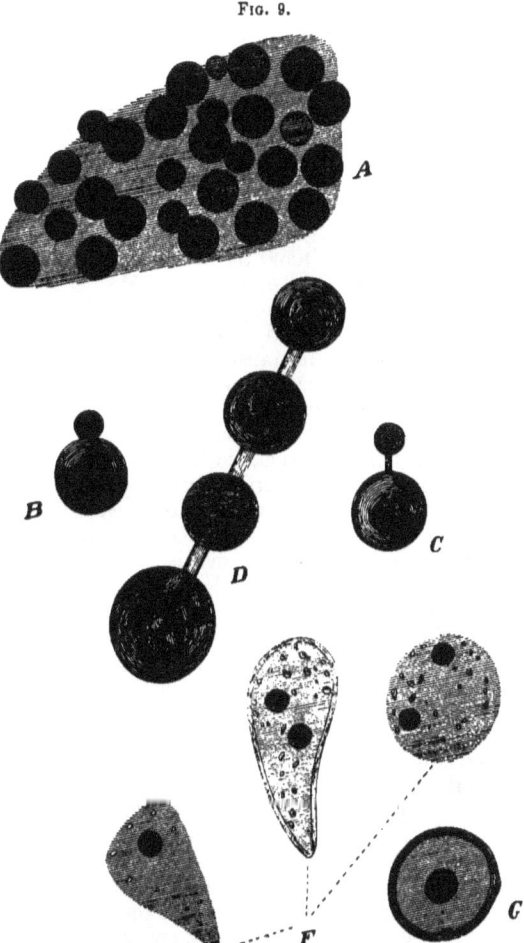

A, mass of fungi: *B*, individual giving off bud; *C*, same, with bud further removed from parent but still attached; *D*, four individuals attached to one another; *F*, small spores in lymph cells and leucocyte; *G*, altered leucocyte containing spore.

pearance when examined under high power and with the assistance of Abbe's sub-stage condensor. The brilliant red or purplish-red color of the "fuchsine bodies" forms a striking contrast to the green and delicate purple of the tissues. He found these bodies grouped in the interior of cells, usually surrounded by a vacuole.

Their intra-cellular location and manner of reproduction is well shown by the accompanying illustration (Fig. 9). He found these bodies invariably present in forty-five specimens of carcinoma taken from so many different individuals. He regards the organism described as a species of fungus which belongs to the sprouting fungi (Sprosspilze of Nägeli), and believes that a direct etiological relationship exists between it and the development of a carcinoma.

At a meeting of the Royal Academy of Medicine in Ireland, Patterson (*British Medical Journal*, December 6, 1890) read a paper on psorospermosis and its relation to cutaneous carcinoma in the light of recent researches, illustrated by drawings and microscopic specimens exhibiting coccidia in proliferating epithelium. In the discussion McKee, while admitting that coccidia might ultimately be shown to be the cause, or at least one cause, of cutaneous and perhaps other forms of carcinoma, led to the inference that the imperfect knowledge available at present rather pointed against than in favor of this view. He maintained that Paget's disease of the nipple resembled eczema, and was apparently inflammatory, yet the same form of coccidium found in it was also credited with the production of the malignant tumor of the breast which sometimes followed it. He thought it more probable that Paget's disease simply acted as one of many irritants, any one of which might give rise to a malignant tumor in a part already predisposed to it.

The strongest proof against the microbic nature of carcinoma is the histological structure of all products of infective processes. The products of all diseases, so far as known, which are caused by microbes and are characterized by local inflammatory processes, consist primarily of granulation tissue. The new tissue surrounding the seat of infection is a homoplastic product which originates from preëxisting tissues. So far, we have no knowledge of a microbe which produces heteroplastic tissue proliferation; that is, the production of cells which do not preëxist in that locality. Many of the lymphomata undoubtedly owe their origin to specific microbes, but these are not true tumors, and should be classified with the inflammatory infective swellings—the granulomata.

There has been, so far, no proof furnished of the existence of a specific bacillus in carcinoma or sarcoma, and the inoculation and implantation experiments have proved so seldom successful, and the experiments which at first appeared successful, have later been shown to be so deceptive, that the microbic origin of malignant tumors has not only not been established, but rendered improbable. The analogy between tumors and infective processes is more apparent than real, and simply consists in the fact that the cells of the former endowed with all the properties of independent growth select the same routes for local and general dissemination as the microbes of the latter.

INDEX OF SUBJECTS.

A BSCESS, description of minute, 88
gonococci in, 236
tubercular, 197
Actinomycosis hominis, 213
clinical history of, 218
fungus of, cultivation of, 215
description of, 214
effect of, on the tissues, 217
inoculation experiment with, 216
staining of, 214
infection through the skin, 220
through carious teeth, 221
of brain, 228
of intestines, 224
of jaw, 221
of lung, 227
of skin, 220
sources of infection, 217
Antagonism among microörganisms, 68
Anthrax, 199
attenuation of virus of, 201
bacillus of, antagonistic to pneumococcus of Friedländer, 68
cultivation of, 200
description of, 200
localization of, 39
staining of, 200
cause of death in, 204
direct transmission of, 21
in man, 202
inoculation of, 201
intensification of virus of, 202
prophylaxis by streptococcus of erysipelas, 68
relation of, to trauma, 38
Arthritis, gonorrhœal, 237
purulent, 100

B ACILLUS ANTHRACIS, 200
cultivation and staining of, 200
coli communis, 77

Bacillus mallei, 208
staining of, 208
of glanders, 208
of smegma, 246
of syphilis, 241
peritonitidus ex-intestinalis cuniculi, 76
pyocyaneus, 84
pyogenes fœtidus, 84
saprogenes, 106
tetani, 143
tuberculosis, cultivation of, 158
description of, 157
in milk, 66
inoculation experiments with, 159
its discovery, 156
staining of, 158
Bacteria, ectogenous, 29
endogenous, 29
Bones, tuberculosis of, 181
always secondary, 181
changes in, 182

C ADAVERIN, 89
Cancer, peritoneal dissemination of, 253
supposed bacillus of, 250
transplantation experiments with, 252
traumatic dissemination of, 253
Chemical irritants as a cause of suppuration, 92
Circulation, force of, in relation to microbic disease, 28
Circumcision, tuberculosis following, 168
Cladothrix, 140
Coccidia, 256
Coli communis, bacillus of, 77
Corpora oryzoidea, 188

INDEX OF SUBJECTS.

ECZEMA, tubercle bacilli in, 175
Elimination of pathogenic microörganisms, 56
  by leucocytes, 56
  in pyæmia, 128
Embolism in pyæmia, 128
  in tuberculosis, 161
Empyema, 98
Epididymis, tuberculosis of, 193
Erysipelas, 129
  cause of death in, 136
  direct transmission of, 20
  history of, 129
  manner of infection in, 133
  microbe of, 130
  relation of, to puerperal fever, 133
    to phlegmonous inflammation and suppuration, 135
  streptococcus of, antagonistic to syphilis, 69
    preventive and curative of anthrax in animals, 69
  therapeutic inoculation of, 131
Erysipeloid, 139

FRIEDLÄNDER, pneumococcus of, 68
Fuchsine bodies, 256

GANGLION, compound, 187
Gangrene, 106
  microbes the cause of, 106
  with emphysema, 107
Genital organs, tuberculosis of, 192
Giant cells (phagocytosis), 56
Glanders, 207
  bacillus of, 208
    cultivation and staining of, 208
  in man, 211
  inoculation experiments, 209
  not transmissible by air, 35
Gonococcus of Neisser, 231
  action of, on peritoneum, 234
    on other tissues, 234
  cultivation of, 230
  description of, 230
  diagnostic value of, 230
  in abscess, 236
  inoculation experiments with, 233
  in purulent ophthalmia, 236

Gonococcus, staining of, 231
Gonorrhœa, 230
Gonorrhœal rheumatism, 237
Granulation tissue in infective inflammation, 73
  in simple inflammation, 71

HEALTH, state of, in relation to microbic disease, 28
Hereditary predisposition, 19
  transmission of microbic disease, 19
    clinical evidences of, 20
    experimental evidences of, 22
Hygroma, 188

INFECTION by inhalation, 32
  mixed, 48
  sources of, 29
Inflammation, 71
  phlegmonous and erysipelatous, 135
  relation to metastatic suppuration, 51
  simple and infective, 71, 72

JOINTS, tuberculosis of, 184

KRYPTO-GENETIC septico-pyæmia, 122

LEUCOCYTES as eliminators of microörganisms, 57
Leucomaïnes, 111
Lupus, 170
  bacillus of tuberculosis in, 173
  identical with tuberculosis, 171
  inoculation with, produces tuberculosis, 159
Lymphatic glands, tuberculosis of, 175
  spontaneous cure by suppuration, 178
  treatment of, 181
Lymphomata, microbic origin of, 258

MACROPHAGI, 60
Malaria infections, urine of, 65
Malignant œdema, 114

Malignant œdema, direct transmission of, 23
Malleus humidus, 207
Mamma, tuberculosis of, 194
Mastitis, suppurative, 72
Metastatic suppuration, 51
Miasma, 29
Microbes, cause of infective inflammation, 72
  encapsulation of, 53
  localization of, in pathological products, 47
Microbic disease in relation to force of circulation, 28
  to state of health, 28
  origin of cancer, 249
  of lymphomata, 258
  of myoma, 255
  of ovarian cysts, 255
Micrococci in urine, 115
Micrococcus pyogenes tenuis, 82
Microörganisms, antagonism among, 68
  excreted in urine, 64
  pathogenic, elimination of, 56
    by phagocytosis, 56
    by kidney and other organs, 63
  in atmospheric air, 31
    detection and estimation of, 32
  in healthy body, 26
  in hospital wards, 34
  localization of, 37
    in relation to trauma, 37
    removed by red blood-corpuscles, 26
  thermal death-point of, 29
Microphagi, 60
Mikrosporon septicum, 110
Milk, bacillus of tuberculosis in, 66
  in puerperal disease, 66
Mixed infection, 48
Mouth, tuberculosis of, 191
Mycosis, 121
  toxic, of the blood, 122

NÄGELI, Sprosspilze of, 258
  Noma, 141

ŒDEMA, malignant, 114
  Ophthalmia, purulent, 236
Osteomyelitis, acute suppurative, 100
  caused by intra-uterine infection, 21
  chronic, 54
  tubercular, 182

PATHOLOGY, its relation to bacteriology, 17
Peritoneum, tolerance of, to pus-microbes, 78
  to unfiltered air, 77
  tuberculosis of, 189
  treatment of, 190
Peritonitis, 97
  experimental, 73
  perforative, 75
  septic, 78
  following inflammation of the intestinal wall, 71
  simple, 78
Phagocytosis, 56, 60.
Placenta, its relation to the passage of microbes, 20-23
Pneumococcus of Friedländer antagonistic to bacillus anthracis, 68
Predisposition, hereditary, 19
Proteus mirabilis, 119
  vulgaris, 119
  Zenkeri, 119
Psorospermia, 256
Ptomaïne of bacillus prodigiosus prevents growth of pus-microbes in culture medium, 70
Ptomaïnes, 119
  diffusion of, prevented by leucocytes, 58
  discovery of, 117
  effect of, on leucocytes, 90
  in relation to suppuration, 89
  prevent coagulation of blood, 91
Puerperal disease, milk in, 66
Pus microbes the cause of suppuration, 84
  classification and description of, 81
  growth of, in culture medium, prevented by ptomaïne of bacillus prodigiosus, 70
  localization of, 43
  their relation to pyæmia, 126

INDEX OF SUBJECTS.

Pus microbes, tolerance of peritoneum to, 77
Pyæmia, 124
    artificial, 124
    changes in vessel-wall in, 127
    chronic, 128
    relation of pus microbes to, 126
    thrombosis and embolism essential to, 128
Pyocyanin, 84

RHEUMATISM, gonorrhœal, 237
    Rice bodies, 188

SAPRÆMIA, 116, 122
    Schizomycetes, 113, 119
Scrofula identical with tuberculosis, 161
Sepsin, 117
Septicæmia, 110
    history of, 110
    progressive, in man, 121
Septic intoxication from introduction of putrid substances, 116
Septico-pyæmia, 122
Serous cavities, suppurative affections of, 97
Skin, actinomycosis of, infection through, 220
    suppurative affections of, 96
    tuberculosis of, 170
Smegma, bacillus of, 246
Spasmatoxin, 151
Staphylococci in suppurative mastitis, 72
    in urine, 67
Staphylococcus cereus albus, 82
    cereus flavus, 82
    flavescens, 82
    pyogenes albus, 81, 85
    pyogenes aureus, 81, 85
        action of, on peritoneum, 74
    pyogenes citreus, 82
Streptococci in suppurative mastitis, 72
Streptococcus erysipelatosus, 129
    antagonistic to syphilis, 69
    preventive and curative of anthrax in animals, 69
    therapeutic inoculation with, 131
    pyogenes, 83, 85
Suppuration, 79
    caused by pus microbes, 84

Suppuration, chemical irritants not a cause of, 87, 88, 92
    in relation to erysipelas, 135
    to ptomaïnes, 89
    history of investigation of, 79
    metastatic, 51
Surgical tuberculosis, 170
Synovitis, experimental, by inoculation with scarlatina-diphtheria cocci, 49
    multiple suppurative, following scarlet fever, 48
Syphilis, 241
    bacillus of, 241
    cultivation and staining of, 242
    effect of erysipelas on, 69
    inoculation experiments, 24, 243
Syphilitic lesions, secondary infection with pus microbes, 247
    favorable soil for growth of bacillus of tuberculosis, 49

TENDO-VAGINITIS granulosa, 187
    synovitis, 187
Testicle, tuberculosis of, 195
Tetanin, 151
Tetanotoxin, 151
Tetanus, 142
    bacillus of, 143
    occurrence of, in earth, 150
    chronic, 153
    incubation period of, 153, 154
    inoculation experiments, 144
    ptomaïnes of bacillus the cause of, 151
Thrombosis in pyæmia, 128
    in tuberculosis, 181
Transmission of microbic disease hereditary, 20
Trauma favors localization of pus microbes, 43
    bacillus tuberculosis, 41
Toxic mycosis of blood, 122
Toxin, muriate of, 151
Trypsin injected into peritoneal cavity causes hemorrhagic peritonitis, 74
Tubercular abscess, 197
Tuberculosis (also bacillus of tuberculosis), 156
    bacillus of, its discovery, 156
    caused by skin-grafts, 167
    diagnosis of, by inoculation, 186

Tuberculosis, entrance of bacillus of, through wounds, 167
  influence of trauma on localization of bacillus of, 41, 164
  inoculation, 163
    experiments, 159
  miliary diffuse, following operative treatment, 42, 181, 182, 185
  of bones, 181
    always secondary, 181
    changes in, 182
  of genital organs, 192
  of joints, 184
  of lymphatic glands, 175
    spontaneous cure due to suppuration, 177
    treatment of, 180
  of mamma, 194
  of mouth, 191
  of peritoneum, 189
    treatment of, 190
  of skin, 170
  of tendon sheaths, 187
  of uterus, 192
    scrofula identical with, 160
    surgical, 170

Tuberculosis, syphilis in relation to, 49
  through skin abrasion, 159, 163
    verrucosa cutis, 170
Tumor albus, 184
Tumors, alleged microbic origin of, 249
  myomatous, microörganisms in, 255

URINE (aseptic), innocuous to peritoneum, 76
  micrococci in, 115
  microörganisms excreted in, 64
  of malaria infections, 65
  of scarlatina infections, 65
  of typhus recurrens infections, 65
  of varicella infections, 65
Uterus, tuberculosis of, 192

VARICELLA infections, urine of, 65
  Vibrion septique, 111, 114
Vulvo-vaginitis, gonococci in, 239

WOUNDS, tubercular infection of, 167

INDEX OF AUTHORS.

A HLFELD, 21
Albarran, 255
Albert, 214
Andromico, 244
Anger, 142
Angerer, 118
Arloing, 107, 161, 202
Aschoff, 52
Aubert, 173

B ABES, 50, 63, 82
Babtchinsky, 132
Bacelli, 155
Baert, 142
Bahrdt, 48, 100
Ballance, 250
Baranski, 215
Barbier, 193
Bary, de, 89, 92, 93
Baumgarten, 251
Becker, 44
Beger, 187
Bender, 173
Bergmann, 117, 118, 237
Besser, 112, 126
Bessnier, 173
Betoli, 142
Beumer, 147, 150
Biedert, 131
Billroth, 129
Biondi, 95
Birch-Hirschfeld, 243
Birch, 118
Bizzozero, 27
Block, 173
Blumberg, 118
Bockhardt, 86, 233
Boeck, 172

Bókai, 233
Bollinger, 22, 66, 161, 199, 204, 213, 228
Bonome, 109, 137, 145, 150
Bordini, 137
Boström, 215
Bouilly, 183
Brauell, 199
Brieger, 48, 49, 119, 151
Bruns, 70, 132
Buchner, 32, 61, 202
Bumm, 49, 66, 232

C ADENE, 35
Canali, 227
Carle, 145
Cavel, 186, 246
Celli, 34
Chamberland, 22
Charrin, 32
Chaveau, 37, 107, 111, 206
Cheyne, Watson, 28, 69, 88, 95, 111, 238
Chiari, 224
Chotzen, 248
Christmas-Dircking-Holmfeld, 60
Ciarrocchi, 108
Clado, 138
Cornet, 34, 159, 192
Cornevin, 202
Cornil, 28, 160, 210
Councilman, 93
Coze, 114
Czerny, 167

D ARIER, 255
Darwin, 113
Davaine, 113, 199
Delafield, 178

INDEX OF AUTHORS.

Delafond, 199
Demme, 172, 174
Disse, 24, 245
Doutrelepont, 160, 174, 242, 243, 244, 247, 248
Dowdeswell, 115
Doyen, 135, 137
Durante, 34

E DELBERG, 118
   Ehrlich, 48, 49, 99, 129
Eiselsberg, 33, 148, 165
Elsenberg, 168
Emmerich, 33, 68
Ernst, 67, 84
Escherich, 66
Eve, 162, 245

F ARDEL, 63
   Fehleisen, 76, 86, 129, 130, 131
Fehling, 190
Feltz, 114
Finger, 170
Firket, 218
Flügge, 144
Fodor, 26, 27
Francke, 250
Fränkel, 44, 48, 76, 98, 119, 179, 239
Frankland, 31
Freudenberg, 48
Friedländer, 171

G AFFKY, 114
   Galippe, 255
Garrè, 70, 86, 97, 100, 102, 198
Gaucher, 32
Gautier, 111
Geppert, J., 30
Giordano, 148
Godwin, 108
Goldschmidt, 149
Gottstein, 242
Grawitz, 77, 89, 92, 94, 97
Griffith, 174
Guarnieri, 34
Gussenbauer, 53, 55
Gusserow, 134
Guttmann, 117

H AAB, 236, 237
   Haberkorn, 245
Habermann, 175
Hahn, 252
Hajeck, 135
Hammonic, 243
Hanau, 226, 263
Hankin, 201
Hanot, 165
Haslund, 237
Hauser, 27, 119
Hess, 60
Hesse, 32
Heubner, 28, 48, 100
Heusinger, 199
Hiller, 118
Hirschberger, 66
Hochenegg, 213, 220, 221, 224
Hochsinger, 145, 247
Hoffa, 100, 121, 204, 237
Hofmokl, 168
Holst, 165
Huber, 38, 49
Hueppe, 64
Hueter, 110, 187

I SRAEL, J., 213, 216, 221
   Israel, O., 215, 219

J AEFE, 108
   Jani, Curt, 20
Janicke, 131
Janowski, 95
Jensen, 217
Johne, 217
Jouin, 194
Jürgensen, 112, 122

K AHLDEN, 138
   Kaloff, 206
Kammerer, 237
Kanzler, 183
Kaposi, 220
Kapper, 219
Karth, 32
Kassowitz, 247
Kaufmann, P., 94
Keller, 229

Kelly, 142
Kircher, 113
Kitasato, 144, 151
Kitt, 209
Klebs, 18, 31, 35, 79, 80, 110, 243, 256
Kleeblatt, 131
Klein, 125
Klemperer, 91, 247
Koch, R., 48, 64, 113, 114, 118, 124, 129, 156, 159
Koch, W., 108, 203
Kocher, 44, 45, 47, 101
Kolisko, 247
König, 162, 168, 182, 188
Konjajeff, 64
Kotschau, 193
Koubassoff, 22
Kranzfeld, 100, 209, 210
Kraske, 103, 196
Krause, 44, 100, 101, 130, 174, 183, 232
Kreis, 232
Kroner, 23
Kümmell, 189
Küttner, 212

LAKER, 228
Lande, 205
Landouzy, 255
Langenbeck, 54
Langhans, 226
Larger, 142
Laruelle, 77
Lebedeff, 20, 194
Lebert, 47
Ledderhose, 84
Legrain, 230
Lehmann, 168
Leistikow, 240
Leloir, 160, 163
Lemaire, 113
Lennander, 166
Leser, 220
Letulle, 183
Letzerich, 129
Leube, 122
Leuwenhoek, 113
Levy, 23
Leyden, 97, 108
Lindfors, 189, 191
Lindt, 228

Lingard, 245
Litten, 122
Löffler, 49, 206
Longard, 96
Loven, 235
Lubarsch, 63
Lübbert, 81, 103
Luecke, 41
Luening, 226
Lukomsky, 129
Lumniczer, 151
Lundgren, 210
Lustgarten, 241
Lustig, 157

MAAS, 119
Magny, 164
Malet, 35
Mangeri, Romeo, 24
Marchand, 21
Marcus, 245
Markuse, 244
Martineau, 243
Mattei, 69
Matterstock, 244
Mazzuschelli, 154
Meisels, 157
Merklen, 165
Metschnikoff, 56, 59
Meyer, 168
Middeldorpff, 167
Mikulicz, 118
Miquel, 31
Mögling, 157, 183
Monpurgo, 64
Moosbrugger, 213, 228
Müller, 102, 162, 182, 183
Muskatblüth, 39

NAEGELI, 36, 113, 199
Nathan, 94
Neelsen, 121, 132
Neisser, 131, 171, 230, 255
Nencki, 209
Nepveau, 54, 129
Netter, 21
Neumann, 32, 64, 69
Nicaise, 188
Nicolaier, 36, 143, 144

Nicolas, 34
Nocard, 159
Noorden, von, 136
Nuttal, 61

OGSTON, 80, 115
Ohlmüller, 149
Okinschitz, 127
Orloff, 52
Orth, 43, 76, 98, 129
Orthmann, 93, 194

PAGENSTECHER, 160
Paget, 249
Paltauf, 170
Panum, 116
Partsch, 213, 220
Passet, 80, 81, 82, 84, 85, 130
Pasteur, 113
Patterson, 258
Pawlowsky, 68, 74, 125, 126
Pernice, 34
Perty, 113
Petri, 33
Pettenkofer, 29
Pfeiffer, 160, 174, 251
Philipowicz, 64
Piana, 217
Pollender, 199
Ponfick, 213
Pott, 239
Poulet, 188

QUINQUAUD, 173

RANKE, 141
Raskina, 83
Rassdnitz, 172
Rattone, 145
Raymond, 164
Recklinghausen, 110
Renken, 183
Renouard, 173
Rheiner, 137
Ribbert, 44, 57, 62, 67, 103

Riedel, 187
Riehl, 170
Rindfleisch, 110
Rinne, 44, 51, 54, 78, 104, 121
Rodet, 102
Rosenbach, 21, 43, 44, 45, 49, 80, 82, 83, 84, 101, 110, 126, 139, 143, 144, 145
Rosenberger, 120
Rosenstein, 63
Rotter, 214, 216
Roux, 159
Ruffer, 62
Ruiys, 95
Russel, 256
Rütimeyer, 228

SACHS, 173
Sahli, 238
Saltzman, 195
Samter, 48
Samuel, 118
Sangalli, 21
Sänger, 44, 238
Scheuerlen, 90, 93, 250
Schlegtendal, 157, 198
Schliferowitsch, 191
Schmidt, 163
Schnell, 181
Schnitzler, 49
Schuchardt, 174, 183
Schueller, 41, 126, 159, 171
Schütz, 206, 242
Schwarz, 236
Schweiger, 65
Schwimmer, 69, 132
Seitz, 64
Selmi, 117
Semmer, 117
Sénásy, 227
Senator, 99
Senger, 251
Shattock, 250
Simone, 137
Sinclair, 235
Sirena, 34
Sjöring, 256
Slessarewskji, 204
Smirnoff, 138, 238
Smith, 115
Socin, 55

INDEX OF AUTHORS.

Soltmann, 217
Sormani, 154
Soyka, 35
Spaeth, 239
Steinschneider, 239
Steinthal, 46
Sternberg, 29, 114, 240
Strauss, 22, 32
Struck, 101
Strümpell, 197
Szuman, 41

TAGUCHI, 24, 245
Tavel (also Cavel), 186, 246
Thiriar, 142
Thoma, 72, 252, 256
Thomas, 202
Tiegel, 121
Tillmanns, 129, 137
Tricomi, 88, 106
Trousseau, 133
Trzebicky, 190
Tschering, 166
Tyndall, 32

ULLMANN, 196
Uskoff, 93

VERCHIÈRE, 42
Verhoogen, 142
Verneuil, 138
Vidal, 112

Vierhoff, 203
Vierordt, 191
Villard, 188
Villemin, 156
Volkmann, 41, 187

WAHL, 167
Waldeyer, 110
Wartmann, 42, 186
Wegner, 77
Wehde, 34
Wehr, 61, 254
Weichselbaum, 77, 98, 157, 177
Weigert, 177
Welander, 231, 233
Wells, Spencer, 190
Werth, 193
Whitney, 131
Wickham, 256
Widmark, 236
Wilde, 129
Williams, Theo., 34
Winckel, 134
Wolf, M., 216
Wolff, Max, 22, 137
Woolbridge, 201
Woronzoff, 30
Wyssokowitsch, 43, 56

ZAHOR, 27
Zeissl, 245
Zemann, 225
Zuckermann, 87
Zweigbaum, 192

LEA BROTHERS & CO.'S
CLASSIFIED CATALOGUE
OF
MEDICAL AND SURGICAL
Publications.

In asking the attention of the profession to the works advertised in the following pages, the publishers would state that no pains are spared to secure a continuance of the confidence earned for the publications of the house by their careful selection and accuracy and finish of execution.

The printed prices are those at which books can generally be supplied by booksellers throughout the United States, who can readily procure for their customers any works not kept in stock. Where access to bookstores is not convenient books will be sent by mail by the publishers postpaid on receipt of the printed price, and as the limit of mailable weight has been removed, no difficulty will be experienced in obtaining through the post-office any work in this catalogue. No risks, however, are assumed either on the money or on the books, and no publications but our own are supplied, so that gentlemen will in most cases find it more convenient to deal with the nearest bookseller.

LEA BROTHERS & CO.
Nos. 706 and 708 SANSOM ST., PHILADELPHIA, March, 1891.

PRACTICAL MEDICAL PERIODICALS.

THE AMERICAN JOURNAL OF THE MEDICAL SCIENCES, Monthly, $4.00 per annum.

THE MEDICAL NEWS, Weekly, $4.00 per annum.

} To one address, post-paid, **$7.50** per annum.

THE MEDICAL NEWS VISITING LIST (4 styles, see page 3), $1.25. With either or both above periodicals, in advance, 75c.

THE YEAR-BOOK OF TREATMENT (see page 17), $1.50. With either JOURNAL or NEWS, or both, 75c. Or JOURNAL, NEWS, VISITING LIST and YEAR-BOOK, $8.50, in advance.

Subscription Price Reduced to $4.00 Per Annum.

THE MEDICAL NEWS.

THAT THE NEWS fulfils the wants of men in active practice is made clear by the steady growth of its subscription list. This increase of readers has rendered possible a reduction in the price of THE NEWS to **Four Dollars** per year, so that it is now by far the cheapest as well as the best large weekly journal published in America.

By means of THE MEDICAL NEWS every physician is now able at a minimum outlay to insure his own receipt of the earliest and most authoritative information on all subjects

THE MEDICAL NEWS—Continued.

of interest to the great medical world. The foremost writers, teachers and practitioners of the day furnish original articles, clinical lectures and notes on practical advances; the latest methods in leading hospitals are promptly reported; a condensed summary of progress is gleaned each week from a large exchange list, comprising the best journals at home and abroad; a special department is assigned to abstracts requiring full treatment for proper presentation; editorial articles are secured from writers able to deal instructively with questions of the day; books are carefully reviewed; society proceedings are represented by the pith alone; regular correspondence is furnished by gentlemen in position to know all occurrences of importance in the district surrounding important medical centres, and minor matters of interest are grouped each week under news items. Everything is presented with such brevity as is compatible with clearness, and in the most attractive manner. In a word THE MEDICAL NEWS is a crisp, fresh, weekly newspaper and as such occupies a well-marked sphere of usefulness, distinct and complementary to the ideal monthly magazinee, THE AMERICAN JOURNAL OF THE MEDICAL SCIENCES.

The American Journal of the Medical Sciences

Published Monthly, at Four Dollars Per Annum

Enters upon its seventy-second year (1891) with assurances of increased usefulness. Encouraged by the emphatic endorsement of the profession, as indicated by a growth in its subscription list of fifty per cent. since its appearance as a monthly at a reduced price, those in charge will spare no effort to maintain its place as the leader of medical periodical literature. Being the medium chosen by the best minds of the profession during the past seventy years for the presentation of their ablest papers, THE AMERICAN JOURNAL has well earned the praise accorded it by an unquestioned authority—"from this file alone, were all other publications of the press for the last fifty years destroyed, it would be possible to reproduce the great majority of the real contributions of the world to medical science during that period." Original Articles, Reviews and Progress, the three main departments into which the contents of THE JOURNAL are divided, will be found to possess still greater interest than in the past. The brightest talent on both sides of the Atlantic is enlisted in its behalf and no effort will be spared to make THE JOURNAL more than ever worthy of its position as the representative of the highest form of medical thought.

COMMUTATION RATE.

Taken together, THE JOURNAL and NEWS form a peculiarly useful combination, and afford their readers the assurance that nothing of value in the progress of medical matters shall escape attention. To lead every reader to prove this personally the commutation rate has been placed at the exceedingly low figure of $7.50.

SPECIAL OFFERS.

The MEDICAL NEWS VISITING LIST (regular price, $1.25, see next page,) or The Year-Book of Treatment (regular price, $1.50, see page 17,) will be furnished to advance-paying subscribers to either or both of these periodicals for 75 cents apiece; or Journal, News, Visiting List and Year-Book, $8.50.

Subscribers can obtain, at the close of each volume, cloth covers for THE JOURNAL *(one annually), and for* THE NEWS *(one annually), free by mail, by remitting Ten Cents for* THE JOURNAL *cover, and Fifteen Cents for* THE NEWS *cover.*

☞ The safest mode of remittance is by bank check or postal money order, drawn to the order of the undersigned; where these are not accessible, remittances for subscriptions may be sent at the risk of the publishers by forwarding in *registered* letters. Address,

LEA BROTHERS & CO., 706 & 708 Sansom Street, Philadelphia.

THE MEDICAL NEWS VISITING LIST FOR 1891

Is published in four styles, Weekly (dated for 30 patients); Monthly (undated, for 120 patients per month); Perpetual (undated, for 30 patients weekly per year); and Perpetual (undated, for 60 patients weekly per year). The 60-patient Perpetual is a novelty for 1891, and consists of 256 pages of assorted blanks. The first three styles contain 32 pages of important data and 176 pages of assorted blanks. Each style is in one wallet-shaped book, leather-bound, with pocket, pencil, rubber, erasable tablet and catheter-scale. Price, each, $1.25.

SPECIAL COMBINATIONS WITH VISITING LIST.

THE AMER. JOURNAL ($4) with VISITING LIST ($1.25), or YEAR-BOOK ($1.50), for $4.75
THE MEDICAL NEWS ($4)   "   "   "   "   "   "   "   "   4.75
THE JOURNAL AND NEWS ($7.50)   "   "   "   "   "   "   "   8.25
THE JOURNAL, NEWS, VISITING LIST and YEAR-BOOK (see page 17), . . 8.50

This list is all that could be desired. It contains a vast amount of useful information, especially for emergencies, and gives good table of doses and therapeutics.—*Canadian Practitioner.*

It is a masterpiece. Some of the features are peculiar to "The Medical News Visiting List," notably the Therapeutic Table, prepared from Dr. T. Lauder Brunton's book, which contains the list of diseases arranged alphabetically, giving under each a list of the prominent drugs employed in the treatment. When ordered, a Ready Reference Thumb-letter Index is furnished. This is a feature peculiar to this Visiting List.—*Physician and Surgeon,* December.

For convenience and elegance it is not surpassable.—*Obstetric Gazette,* November.

THE MEDICAL NEWS PHYSICIANS' LEDGER.

Containing 300 pages of fine linen "ledger" paper, ruled so that all the accounts of a large practice may be conveniently kept in it, either by single or double entry, for a long period. Strongly bound in leather, with cloth sides, and with a patent flexible back, which permits it to lie perfectly flat when opened at any place. Price, $4.00.

HARTSHORNE, HENRY, A. M., M. D., LL. D.,

Lately Professor of Hygiene in the University of Pennsylvania.

A Conspectus of the Medical Sciences; Containing Handbooks on Anatomy, Physiology, Chemistry, Materia Medica, Practice of Medicine, Surgery and Obstetrics. Second edition, thoroughly revised and greatly improved. In one large royal 12mo. volume of 1028 pages, with 477 illustrations. Cloth, $4.25; leather, $5.00.

The object of this manual is to afford a convenient work of reference to students during the brief moments at their command while in attendance upon medical lectures. It is a favorable sign that it has been found necessary, in a short space of time, to issue a new and carefully revised edition. The illustrations are very numerous and unusually clear, and each part seems to have received its due share of attention. We can conceive such a work to be useful, not only to students, but to practitioners as well. It reflects credit upon the industry and energy of its able editor.—*Boston Medical and Surgical Journal,* Sept. 3, 1874.

We can say with the strictest truth that it is the best work of the kind with which we are acquainted. It embodies in a condensed form all recent contributions to practical medicine, and is therefore useful to every busy practitioner throughout our country, besides being admirably adapted to the use of students of medicine. The book is faithfully and ably executed.—*Charleston Medical Journal,* April, 1875.

NEILL, JOHN, M. D., and SMITH, F. G., M. D.,

Late Surgeon to the Penna. Hospital.   *Prof. of the Institutes of Med. in the Univ. of Penna.*

An Analytical Compendium of the Various Branches of Medical Science, for the use and examination of Students. A new edition, revised and improved. In one large royal 12mo. volume of 974 pages, with 374 woodcuts. Cloth, $4; leather, $4.75.

LUDLOW, J. L., M. D.,

Consulting Physician to the Philadelphia Hospital, etc.

A Manual of Examinations upon Anatomy, Physiology, Surgery, Practice of Medicine, Obstetrics, Materia Medica, Chemistry, Pharmacy and Therapeutics. To which is added a Medical Formulary. Third edition, thoroughly revised, and greatly enlarged. In one 12mo. volume of 816 pages, with 370 illustrations. Cloth, $3.25; leather, $3.75.

The arrangement of this volume in the form of question and answer renders it especially suitable for the office examination of students, and for those preparing for graduation

HOBLYN, RICHARD D., M. D.

A Dictionary of the Terms Used in Medicine and the Collateral Sciences. Revised, with numerous additions, by ISAAC HAYS, M. D., late editor of The American Journal of the Medical Sciences. In one large royal 12mo. volume of 520 double-columned pages. Cloth, $1.50; leather, $2.00.

It is the best book of definitions we have, and ought always to be upon the student's table.—*Southern Medical and Surgical Journal.*

JUST READY.

THE NATIONAL MEDICAL DICTIONARY

INCLUDING

English, French, German, Italian and Latin Technical Terms used in Medicine and the Collateral Sciences, and a Series of Tables of Useful Data.

BY

John S. Billings, M.D., LL.D., Edin. and Harv., D.C.L., Oxon.

Member of the National Academy of Sciences, Surgeon U. S. A., etc.

WITH THE COLLABORATION OF

PROF. W. O. ATWATER,	JAMES M. FLINT, M. D.,	WASHINGTON MATTHEWS, M. D.,
FRANK BAKER, M. D.,	J. H. KIDDER, M. D.,	C. S. MINOT, M. D.
S. M. BURNETT, M. D.,	WILLIAM LEE, M. D.,	H. C. YARROW, M. D.,
W. T. COUNCILMAN, M. D.,	R. LORINI, M. D.,	

In two very handsome royal octavo volumes containing 1574 pages, with two colored plates.

Per Volume—Cloth, $6; Leather, $7; Half Morocco, Marbled Edges, $8.50. For Sale by Subscription only. Specimen pages on application. Address the Publishers.

The publishers have great pleasure in presenting to the profession a new practical working dictionary embracing in one alphabet all current terms used in every department of medicine in the five great languages constituting modern medical literature.

For the vast and complex labor involved in such an undertaking no one better qualified than Dr. Billings could have been selected. He has planned the work, chosen the most accomplished men to assist him in special departments, and personally supervised and combined their work into a consistent and uniform whole.

Special care has been taken to render the definitions clear, sharp and concise. They are given in English, with synonyms in French, German and Italian of the more important words in English and Latin.

Regarded as a dictionary, therefore, this standard work supplies the physician, surgeon and specialist with all information concerning medical words, simple and compound, found in English, giving correct spelling, clear, sharp definitions and accentuation, and furthermore it enables him to consult foreign works and to understand the large and increasing number of foreign words used in medical English. It is especially full in phrases comprising two, three or more words used in special senses in the various departments of medicine.

The work is, however, far more than a dictionary, and partakes of the nature of an encyclopædia, as it gives in its body a large amount of valuable therapeutical and chemical information, and groups in its tables, in a condensed and convenient form, a vast amount of important data which will be consulted daily by all in active practice.

The completeness of the work is made evident by the fact that it defines 84,844 separate words and phrases.

The type has been most carefully selected for boldness and clearness, and everything has been done to secure ease, rapidity and durability in use.

Its scope is one which will at once satisfy the student and meet all the requirements of the medical practitioner. Clear and comprehensive definitions of words should form the prime feature of any dictionary, and in this one the chief aim seems to be to give the exact signification and the different meanings of terms in use in medicine and the collateral sciences in language as terse as is compatible with lucidity. The utmost brevity and conciseness have been kept in view. The work is remarkable, too, for its fulness. The enumerations and subdivisions under each word heading are strikingly complete, as regards alike the English tongue and the languages chiefly employed by ancient and modern science. It is impossible to do justice to the dictionary by any casual illustration. It presents to the English reader a thoroughly scientific mode of acquiring a rich vocabulary and offers an accurate and ready means of reference in consulting works in any of the three modern continental languages which are richest in medical literature. To add to its usefulness as a work of reference some valuable tables are given. Another feature of the work is the accuracy of its definitions, all of which have been checked by comparison with many other standard works in the different languages it deals with. Apart from the boundless stores of information which may be gained by the study of a good dictionary, one is enabled by the work under notice to read intelligently any technical treatise in any of the four chief modern languages. There cannot be two opinions as to the great value and usefulness of this dictionary as a book of ready reference for all sorts and conditions of medical men. So far as we have been able to see, no subject has been omitted, and in respect of completeness it will be found distinctly superior to any medical lexicon yet published.—*The London Lancet*, April 5, 1890.

GRAY, HENRY, F. R. S.,
Lecturer on Anatomy at St. George's Hospital, London.

Anatomy, Descriptive and Surgical. Edited by T. PICKERING PICK, F. R. C. S., Surgeon to and Lecturer on Anatomy at St. George's Hospital, London, Examiner in Anatomy, Royal College of Surgeons of England. A new American from the eleventh enlarged and improved London edition, thoroughly revised and re-edited by WILLIAM W. KEEN, M. D., Professor of Surgery in the Jefferson Medical College of Philadelphia. To which is added the second American from the latest English edition of LANDMARKS, MEDICAL AND SURGICAL, by LUTHER HOLDEN, F. R. C. S. In one imperial octavo volume of 1098 pages, with 685 large and elaborate engravings on wood. Price of edition in black: Cloth, $6; leather, $7; half Russia, $7.50. Price of edition in colors (see below): Cloth, $7.25; leather, $8.25; half Russia, $8.75.

This work covers a more extended range of subjects than is customary in the ordinary text-books, giving not only the details necessary for the student, but also the application to those details to the practice of medicine and surgery. It thus forms both a guide for the learner and an admirable work of reference for the active practitioner. The engraving form a special feature in the work, many of them being the size of nature, nearly all original, and having the names of the various parts printed on the body of the cut, in place of figures of reference with descriptions at the foot. In this edition a new departure has been taken by the issue of the work with the arteries, veins and nerves distinguished by different colors. The engravings thus form a complete and splendid series, which will greatly assist the student in forming a clear idea of Anatomy, and will also serve to refresh the memory of those who may find in the exigencies of practice the necessity of recalling the details of the dissecting-room. Combining, as it does, a complete Atlas of Anatomy with a thorough treatise on systematic, descriptive and applied Anatomy, the work will be found of great service to all physicians who receive students in their offices, relieving both preceptor and pupil of much labor in laying the groundwork of a thorough medical education.

For the convenience of those who prefer not to pay the slight increase in cost necessitated by the use of colors, the volume is published also in black alone, and maintained in this style at the price of former editions, notwithstanding its largely increased size.

Landmarks, Medical and Surgical, by the distinguished Anatomist, Mr. Luther Holden, has been appended to the present edition as it was to the previous one. This work gives in a clear, condensed and systematic way all the information by which the practitioner can determine from the external surface of the body the position of internal parts. Thus complete, the work will furnish all the assistance that can be rendered by type and illustration in anatomical study.

The most popular work on anatomy ever written. It is sufficient to say of it that this edition, thanks to its American editor, surpasses all other editions.—*Jour. of the Amer. Med. Ass'n,* Dec. 31, 1887.

A work which for more than twenty years has had the lead of all other text-books on anatomy throughout the civilized world comes to hand in such beauty of execution and accuracy of text and illustration as more than to make good the large promise of the prospectus. It would be indeed difficult to name a feature wherein the present American edition of Gray could be mended or bettered, and it needs no prophet to see that the royal work is destined for many years to come to hold the first place among anatomical text-books. The work is published with black and colored plates. It is a marvel of book-making.—*American Practitioner and News,* Jan. 21, 1888.

Gray's *Anatomy* is the most magnificent work upon anatomy which has ever been published in the English or any other language.—*Cincinnati Medical News,* Nov. 1887.

As the book now goes to the purchaser he is receiving the best work on anatomy that is published in any language.—*Virginia Med. Monthly,* Dec. 1887.

Gray's standard *Anatomy* has been and will be for years the text-book for students. The book needs only to be examined to be perfectly understood.—*Medical Press of Western New York,* Jan. 1888.

ALSO FOR SALE SEPARATE—

HOLDEN, LUTHER, F. R. C. S.,
Surgeon to St. Bartholomew's and the Foundling Hospitals, London.

Landmarks, Medical and Surgical. Second American from the latest revised English edition, with additions by W. W. KEEN, M. D., Professor of Artistic Anatomy in the Penna. Academy of Fine Arts. In one 12mo. volume of 148 pages. Cloth, $1.00.

DUNGLISON, ROBLEY, M. D.,
Late Professor of Institutes of Medicine in the Jefferson Medical College of Philadelphia.

MEDICAL LEXICON; A Dictionary of Medical Science: Containing a concise Explanation of the various Subjects and Terms of Anatomy, Physiology, Pathology, Hygiene, Therapeutics, Pharmacology, Pharmacy, Surgery, Obstetrics, Medical Jurisprudence and Dentistry, Notices of Climate and of Mineral Waters, Formulæ for Officinal, Empirical and Dietetic Preparations, with the Accentuation and Etymology of the Terms, and the French and other Synonymes, so as to constitute a French as well as an English Medical Lexicon. Edited by RICHARD J. DUNGLISON, M. D. In one very large and handsome royal octavo volume of 1139 pages. Cloth, $6.50; leather, raised bands, $7.50; very handsome half Russia, raised bands, $8.

It has the rare merit that it certainly has no rival in the English language for accuracy and extent of references.—*London Medical Gazette.*

ALLEN, HARRISON, M. D.,
Professor of Physiology in the University of Pennsylvania.

A System of Human Anatomy, Including Its Medical and Surgical Relations. For the use of Practitioners and Students of Medicine. With an Introductory Section on Histology. By E. O. SHAKESPEARE, M. D., Ophthalmologist to the Philadelphia Hospital. Comprising 813 double-columned quarto pages, with 380 illustrations on 109 full page lithographic plates, many of which are in colors, and 241 engravings in the text. In six Sections, each in a portfolio. Section I. HISTOLOGY. Section II. BONES AND JOINTS. Section III. MUSCLES AND FASCIÆ. Section IV. ARTERIES, VEINS AND LYMPHATICS. Section V. NERVOUS SYSTEM. Section VI. ORGANS OF SENSE, OF DIGESTION AND GENITO-URINARY ORGANS, EMBRYOLOGY, DEVELOPMENT, TERATOLOGY, SUPERFICIAL ANATOMY, POST-MORTEM EXAMINATIONS, AND GENERAL AND CLINICAL INDEXES. Price per Section, $3.50; also bound in one volume, cloth, $23.00; very handsome half Russia, raised bands and open back, $25.00. *For sale by subscription only. Apply to the Publishers.*

It is to be considered a study of applied anatomy in its widest sense—a systematic presentation of such anatomical facts as can be applied to the practice of medicine as well as of surgery. Our author is concise, accurate and practical in his statements, and succeeds admirably in infusing an interest into the study of what is generally considered a dry subject. The department of Histology is treated in a masterly manner, and the ground is travelled over by one thoroughly familiar with it. The illustrations are made with great care, and are simply superb. There is as much of practical application of anatomical points to the every-day wants of the medical clinician as to those of the operating surgeon. In fact, few general practitioners will read the work without a feeling of surprised gratification that so many points, concerning which they may never have thought before are so well presented for their consideration. It is a work which is destined to be the best of its kind in any language.—*Medical Record*, Nov. 25, 1882.

CLARKE, W. B., F.R.C.S. & LOCKWOOD, C. B., F.R.C.S.
Demonstrators of Anatomy at St. Bartholomew's Hospital Medical School, London.

The Dissector's Manual. In one pocket-size 12mo. volume of 396 pages, with 49 illustrations. Limp cloth, red edges, $1.50. See *Students' Series of Manuals*, page 31.

Messrs. Clarke and Lockwood have written a book that can hardly be rivalled as a practical aid to the dissector. Their purpose, which is "how to describe the best way to display the anatomical structure," has been fully attained. They excel in a lucidity of demonstration and graphic terseness of expression, which only a long training and intimate association with students could have given. With such a guide as this, accompanied by so attractive a commentary as Treves' *Surgical Applied Anatomy* (same series), no student could fail to be deeply and absorbingly interested in the study of anatomy.—*New Orleans Medical and Surgical Journal*, April, 1884.

TREVES, FREDERICK, F. R. C. S.,
Senior Demonstrator of Anatomy and Assistant Surgeon at the London Hospital.

Surgical Applied Anatomy. In one pocket-size 12mo. volume of 540 pages, with 61 illustrations. Limp cloth, red edges, $2.00. See *Students' Series of Manuals*, page 31.

He has produced a work which will command a larger circle of readers than the class for which it was written. This union of a thorough, practical acquaintance with these fundamental branches, quickened by daily use as a teacher and practitioner, has enabled our author to prepare a work which it would be a most difficult task to excel.— *The American Practitioner*, Feb. 1884. This number of the "Manuals for Students" is most excellent, giving just such practical knowledge as will be required for application in relieving the injuries to which the living body is liable. The book is intended mainly for students, but it will also be of great use to practitioners. The illustrations are well executed and fully elucidate the text.—*Southern Practitioner*, Feb. 1884.

BELLAMY, EDWARD, F. R. C. S.,
Senior Assistant-Surgeon to the Charing-Cross Hospital, London.

The Student's Guide to Surgical Anatomy: Being a Description of the most Important Surgical Regions of the Human Body, and intended as an Introduction to operative Surgery. In one 12mo. volume of 300 pages, with 50 illustrations. Cloth, $2.25.

WILSON, ERASMUS, F. R. S.

A System of Human Anatomy, General and Special. Edited by W. H. GOBRECHT, M. D., Professor of General and Surgical Anatomy in the Medical College of Ohio. In one large and handsome octavo volume of 616 pages, with 397 illustrations. Cloth, $4.00; leather, $5.00.

CLELAND, JOHN, M. D., F. R. S.,
Professor of Anatomy and Physiology in Queen's College, Galway.

A Directory for the Dissection of the Human Body. In one 12mo. volume of 178 pages. Cloth, $1.25.

HARTSHORNE'S HANDBOOK OF ANATOMY AND PHYSIOLOGY. Second edition, revised. In one royal 12mo. volume of 310 pages, with 220 woodcuts. Cloth, $1.75.

HORNER'S SPECIAL ANATOMY AND HISTOLOGY. Eighth edition, extensively revised and modified. In two octavo volumes of 1007 pages, with 320 woodcuts. Cloth, $6.00.

DRAPER, JOHN C., M. D., LL. D.,
Professor of Chemistry in the University of the City of New York.

Medical Physics. A Text-book for Students and Practitioners of Medicine. In one octavo volume of 734 pages, with 376 woodcuts, mostly original. Cloth, $4.

FROM THE PREFACE.

The fact that a knowledge of Physics is indispensable to a thorough understanding of Medicine has not been as fully realized in this country as in Europe, where the admirable works of Desplats and Gariel, of Robertson and of numerous German writers constitute a branch of educational literature to which we can show no parallel. A full appreciation of this the author trusts will be sufficient justification for placing in book form the substance of his lectures on this department of science, delivered during many years at the University of the City of New York.

Broadly speaking, this work aims to impart a knowledge of the relations existing between Physics and Medicine in their latest state of development, and to embody in the pursuit of this object whatever experience the author has gained during a long period of teaching this special branch of applied science.

While all enlightened physicians will agree that a knowledge of physics is desirable for the medical student, only those actually engaged in the teaching of the primary subjects can be fully aware of the difficulties encountered by students who attempt the study of these subjects without a knowledge of either physics or chemistry. These are especially felt by the teacher of physiology.

It is, however, impossible for him to impart a knowledge of the main facts of his subject and establish them by reasons and experimental demonstration, and at the same time undertake to teach *ab initio* the principles of chemistry or physics. Hence the desirability, we may say the necessity, for some such work as the present one.

No man in America was better fitted than Dr. Draper for the task he undertook, and he has provided the student and practitioner of medicine with a volume at once readable and thorough. Even to the student who has some knowledge of physics this book is useful, as it shows him its applications to the profession that he has chosen. Dr. Draper, as an old teacher, knew well the difficulties to be encountered in bringing his subject within the grasp of the average student, and that he has succeeded so well proves once more that the man to write for and examine students is the one who has taught and is teaching them. The book is well printed and fully illustrated, and in every way deserves grateful recognition.—*The Montreal Medical Journal*, July, 1890.

ROBERTSON, J. McGREGOR, M. A., M. B.,
Muirhead Demonstrator of Physiology, University of Glasgow.

Physiological Physics. In one 12mo. volume of 537 pages, with 219 illustrations. Limp cloth, $2.00. See *Students' Series of Manuals*, page 31.

The title of this work sufficiently explains the nature of its contents. It is designed as a manual for the student of medicine, an auxiliary to his text-book in physiology, and it would be particularly useful as a guide to his laboratory experiments. It will be found of great value to the practitioner. It is a carefully prepared book of reference, concise and accurate, and as such we heartily recommend it.—*Journal of the American Medical Association*, Dec. 6, 1884.

DALTON, JOHN C., M. D.,
Professor Emeritus of Physiology in the College of Physicians and Surgeons, New York.

Doctrines of the Circulation of the Blood. A History of Physiological Opinion and Discovery in regard to the Circulation of the Blood. In one handsome 12mo. volume of 293 pages. Cloth, $2.

Dr. Dalton's work is the fruit of the deep research of a cultured mind, and to the busy practitioner it cannot fail to be a source of instruction. It will inspire him with a feeling of gratitude and admiration for those plodding workers of olden times, who laid the foundation of the magnificent temple of medical science as it now stands.—*New Orleans Medical and Surgical Journal*, Aug. 1885.

In the progress of physiological study no fact was of greater moment, none more completely revolutionized the theories of teachers, than the discovery of the circulation of the blood. This explains the extraordinary interest it has to all medical historians. The volume before us is one of three or four which have been written within a few years by American physicians. It is in several respects the most complete. The volume, though small in size, is one of the most creditable contributions from an American pen to medical history that has appeared.—*Med. & Surg. Rep.*, Dec. 6, 1884.

BELL, F. JEFFREY, M. A.,
Professor of Comparative Anatomy at King's College, London.

Comparative Physiology and Anatomy. In one 12mo. volume of 561 pages, with 229 illustrations. Limp cloth, $2.00. See *Students' Series of Manuals*, page 31.

The manual is preëminently a student's book—clear and simple in language and arrangement. It is well and abundantly illustrated, and is readable and interesting. On the whole we consider it the best work in existence in the English language to place in the hands of the medical student.—*Bristol Medico-Chirurgical Journal*, Mar. 1886.

ELLIS, GEORGE VINER,
Emeritus Professor of Anatomy in University College, London.

Demonstrations of Anatomy. Being a Guide to the Knowledge of the Human Body by Dissection. From the eighth and revised London edition. In one very handsome octavo volume of 716 pages, with 249 illustrations. Cloth, $4.25; leather, $5.25.

ROBERTS, JOHN B., A. M., M. D.,
Lecturer in Anatomy in the University of Pennsylvania.

The Compend of Anatomy. For use in the dissecting-room and in preparing for examinations. In one 16mo. volume of 196 pages. Limp cloth, 75 cents.

8   LEA BROTHERS & CO.'S PUBLICATIONS—**Physiology, Chemistry.**

CHAPMAN, HENRY C., M. D.,
Professor of Institutes of Medicine and Medical Juris. in the Jefferson Med. Coll. of Philadelphia.

A Treatise on Human Physiology. In one handsome octavo volume of 925 pages, with 605 fine engravings. Cloth, $5.50; leather, $6.50.

It represents very fully the existing state of physiology. The present work has a special value to the student and practitioner as devoted more to the practical application of well-known truths which the advance of science has given to the profession in this department, which may be considered the foundation of rational medicine.—*Buffalo Medical and Surgical Journal*, Dec. 1887.

Matters which have a practical bearing on the practice of medicine are lucidly expressed; technical matters are given in minute detail; elaborate directions are stated for the guidance of students in the laboratory. In every respect the work fulfils its promise, whether as a complete treatise for the student or for the physician; for the former it is so complete that he need look no farther, and the latter will find entertainment and instruction in an admirable book of reference.—*North Carolina Medical Journal*, Nov. 1887.

The work certainly commends itself to both student and practitioner. What is most demanded by the progressive physician of to-day is an adaptation of physiology to practical therapeutics, and this work is a decided improvement in this respect over other works in the market. It will certainly take place among the most valuable text-books.—*Medical Age*, Nov. 25, 1887.

It is the production of an author delighted with his work, and able to inspire students with an enthusiasm akin to his own.—*American Practitioner and News*, Nov. 12, 1887.

DALTON, JOHN C., M. D.,
Professor of Physiology in the College of Physicians and Surgeons, New York, etc.

A Treatise on Human Physiology. Designed for the use of Students and Practitioners of Medicine. Seventh edition, thoroughly revised and rewritten. In one very handsome octavo volume of 722 pages, with 252 beautiful engravings on wood. Cloth, $5.00; leather, $6.00.

From the first appearance of the book it has been a favorite, owing as well to the author's renown as an oral teacher as to the charm of simplicity with which, as a writer, he always succeeds in investing even intricate subjects. It must be gratifying to him to observe the frequency with which his work, written for students and practitioners, is quoted by other writers on physiology. This fact attests its value, and, in great measure, its originality. It now needs no such seal of approbation, however, for the thousands who have studied it in its various editions have never been in any doubt as to its sterling worth.—*N. Y. Medical Journal*, Oct. 1882.

Professor Dalton's well-known and deservedly-appreciated work has long passed the stage at which it could be reviewed in the ordinary sense. The work is eminently one for the medical practitioner, since it treats most fully of those branches of physiology which have a direct bearing on the diagnosis and treatment of disease. The work is one which we can highly recommend to all our readers.—*Dublin Journal of Medical Science*, Feb.'83.

FOSTER, MICHAEL, M. D., F. R. S.,
Prelector in Physiology and Fellow of Trinity College, Cambridge, England.

Text-Book of Physiology. New (fourth) and enlarged American from the fifth and revised English edition, with notes and additions. *Preparing.*

A REVIEW OF THE FIFTH ENGLISH EDITION IS APPENDED.

It is delightful to meet a book which deserves only unqualified praise. Such a book is now before us. It is in all respects an ideal text-book. With a complete, accurate and detailed knowledge of his subject, the author has succeeded in giving a thoroughly consecutive and philosophic account of the science. A student's attention is kept throughout fixed on the great and salient questions, and his energies are not frittered away and degenerated on petty and trivial details. Reviewing this volume as a whole we are justified in saying that it is the only thoroughly good text-book of physiology in the English language, and that it is probably the best text-book in any language.—*Edinburgh Medical Journal*, December 1889.

POWER, HENRY, M. B., F. R. C. S.,
Examiner in Physiology, Royal College of Surgeons of England.

Human Physiology. Second edition. In one handsome pocket-size 12mo. volume of 509 pp., with 68 illustrations. Cloth, $1.50. See *Students' Series of Manuals*, p. 31.

SIMON, W., Ph. D., M. D.,
Professor of Chemistry and Toxicology in the College of Physicians and Surgeons, Baltimore, and Professor of Chemistry in the Maryland College of Pharmacy.

Manual of Chemistry. A Guide to Lectures and Laboratory work for Beginners in Chemistry. A Text-book, specially adapted for Students of Pharmacy and Medicine. New (second) edition. In one 8vo. vol. of 478 pp., with 44 woodcuts and 7 colored plates illustrating 56 of the most important chemical tests. Cloth, $3.25.

In this book the author has endeavored to meet the wants of the student of medicine or pharmacy in regard to his chemical studies, and he has succeeded in presenting his subject so clearly that no one who really wishes to acquire a fair knowledge of chemistry can fail to do so with the help of this work. The largest section of the book is naturally that devoted to the consideration of the carbon compounds, or organic chemistry. An excellent feature is the introduction of a number of plates showing the various colors of the most important chemical reactions of the metallic salts, of some of the alkaloids, and of the urinary tests. In the part treating of physiological chemistry the section on analysis of the urine will be found very practical, and well suited to the needs of the practitioner of medicine.—*The Medical Record*, May 25, 1889.

Wöhler's Outlines of Organic Chemistry. Edited by FITTIG. Translated by IRA REMSEN, M. D., Ph. D. In one 12mo. volume of 550 pages. Cloth, $3.

LEHMANN'S MANUAL OF CHEMICAL PHYSIOLOGY. In one octavo volume of 327 pages, with 41 illustrations. Cloth, $2.25.
CARPENTER'S HUMAN PHYSIOLOGY. Edited by HENRY POWER. In one octavo volume.

CARPENTER'S PRIZE ESSAY ON THE USE AND ABUSE OF ALCOHOLIC LIQUORS IN HEALTH AND DISEASE. With explanations of scientific words. Small 12mo. 178 pages. Cloth, 60 cents.

FRANKLAND, E., D. C. L., F.R.S., & JAPP, F. R., F. I. C.,
Professor of Chemistry in the Normal School of Science, London.   *Assist. Prof. of Chemistry in the Normal School of Science, London.*

Inorganic Chemistry. In one handsome octavo volume of 677 pages with 51 woodcuts and 2 plates. Cloth, $3.75; leather, $4.75.

This work should supersede other works of its class in the medical colleges. It is certainly better adapted than any work upon chemistry, with which we are acquainted, to impart that clear and full knowledge of the science which students of medicine should have. Physicians who feel that their chemical knowledge is behind the times, would do well to devote some of their leisure time to the study of this work. The descriptions and demonstrations are made so plain that there is no difficulty in understanding them.—*Cincinnati Medical News*, January, 1886.

This excellent treatise will not fail to take its place as one of the very best on the subject of which it treats. We have been much pleased with the comprehensive and lucid manner in which the difficulties of chemical notation and nomenclature have been cleared up by the writers. It shows on every page that the problem of rendering the obscurities of this science easy of comprehension has long and successfully engaged the attention of the authors.—*Medical and Surgical Reporter*, October 31, 1885.

FOWNES, GEORGE, Ph. D.

A Manual of Elementary Chemistry; Theoretical and Practical. Embodying WATTS' *Physical and Inorganic Chemistry.* New American, from the twelfth English edition. In one large royal 12mo. volume of 1061 pages, with 168 illustrations on wood and a colored plate. Cloth, $2.75; leather, $3.25.

Fownes' Chemistry has been a standard textbook upon chemistry for many years. Its merits are very fully known by chemists and physicians everywhere in this country and in England. As the science has advanced by the making of new discoveries, the work has been revised so as to keep it abreast of the times. It has steadily maintained its position as a text-book with medical students. In this work are treated fully: Heat, Light and Electricity, including Magnetism. The influence exerted by these forces in chemical action upon health and disease, etc., is of the most important kind, and should be familiar to every medical practitioner. We can commend the

work as one of the very best text-books upon chemistry extant.—*Cincinnati Med. News*, Oct. '85. Of all the works on chemistry intended for the use of medical students, Fownes' Chemistry is perhaps the most widely used. Its popularity is based upon its excellence. This last edition contains all of the material found in the previous, and it is also enriched by the addition of Watts' *Physical and Inorganic Chemistry*. All of the matter is brought to the present standpoint of chemical knowledge. We may safely predict for this work a continuance of the fame and favor it enjoys among medical students.—*New Orleans Medical and Surgical Journal*, March, 1886.

ATTFIELD, JOHN, M. A., Ph. D., F. I. C., F. R. S., Etc.
Professor of Practical Chemistry to the Pharmaceutical Society of Great Britain, etc.

Chemistry, General, Medical and Pharmaceutical; Including the Chemistry of the U. S. Pharmacopœia. A Manual of the General Principles of the Science, and their Application to Medicine and Pharmacy. A new American, from the twelfth English edition, specially revised by the Author for America. In one handsome royal 12mo. volume of 782 pages, with 88 illustrations. Cloth, $2.75; leather, $3.25.

Attfield's Chemistry is the most popular book among students of medicine and pharmacy. This popularity has a good, substantial basis. It rests upon real merits. Attfield's work combines in the happiest manner a clear exposition of the theory of chemistry with the practical application of his knowledge to the everyday dealings of the physician and pharmacist. His discernment is shown not only in what he puts into his work, but also in what he leaves out. His book is precisely what the title claims for it. The admirable arrangement of the text enables a reader to get a good idea of chemistry without the aid of experiments, and

again it is a good laboratory guide, and finally it contains such a mass of well-arranged information that it will always serve as a handy book of reference. He does not allow any unutilizable knowledge to slip into his book; his long years of experience have produced a work which is both scientific and practical, and which shuts out everything in the nature of a superfluity, and therein lies the secret of its success. This last edition shows the marks of the latest progress made in chemistry and chemical teaching.—*New Orleans Medical and Surgical Journal*, Nov. 1889.

BLOXAM, CHARLES L.,
Professor of Chemistry in King's College, London.

Chemistry, Inorganic and Organic. New American from the fifth London edition, thoroughly revised and much improved. In one very handsome octavo volume of 727 pages, with 292 illustrations. Cloth, $2.00; leather, $3.00.

Comment from us on this standard work is almost superfluous. It differs widely in scope and aim from that of Attfield, and in its way is equally beyond criticism. It adopts the most direct methods in stating the principles, hypotheses and facts of the science. Its language is so terse and lucid, and its arrangement of matter so logical in sequence that the student never has occasion to complain that chemistry is a hard study. Much attention is paid to experimental illustrations of chemical principles and phenomena, and the mode of conducting these experiments. The book maintains the position it has always held as one of

the best manuals of general chemistry in the English language.—*Detroit Lancet*, Feb. 1884. We know of no treatise on chemistry which contains so much practical information in the same number of pages. The book can be readily adapted not only to the needs of those who desire a tolerably complete course of chemistry, but also to the needs of those who desire only a general knowledge of the subject. We take pleasure in recommending this work both as a satisfactory text-book, and as a useful book of reference.—*Boston Medical and Surgical Journal*, June 19, 1884.

GREENE, WILLIAM H., M. D.,
Demonstrator of Chemistry in the Medical Department of the University of Pennsylvania.

A Manual of Medical Chemistry. For the use of Students. Based upon Bowman's Medical Chemistry. In one 12mo. volume of 310 pages, with 74 illus. Cloth, $1.75.

It is a concise manual of three hundred pages, giving an excellent summary of the best methods of analyzing the liquids and solids of the body, both for the estimation of their normal constituent and

the recognition of compounds due to pathological conditions. The detection of poisons is treated with sufficient fulness for the purpose of the student or practitioner.—*Boston Jl. of Chem.* June, '80.

REMSEN, IRA, M. D., Ph. D.,
Professor of Chemistry in the Johns Hopkins University, Baltimore.

Principles of Theoretical Chemistry, with special reference to the Constitution of Chemical Compounds. New (third) and thoroughly revised edition. In one handsome royal 12mo. volume of 316 pages. Cloth, $2.00

This work of Dr. Remsen is the very text-book needed, and the medical student who has it at his fingers' ends, so to speak, can, if he chooses, make himself familiar with any branch of chemistry which he may desire to pursue. It would be difficult indeed to find a more lucid, full, and at the same time compact explication of the philosophy of chemistry, than the book before us, and we recommend it to the careful and impartial examination of college faculties as the text-book of chemical instruction.—*St. Louis Medical and Surgical Journal*, January, 1888.

It is a healthful sign when we see a demand for a third edition of such a book as this. This edition is larger than the last by about seventy-five pages, and much of it has been rewritten, thus bringing it fully abreast of the latest investigations.—*N. Y. Medical Journal*, Dec. 31, 1887.

CHARLES, T. CRANSTOUN, M. D., F. C. S., M. S.,
Formerly Asst. Prof. and Demonst. of Chemistry and Chemical Physics, Queen's College, Belfast.

The Elements of Physiological and Pathological Chemistry. A Handbook for Medical Students and Practitioners. Containing a general account of Nutrition, Foods and Digestion, and the Chemistry of the Tissues, Organs, Secretions and Excretions of the Body in Health and in Disease. Together with the methods for preparing or separating their chief constituents, as also for their examination in detail, and an outline syllabus of a practical course of instruction for students. In one handsome octavo volume of 463 pages, with 38 woodcuts and 1 colored plate. Cloth, $3.50.

Dr. Charles is fully impressed with the importance and practical reach of his subject, and he has treated it in a competent and instructive manner. We cannot recommend a better book than the present. In fact, it fills a gap in medical text-books, and that is a thing which can rarely be said nowadays. Dr. Charles has devoted much space to the elucidation of urinary mysteries. He does this with much detail, and yet in a practical and intelligible manner. In fact, the author has filled his book with many practical hints.—*Medical Record*, December 20, 1884.

HOFFMANN, F., A.M., Ph.D., & POWER, F.B., Ph.D.,
Public Analyst to the State of New York.   Prof. of Anal. Chem. in the Phil. Coll. of Pharmacy.

A Manual of Chemical Analysis, as applied to the Examination of Medicinal Chemicals and their Preparations. Being a Guide for the Determination of their Identity and Quality, and for the Detection of Impurities and Adulterations. For the use of Pharmacists, Physicians, Druggists and Manufacturing Chemists, and Pharmaceutical and Medical Students. Third edition, entirely rewritten and much enlarged. In one very handsome octavo volume of 621 pages, with 179 illustrations. Cloth, $4.25.

We congratulate the author on the appearance of the third edition of this work, published for the first time in this country also. It is admirable for the information it undertakes to supply is both extensive and trustworthy. The selection of processes for determining the purity of the substances of which it treats is excellent and the description of them singularly explicit. Moreover, it is exceptionally free from typographical errors. We have no hesitation in recommending it to those who are engaged either in the manufacture or the testing of medicinal chemicals.—*London Pharmaceutical Journal and Transactions*, 1883.

CLOWES, FRANK, D. Sc., London,
Senior Science-Master at the High School, Newcastle-under-Lyme, etc.

An Elementary Treatise on Practical Chemistry and Qualitative Inorganic Analysis. Specially adapted for use in the Laboratories of Schools and Colleges and by Beginners. Third American from the fourth and revised English edition. In one very handsome royal 12mo. volume of 387 pages, with 55 illustrations. Cloth, $2.50.

This work has long been a favorite with laboratory instructors on account of its systematic plan, carrying the student step by step from the simplest questions of chemical analysis, to the more recondite problems. Features quite as commendable are the regularity and system demanded of the student in the performance of each analysis. These characteristics are preserved in the present edition, which we can heartily recommend as a satisfactory guide for the student of inorganic chemical analysis.—*New York Medical Journal*, Oct. 9, 1886.

RALFE, CHARLES H., M. D., F. R. C. P.,
Assistant Physician at the London Hospital.

Clinical Chemistry. In one pocket-size 12mo. volume of 314 pages, with 16 illustrations. Limp cloth, red edges, $1.50. See *Students' Series of Manuals*, page 31.

This is one of the most instructive little works that we have met with in a long time. The author is a physician and physiologist, as well as a chemist, consequently the book is unqualifiedly practical, telling the physician just what he ought to know, of the applications of chemistry in medicine. Dr. Ralfe is thoroughly acquainted with the latest contributions to his science, and it is quite refreshing to find the subject dealt with so clearly and simply, yet in such evident harmony with the modern scientific methods and spirit.—*Medical Record*, February 2, 1884.

CLASSEN, ALEXANDER,
Professor in the Royal Polytechnic School, Aix-la-Chapelle.

Elementary Quantitative Analysis. Translated, with notes and additions, by EDGAR F. SMITH, Ph. D., Assistant Professor of Chemistry in the Towne Scientific School, University of Penna. In one 12mo. volume of 324 pages, with 36 illus. Cloth, $2.00.

It is probably the best manual of an elementary nature extant, insomuch as its methods are the best. It teaches by examples, commencing with single determinations, followed by separations, and then advancing to the analysis of minerals and such products as are met with in applied chemistry. It is an indispensable book for students in chemistry.—*Boston Journal of Chemistry*, Oct. 1878.

LEA BROTHERS & CO.'S PUBLICATIONS—Pharm., Mat. Med., Therap. 11

HARE, HOBART AMORY, B. Sc., M. D.,
Clinical Professor of Diseases of Children and Demonstrator of Therapeutics in the University of Pennsylvania; Secretary of the Convention for the Revision of the United States Pharmacopœia of 1890.

A Text-Book of Practical Therapeutics; With Especial Reference to the Application of Remedial Measures to Disease and their Employment upon a Rational Basis. With special chapters by DRS. G. E. DE SCHWEINITZ, EDWARD MARTIN, J. HOWARD REEVES and BARTON C. HIRST. In one handsome octavo volume of 622 pages. Cloth, $3.75; leather, $4.75.

That the student is too often required to perform acrobatic feats of memory and invention in associating and reconciling widely separated statements is certain, and disgust over his failure is apt to develop him into a physician, without faith in the reasonableness of his art. Dr. Hare has obviated this difficulty by comprising in one cover a work on therapeutics and on treatment, each part being so interwoven with the other by references that there will be the least possible difficulty in learning and remembering the nature of therapeutic resources, and in using them to the best advantage. The portion devoted to treatment occupies at least one-half of the work, with clear directions for the therapeutic measures to be employed, together with the reasons for the choice of drugs, according to the varying stages and symptoms.—*Medical Age*, September 25, 1890. We may say without exaggeration that the present volume is in many respects unique and a great credit to the author. Dr. Hare is already well known as an able experimental, didactic and clinical therapeutist, a happy combination, which has eminently fitted him for the preparation of the present work. He is thoroughly acquainted with the latest contributions to therapeutical science, and his book represents the actual state of the science. It is a model of concise, clear and forcible description, an exponent of plain facts, and a thoroughly practical guide to the rational treatment of disease. Books like this make a lasting impression. We heartily commend the present volume to the student, the scientific therapeutist and the general practitioner, not only as a most satisfactory text-book, but also as a highly valuable work of reference. We bespeak for Dr. Hare's "Practical Therapeutics" the greatest success in every way.--*University Medical Magazine*, Nov. 1890.

BRUNTON, T. LAUDER, M.D., D.Sc., F.R.S., F.R.C.P.,
Lecturer on Materia Medica and Therapeutics at St. Bartholomew's Hospital, London, etc.

A Text-Book of Pharmacology, Therapeutics and Materia Medica; Including the Pharmacy, the Physiological Action and the Therapeutical Uses of Drugs. Third edition. Octavo, 1305 pages, 230 illustrations. Cloth, $5.50; leather, $6.50.

No words of praise are needed for this work, for it has already spoken for itself in former editions. It was by unanimous consent placed among the foremost books on the subject ever published in any language, and the better it is known and studied the more highly it is appreciated. The present edition contains much new matter, the insertion of which has been necessitated by the advances made in various directions in the art of therapeutics, and it now stands unrivalled in its thoroughly scientific presentation of the modes of drug action. No one who wishes to be fully up to the times in this science can afford to neglect the study of Dr. Brunton's work. The indexes are excellent, and add not a little to the practical value of the book. —*Medical Record*, May 25, 1889.

MAISCH, JOHN M., Phar. D.,
Professor of Materia Medica and Botany in the Philadelphia College of Pharmacy.

A Manual of Organic Materia Medica; Being a Guide to Materia Medica of the Vegetable and Animal Kingdoms. For the use of Students, Druggists, Pharmacists and Physicians. New (4th) edition, thoroughly revised. In one handsome royal 12mo. volume of 529 pages, with 258 illustrations. Cloth, $3.

For everyone interested in materia medica, Maisch's Manual, first published in 1882, and now in its fourth edition, is an indispensable book. For the American pharmaceutical student it is the work which will give him the necessary knowledge in the easiest way, partly because the text is brief, concise, and free from unnecessary matter, and partly because of the numerous illustrations, which bring facts worth knowing immediately before his eyes. That it answers its purposes in this respect the rapid succession of editions is the best evidence. It is the favorite book of the American student even outside of Maisch's several hundred personal students. The arrangement of its contents shows the practical tendency of the book. Maisch's system of classification is easy and comprehensive.—*Pharmaceutische Zeitung*, Germany, 1890.

PARRISH, EDWARD,
Late Professor of the Theory and Practice of Pharmacy in the Philadelphia College of Pharmacy.

A Treatise on Pharmacy: Designed as a Text-book for the Student, and as a Guide for the Physician and Pharmaceutist. With many Formulæ and Prescriptions. Fifth edition, thoroughly revised, by THOMAS S. WIEGAND, Ph. G. In one handsome octavo volume of 1093 pages, with 256 illustrations. Cloth, $5; leather, $6.

No thorough-going pharmacist will fail to possess himself of so useful a guide to practice, and no physician who properly estimates the value of an accurate knowledge of the remedial agents employed by him in daily practice, so far as their miscibility, compatibility and most effective methods of combination are concerned, can afford to leave this work out of the list of their works of reference. The country practitioner, who must always be in a measure his own pharmacist, will find it indispensable.—*Louisville Medical News*, March 29, 1884.

HERMANN, Dr. L.,
Professor of Physiology in the University of Zurich.

Experimental Pharmacology. A Handbook of Methods for Determining the Physiological Actions of Drugs. Translated, with the Author's permission, and with extensive additions, by ROBERT MEADE SMITH, M. D., Demonstrator of Physiology in the University of Pennsylvania. 12mo., 199 pages, with 32 illustrations Cloth, $1.50

STILLÉ, ALFRED, M. D., LL. D.,
Professor of Theory and Practice of Med. and of Clinical Med. in the Univ. of Penna.

Therapeutics and Materia Medica. A Systematic Treatise on the Action and Uses of Medicinal Agents, including their Description and History. Fourth edition, revised and enlarged. In two large and handsome octavo volumes, containing 1936 pages. Cloth, $10.00; leather, $12.00.

STILLÉ, A., M. D., LL. D., & MAISCH, J. M., Phar. D.,
Professor Emeritus of the Theory and Practice of Medicine and of Clinical Medicine in the University of Pennsylvania.
Prof. of Mat. Med. and Botany in Phila. College of Pharmacy, Sec'y to the American Pharmaceutical Association.

The National Dispensatory.

CONTAINING THE NATURAL HISTORY, CHEMISTRY, PHARMACY, ACTIONS AND USES OF MEDICINES, INCLUDING THOSE RECOGNIZED IN THE PHARMACOPŒIAS OF THE UNITED STATES, GREAT BRITAIN AND GERMANY, WITH NUMEROUS REFERENCES TO THE FRENCH CODEX.

Fourth edition revised, and covering the new British Pharmacopœia. In one magnificent imperial octavo volume of 1794 pages, with 311 elaborate engravings. Price in cloth, $7.25; leather, raised bands, $8.00. ***This work will be furnished with Patent Ready Reference Thumb-letter Index for $1.00 in addition to the price in any style of binding.*

In this new edition of THE NATIONAL DISPENSATORY, all important changes in the recent British Pharmacopœia have been incorporated throughout the volume, while in the Addenda will be found, grouped in a convenient section of 24 pages, all therapeutical novelties which have been established in professional favor since the publication of the third edition two years ago. Since its first publication, THE NATIONAL DISPENSATORY has been the most accurate work of its kind, and in this edition, as always before, it may be said to be the representative of the most recent state of American, English, German and French Pharmacology, Therapeutics and Materia Medica.

It is with much pleasure that the fourth edition of this magnificent work is received. The authors and publishers have reason to feel proud of this, the most comprehensive, elaborate and accurate work of the kind ever printed in this country. It is no wonder that it has become the standard authority for both the medical and pharmaceutical profession, and that four editions have been required to supply the constant and increasing demand since its first appearance in 1879. The entire field has been gone over and the various articles revised in accordance with the latest developments regarding the attributes and therapeutical action of drugs. The remedies of recent discovery have received due attention.—*Kansas City Medical Index,* Nov. 1887.

We think it a matter for congratulation that the profession of medicine and that of pharmacy have shown such appreciation of this great work as to call for four editions within the comparatively brief period of eight years. The matters with which it deals are of so practical a nature that neither the physician nor the pharmacist can do without the latest text-books on them, especially those that are so accurate and comprehensive as this one. The book is in every way creditable both to the authors and to the publishers.—*New York Medical Journal,* May 21, 1887.

FARQUHARSON, ROBERT, M. D., F. R. C. P., LL. D.,
Lecturer on Materia Medica at St. Mary's Hospital Medical School, London.

A Guide to Therapeutics and Materia Medica. New (fourth) American, from the fourth English edition. Enlarged and adapted to the U. S. Pharmacopœia. By FRANK WOODBURY, M. D., Professor of Materia Medica and Therapeutics and Clinical Medicine in the Medico-Chirurgical College of Philadelphia. In one handsome 12mo. volume of 581 pages. Cloth, $2.50.

It may correctly be regarded as the most modern work of its kind. It is concise, yet complete. Containing an account of all remedies that have a place in the British and United States Pharmacopœias, as well as considering all non-official but important new drugs, it becomes in fact a miniature dispensatory.—*Pacific Medical Journal,* June, 1889. An especially attractive feature is an arrangement by which the physiological and therapeutical actions of various remedies are shown in parallel columns. This aids greatly in fixing attention and facilitates study. The American editor has enlarged the work so as to make it include all the remedies and preparations in the U. S. Pharmacopœia. The book is a most valuable addition to the list of treatises on this most important subject. —*American Practitioner and News,* Nov. 9th, 1889.

EDES, ROBERT T., M. D.,
Jackson Professor of Clinical Medicine in Harvard University, Medical Department.

A Text-Book of Therapeutics and Materia Medica. Intended for the Use of Students and Practitioners. Octavo, 544 pages. Cloth, $3.50; leather, $4.50.

The present work seems destined to take a prominent place as a text-book on the subjects of which it treats. It possesses all the essentials which we expect in a book of its kind, such as conciseness, clearness, a judicious classification, and a reasonable degree of dogmatism. All the newest drugs of promise are treated of. The clinical index at the end will be found very useful. We heartily commend the book and congratulate the author on having produced so good a one.—*N. Y. Medical Journal,* Feb. 18, 1888.

Dr. Edes' book represents better than any older book the practical therapeutics of the present day. The book is a thoroughly practical one. The classification of remedies has reference to their therapeutic action.—*Pharmaceutical Era,* Jan. 1888.

BRUCE, J. MITCHELL, M. D., F. R. C. P.,
Physician and Lecturer on Materia Medica and Therapeutics at Charing Cross Hospital, London.

Materia Medica and Therapeutics. An Introduction to Rational Treatment. Fourth edition. 12mo., 591 pages. Cloth, $1.50. See *Students' Series of Manuals,* page 31.

GRIFFITH, ROBERT EGLESFIELD, M. D.

A Universal Formulary, containing the Methods of Preparing and Administering Officinal and other Medicines. The whole adapted to Physicians and Pharmaceutists. Third edition, thoroughly revised, with numerous additions, by JOHN M. MAISCH, Phar. D., Professor of Materia Medica and Botany in the Philadelphia College of Pharmacy. In one octavo volume of 775 pages, with 38 illustrations. Cloth, $4.50; leather, $5.50.

GREEN, T. HENRY, M. D.,
Lecturer on Pathology and Morbid Anatomy at Charing-Cross Hospital Medical School, London.

Pathology and Morbid Anatomy. New (sixth) American from the seventh revised English edition. Octavo, 539 pp., with 167 engravings. Cloth, $2.75. *Just ready.*

The Pathology and Morbid Anatomy of Dr. Green is too well known by members of the medical profession to need any commendation. There is scarcely an intelligent physician anywhere who has not the work in his library, for it is almost an essential. In fact it is better adapted to the wants of general practitioners than any work of the kind with which we are acquainted. The works of German authors upon pathology, which have been translated into English, are too abstruse for the physician. Dr. Green's work precisely meets his wishes. The cuts exhibit the appearances of pathological structures just as they are seen through the microscope. The fact that it is so generally employed as a text-book by medical students is evidence that we have not spoken too much in its favor.—*Cincinnati Medical News,* Oct. 1889.

PAYNE, JOSEPH F., M. D., F. R. C. P.,
Senior Assistant Physician and Lecturer on Pathological Anatomy, St. Thomas' Hospital, London.

A Manual of General Pathology. Designed as an Introduction to the Practice of Medicine. Octavo of 524 pages, with 152 illus. and a colored plate. Cloth, $3.50.

Knowing, as a teacher and examiner, the exact needs of medical students, the author has in the work before us prepared for their especial use what we do not hesitate to say is the best introduction to general pathology that we have yet examined. A departure which our author has taken is the greater attention paid to the causation of disease, and more especially to the etiological factors in those diseases now with reasonable certainty ascribed to pathogenetic microbes. In this department he has been very full and explicit, not only in a descriptive manner, but in the technique of investigation. The Appendix, giving methods of research, is alone worth the price of the book, several times over, to every student of pathology.—*St. Louis Med. and Surg. Jour.,* Jan.'89.

SENN, NICHOLAS, M.D., Ph.D.,
Professor of Principles of Surgery and Surgical Pathology in Rush Medical College, Chicago.

Surgical Bacteriology. New (second) edition. In one handsome octavo of about 250 pages, with 13 plates, of which 9 are colored. *In press.*

COATS, JOSEPH, M. D., F. F. P. S.,
Pathologist to the Glasgow Western Infirmary.

A Treatise on Pathology. In one very handsome octavo volume of 829 pages, with 339 beautiful illustrations. Cloth, $5.50; leather, $6.50.

Medical students as well as physicians, who desire a work for study or reference, that treats the subjects in the various departments in a very thorough manner, but without prolixity, will certainly give this one the preference to any with which we are acquainted. It sets forth the most recent discoveries, exhibits, in an interesting manner, the changes from a normal condition effected in structures by disease, and points out the characteristics of various morbid agencies, so that they can be easily recognised. But, not limited to morbid anatomy, it explains fully how the functions of organs are disturbed by abnormal conditions.—*Cincinnati Medical News,* Oct. 1883.

GIBBES, HENEAGE, M. D.,
Professor of Pathology in the University of Michigan, Medical Department.

Practical Pathology. In one very handsome octavo volume of about 400 pages, with about 75 illustrations. *In press.*

WOODHEAD, G. SIMS, M. D., F. R. C. P., E.,
Demonstrator of Pathology in the University of Edinburgh.

Practical Pathology. A Manual for Students and Practitioners. In one beautiful octavo volume of 497 pages, with 136 exquisitely colored illustrations. Cloth, $6.00.

SCHÄFER, EDWARD A., F. R. S.,
Jodrell Professor of Physiology in University College, London.

The Essentials of Histology. In one octavo volume of 246 pages, with 281 illustrations. Cloth, $2.25.

This admirable work was greatly needed. It has been written with the object of supplying the student with directions for the microscopical examination of the tissues, which are given in a clear and understandable way. Although especially adapted for laboratory work, at the same time it is intended to serve as an elementary text-book of histology, comprising all the essential facts of the science.—*The Physician and Surgeon,* July, 1887.

KLEIN, E., M. D., F. R. S.,
Joint Lecturer on General Anat. and Phys. in the Med. School of St. Bartholomew's Hosp., London.

Elements of Histology. Fourth edition. In one 12mo. volume of 376 pages, with 194 illus. Limp cloth, $1.75. See *Students' Series of Manuals,* page 31.

Considered with regard to its contents, it can only be looked on as a large and comprehensive volume. New and original illustrations have been added, with the help of which the structure of each tissue becomes clear to the reader. A copious index affords a ready reference to the histology of every tissue and organ, and presents, at the same time, a complete glossary of the scientific terms.—*Provincial Medical Journal,* May 1, 1889.

PEPPER, A. J., M. B., M. S., F. R. C. S.,
Surgeon and Lecturer at St. Mary's Hospital, London.

Surgical Pathology. In one pocket-size 12mo. volume of 511 pages, with 81 illustrations. Limp cloth, red edges, $2.00. See *Students' Series of Manuals,* page 31.

FLINT, AUSTIN, M. D., LL. D.,
Prof. of the Principles and Practice of Med. and of Clin. Med. in Bellevue Hospital Medical College, N. Y.

A Treatise on the Principles and Practice of Medicine. Designed for the use of Students and Practitioners of Medicine. New (sixth) edition, thoroughly revised and rewritten by the Author, assisted by WILLIAM H. WELCH, M. D., Professor of Pathology, Johns Hopkins University, Baltimore, and AUSTIN FLINT, JR., M. D., LL. D., Professor of Physiology, Bellevue Hospital Medical College, N. Y. In one very handsome octavo volume of 1160 pages, with illustrations. Cloth, $5.50; leather, $6.50.

No text-book on the principles and practice of medicine has ever met in this country with such general approval by medical students and practitioners as the work of Professor Flint. In all the medical colleges of the United States it is the favorite work upon Practice; and, as we have stated before in alluding to it, there is no other medical work that can be so generally found in the libraries of physicians. In every state and territory of this vast country the book that will be most likely to be found in the office of a medical man, whether in city, town, village, or at some cross-roads, is Flint's *Practice*. We make this statement to a considerable extent from personal observation, and it is the testimony also of others. An examination shows that very considerable changes have been made in the sixth edition. The work may undoubtedly be regarded as fairly representing the present state of the science of medicine, and as reflecting the views of those who exemplify in their practice the present stage of progress of medical art.—*Cincinnati Medical News*, Oct. 1886.

BRISTOWE, JOHN SYER, M. D., LL. D., F. R. S.,
Senior Physician to and Lecturer on Medicine at St. Thomas' Hospital, London.

A Treatise on the Science and Practice of Medicine. Seventh edition. In one large octavo volume of 1325 pages. Cloth, $6.50; leather, $7.50. *Just ready.*

The remarkable regularity with which new editions of this text-book make their appearance is striking testimony to its excellence and value. This, too, in spite of the numerous rivals for the favor of the student which have been put forth within the sixteen years since Bristowe's "Medicine" first appeared. Nor can it be said that the author himself has failed to keep his manual abreast of advancing knowledge, arduous as that task must prove. So long as there is shown such care and circumspection in the inclusion of all new matter that has stood the test of criticism, so long will this work retain the favor which it has always met. For it is a work that is built on a stable foundation, systematic, scientific and practical, containing the matured experience of a physician who has every claim to be considered an authority, and composed in a style which attracts the practitioner as much as the student. No one can say that this book has obtained a success which was undeserved, and we trust that its author will long continue to supervise the production of fresh editions for the advantage of the coming generation of medical students.—*The Lancet*, July 12, 1890.

Dr. Bristowe's now famous treatise appears in its seventh edition. It has long passed the stage in which it requires critical examination or commendation, and has thoroughly established itself as among the most complete and useful of text-books.—*British Medical Journal*, September 27, 1890.

HARTSHORNE, HENRY, M. D., LL. D.,
Lately Professor of Hygiene in the University of Pennsylvania.

Essentials of the Principles and Practice of Medicine. A Handbook for Students and Practitioners. Fifth edition, thoroughly revised and rewritten. In one royal 12mo. volume of 669 pages, with 144 illustrations. Cloth, $2.75; half bound, $3.00.

Within the compass of 600 pages it treats of the history of medicine, general pathology, general symptomatology, and physical diagnosis (including laryngoscope, ophthalmoscope, etc.), general therapeutics, nosology, and special pathology and practice. There is a wonderful amount of information contained in this work, and it is one of the best of its kind that we have seen.—*Glasgow Medical Journal*, Nov. 1882.

An indispensable book. No work ever exhibited a better average of actual practical treatment than this one; and probably not one writer in our day had a better opportunity than Dr. Hartshorne for condensing all the views of eminent practitioners into a 12mo. The numerous illustrations will be very useful to students especially. These essentials, as the name suggests, are not intended to supersede the text-books of Flint and Bartholow, but they are the most valuable in affording the means to see at a glance the whole literature of any disease, and the most valuable treatment.—*Chicago Medical Journal and Examiner*, April, 1882.

REYNOLDS, J. RUSSELL, M. D.,
Professor of the Principles and Practice of Medicine in University College, London.

A System of Medicine. With notes and additions by HENRY HARTSHORNE, A. M., M. D., late Professor of Hygiene in the University of Pennsylvania. In three large and handsome octavo volumes, containing 3056 double-columned pages, with 317 illustrations. Price per volume, cloth, $5.00; sheep, $6.00; very handsome half Russia, raised bands, $6.50. Per set, cloth, $15; leather, $18. *Sold only by subscription.*

STILLÉ, ALFRED, M. D., LL. D.,
Professor Emeritus of the Theory and Practice of Med. and of Clinical Med. in the Univ. of Penna.

Cholera: Its Origin, History, Causation, Symptoms, Lesions, Prevention and Treatment. In one handsome 12mo. volume of 163 pages, with a chart. Cloth, $1.25.

WATSON, SIR THOMAS, M. D.,
Late Physician in Ordinary to the Queen.

Lectures on the Principles and Practice of Physic. A new American from the fifth English edition. Edited, with additions, and 190 illustrations, by HENRY HARTSHORNE, A. M., M. D., late Professor of Hygiene in the University of Pennsylvania. In two large octavo volumes of 1840 pages. Cloth, $9.00; leather, $11.00.

For Sale by Subscription Only.

A System of Practical Medicine.
BY AMERICAN AUTHORS.
EDITED BY WILLIAM PEPPER, M. D., LL. D.,
PROVOST AND PROFESSOR OF THE THEORY AND PRACTICE OF MEDICINE AND OF CLINICAL MEDICINE IN THE UNIVERSITY OF PENNSYLVANIA,

Assisted by LOUIS STARR, M. D., Clinical Professor of the Diseases of Children in the Hospital of the University of Pennsylvania.

The complete work, in five volumes, containing 5573 pages, with 198 illustrations, is now ready. Price per volume, cloth, $5; leather, $6; half Russia, raised bands and open back, $7.

In this great work American medicine is for the first time reflected by its worthiest teachers, and presented in the full development of the practical utility which is its preëminent characteristic. The most able men—from the East and the West, from the North and the South, from all the prominent centres of education, and from all the hospitals which afford special opportunities for study and practice—have united in generous rivalry to bring together this vast aggregate of specialized experience.

The distinguished editor has so apportioned the work that to each author has been assigned the subject which he is peculiarly fitted to discuss, and in which his views will be accepted as the latest expression of scientific and practical knowledge. The practitioner will therefore find these volumes a complete, authoritative and unfailing work of reference, to which he may at all times turn with full certainty of finding what he needs in its most recent aspect, whether he seeks information on the general principles of medicine, or minute guidance in the treatment of special disease. So wide is the scope of the work that, with the exception of midwifery and matters strictly surgical, it embraces the whole domain of medicine, including the departments for which the physician is accustomed to rely on special treatises, such as diseases of women and children, of the genito-urinary organs, of the skin, of the nerves, hygiene and sanitary science, and medical ophthalmology and otology. Moreover, authors have inserted the formulas which they have found most efficient in the treatment of the various affections. It may thus be truly regarded as a COMPLETE LIBRARY OF PRACTICAL MEDICINE, and the general practitioner possessing it may feel secure that he will require little else in the daily round of professional duties.

In spite of every effort to condense the vast amount of practical information furnished, it has been impossible to present it in less than 5 large octavo volumes, containing about 5600 beautifully printed pages, and embodying the matter of about 15 ordinary octavos. Illustrations are introduced wherever requisite to elucidate the text.

A detailed prospectus will be sent to any address on application to the publishers.

These two volumes bring this admirable work to a close, and fully sustain the high standard reached by the earlier volumes; we have only therefore to echo the eulogium pronounced upon them. We would warmly congratulate the editor and his collaborators at the conclusion of their laborious task on the admirable manner in which, from first to last, they have performed their several duties. They have succeeded in producing a work which will long remain a standard work of reference, to which practitioners will look for guidance, and authors will resort for facts. From a literary point of view, the work is without any serious blemish, and in respect of production, it has the beautiful finish that Americans always give their works.—*Edinburgh Medical Journal*, Jan. 1887.

* * The greatest distinctively American work on the practice of medicine, and, indeed, the superlative adjective would not be inappropriate when even all other productions placed in comparison. An examination of the five volumes is sufficient to convince one of the magnitude of the enterprise, and of the success which has attended its fulfilment.—*The Medical Age*, July 26, 1886.

This huge volume forms a fitting close to the great system of medicine which in so short a time has won so high a place in medical literature, and has done such credit to the profession in this country. Among the twenty-three contributors are the names of the leading neurologists in America, and most of the work in the volume is of the highest order.—*Boston Medical and Surgical Journal*, July 21, 1887.

We consider it one of the grandest works on Practical Medicine in the English language. It is a work of which the profession of this country can feel proud. Written exclusively by American physicians who are acquainted with all the varieties of climate in the United States, the character of the soil, the manners and customs of the people, etc., it is peculiarly adapted to the wants of American practitioners of medicine, and it seems to us that every one of them would desire to have it. It has been truly called a "Complete Library of Practical Medicine," and the general practitioner will require little else in his round of professional duties.—*Cincinnati Medical News*, March, 1886.

Each of the volumes is provided with a most copious index, and the work altogether promises to be one which will add much to the medical literature of the present century, and reflect great credit upon the scholarship and practical acumen of its authors.—*The London Lancet*, Oct. 3, 1885.

The feeling of proud satisfaction with which the American profession sees this, its representative system of practical medicine issued to the medical world, is fully justified by the character of the work. The entire caste of the system is in keeping with the best thoughts of the leaders and followers of our home school of medicine, and the combination of the scientific study of disease and the practical application of exact and experimental knowledge to the treatment of human maladies, makes a place every one of us share in the pride of the prolixity that wearies the readers of the German school, the articles glean these same fields for all that is valuable. It is the outcome of American brains, and is marked throughout by much of the sturdy independence of thought and originality that is a national characteristic. Yet nowhere is there lack of study of the most advanced views of the day.—*North Carolina Medical Journal*, Sept. 1886.

FOTHERGILL, J. M., M. D., Edin., M. R. C. P., Lond.,
Physician to the City of London Hospital for Diseases of the Chest.
The Practitioner's Handbook of Treatment; Or, The Principles of Therapeutics. New (third) edition. In one 8vo. vol. of 661 pages. Cloth, $3.75; leather, $4.75.

To have a description of the normal physiological processes of an organ and of the methods of treatment of its morbid conditions brought together in a single chapter, and the relations between the two clearly stated, cannot fail to prove a great convenience to many thoughtful but busy physicians. The practical value of the volume is greatly increased by the introduction of many prescriptions. That the profession appreciates that the author has undertaken an important work and has accomplished it is shown by the demand for this third edition.—*N. Y. Med. Jour.*, June 11, '87.

This is a wonderful book. If there be such a thing as "medicine made easy," this is the work to accomplish this result.—*Va. Med. Month.*, June, '87.
It is an excellent, practical work on therapeutics, well arranged and clearly expressed, useful to the student and young practitioner, perhaps even to the old.—*Dublin Journal of Medical Science*, March, 1888.
We do not know a more readable, practical and useful work on the treatment of disease than the one we have now before us.—*Pacific Medical and Surgical Journal*, October, 1887.

VAUGHAN, VICTOR C., Ph. D., M. D.,
Prof. of Phys. and Path. Chem. and Assoc. Prof. of Therap. and Mat. Med. in the Univ. of Mich.
and NOVY, FREDERICK G., M. D.
Instructor in Hygiene and Phys. Chem. in the Univ. of Mich.
Ptomaines and Leucomaines, or Putrefactive and Physiological Alkaloids. New Edition. In one handsome 12mo. vol. of about 300 pages. *Preparing.*

FINLAYSON, JAMES, M. D., Editor,
Physician and Lecturer on Clinical Medicine in the Glasgow Western Infirmary, etc.
Clinical Manual for the Study of Medical Cases. With Chapters by Prof. Gairdner on the Physiognomy of Disease; Prof. Stephenson on Diseases of the Female Organs; Dr. Robertson on Insanity; Dr. Gemmell on Physical Diagnosis; Dr. Coats on Laryngoscopy and Post-Mortem Examinations, and by the Editor on Casetaking, Family History and Symptoms of Disorder in the Various Systems. New edition. In one 12mo. volume of 682 pages, with 158 illustrations. Cloth, $2.50.

The profession cannot but welcome the second edition of this very valuable work of Finlayson and his collaborators. The size of the book has been increased and the number of illustrations nearly doubled. The manner in which the subject is treated is a most practical one. Symptoms alone and their diagnostic indications form the basis of discussion. The text explains clearly and fully the methods of examinations and the conclusions to be drawn from the physical signs.—*The Medical News*, April 23, 1887.
We are pleased to see a second edition of this admirable book. It is essentially a practical treatise on medical diagnosis, in which every sign and symptom of disease is carefully analyzed, and their relative significance in the different affections in which they occur pointed out. From their synthesis the student can accurately determine the disease with which he has to deal. The book has no competitor, nor is it likely to have as long as future editions maintain its present standard of excellence. The general practitioner will find many practical hints in its pages, while a careful study of the work will save him from many pitfalls in diagnosis.—*Liverpool Medico-Chirurgical Journal*, January, 1887.

BROADBENT, W. H., M. D., F. R. C. P.,
Physician to and Lecturer on Medicine at St. Mary's Hospital, London.
The Pulse. In one 12mo. volume of 312 pages. Cloth, $1.75. See *Series of Clinical Manuals*, page 31

This little book probably represents the best practical thought on this subject in the English language. A correct interpretation of the pulse, with its almost infinite modifications, brought about by almost unlimited bodily variations, can only be achieved by experience, and, as an aid toward attaining this goal, nothing will be of more service than this brochure on the study of the pulse.—*The American Journal of Medical Sciences*, September, 1890.

HABERSHON, S. O., M. D.,
Senior Physician to and late Lect. on Principles and Practice of Med. at Guy's Hospital, London.
On the Diseases of the Abdomen; Comprising those of the Stomach, and other parts of the Alimentary Canal, Œsophagus, Cæcum, Intestines and Peritoneum. Second American from third enlarged and revised English edition. In one handsome octavo volume of 554 pages, with illustrations. Cloth, $3.50.

This valuable treatise on diseases of the stomach and abdomen will be found a cyclopædia of information, systematically arranged, on all diseases of the alimentary tract, from the mouth to the rectum. A fair proportion of each chapter is devoted to symptoms, pathology, and therapeutics. The present edition is fuller than former ones in many particulars, and has been thoroughly revised and amended by the author. Several new chapters have been added, bringing the work fully up to the times, and making it a volume of interest to the practitioner in every field of medicine and surgery. Perverted nutrition is in some form associated with all diseases we have to combat, and we need all the light that can be obtained on a subject so broad and general. Dr. Habershon's work is one that every practitioner should read and study for himself.—*N. Y. Medical Journal*, April, 1879.

TANNER, THOMAS HAWKES, M. D.
A Manual of Clinical Medicine and Physical Diagnosis. Third American from the second London edition. Revised and enlarged by TILBURY FOX, M. D. In one small 12mo. volume of 362 pages, with illustrations. Cloth, $1.50.

LECTURES ON THE STUDY OF FEVER. By A. HUDSON, M. D., M. R. I. A. In one octavo volume of 308 pages. Cloth, $2.50.
A TREATISE ON FEVER. By ROBERT D. LYONS, K. C. C. In one 8vo. vol. of 354 pp. Cloth, $2.25.

LA ROCHE ON YELLOW FEVER, considered in its Historical, Pathological, Etiological and Therapeutical Relations. In two large and handsome octavo volumes of 1468 pp. Cloth, $7.00.

BARTHOLOW, ROBERTS, A. M., M. D., LL. D.,
Prof. of Materia Medica and General Therapeutics in the Jefferson Med. Coll. of Phila., etc.
Medical Electricity. A Practical Treatise on the Applications of Electricity to Medicine and Surgery. New (third) edition. In one very handsome octavo volume of 308 pages, with 110 illustrations. Cloth, $2.50.

The fact that this work has reached its third edition in six years, and that it has been kept fully abreast with the increasing use and knowledge of electricity, demonstrates its claim to be considered a practical treatise of tried value to the profession. The matter added to the present edition embraces the most recent advances in electrical treatment. The illustrations are abundant and clear, and the work constitutes a full, clear and concise manual well adapted to the needs of both student and practitioner.—*The Medical News,* May 14, 1887.

YEO, I. BURNEY, M. D., F. R. C. P.,
Professor of Clinical Therapeutics in King's College, London, and Physician to King's College Hospital.
Food in Health and Disease. In one 12mo. volume of 590 pages. Cloth, $2. *See Series of Clinical Manuals,* page 31.

Dr. Yeo supplies in a compact form nearly all that the practitioner requires to know on the subject of diet. The work is divided into two parts—food in health and food in disease. Dr. Yeo has gathered together from all quarters an immense amount of useful information within a comparatively small compass, and he has arranged and digested his materials with skill for the use of the practitioner. We have seldom seen a book which more thoroughly realizes the object for which it was written than this little work of Dr. Yeo.—*British Medical Journal,* Feb. 8, 1890.

RICHARDSON, B. W., M. D., LL. D., F. R. S.,
Fellow of the Royal College of Physicians, London.
Preventive Medicine. In one octavo volume of 729 pages. Cloth, $4; leather. $5.

Dr. Richardson has succeeded in producing a work which is elevated in conception, comprehensive in scope, scientific in character, systematic in arrangement, and which is written in a clear, concise and pleasant manner. He evinces the happy faculty of extracting the pith of what is known on the subject, and of presenting it in a most simple, intelligent and practical form. There is perhaps no similar work written for the general public that contains such a complete, reliable and instructive collection of data upon the diseases common to the race, their origins, causes, and the measures for their prevention. The descriptions of diseases are clear, chaste and scholarly; the discussion of the question of disease is comprehensive, masterly and fully abreast with the latest and best knowledge on the subject, and the preventive measures advised are accurate, explicit and reliable.—*The American Journal of the Medical Sciences,* April, 1884.

THE YEAR-BOOK OF TREATMENT FOR 1891.
A Comprehensive and Critical Review for Practitioners of Medicine. In one 12mo. volume of 484 pages. Cloth, $1.50. *Just ready.*

** For special commutations with periodicals see pages 1 and 2.

Heart and Circulation. By J. MITCHELL BRUCE, M. D., F. R. C. P.——Lungs and Organs of Respiration. by E. MARKHAM SKERRITT, M. D., F. R. C. P.——Nervous System. By JAMES ROSS, M. D., LL. D., F. R. C. P., and ERNEST SEPTIMUS REYNOLDS, M. D., M. R. C. P.——Stomach, Intestines, Liver, etc. By ROBERT MAGUIRE, M. D., F. R. C. P.——Kidney, Diabetes, etc. By CHARLES H. RALFE, M. A., M. D., Cantab., F. R. C. P.——Gout, Rheumatism, and Rheumatoid Arthritis. By ARCHIBALD E. GARROD, M. A., M. D., Oxon., M. R. C. P.——Infectious Fevers. By SIDNEY PHILLIPS, M. D., M. R. C. P.——General Surgery. By STANLEY BOYD, B. S., Lond., F. R. C. S.——Orthopædic Surgery. By W. J. WALSHAM, F.R.C.S.——Surgical Diseases of Children. By EDMUND OWEN, M. B., F. R. C. S.——Diseases of the Genito-Urinary System. By REGINALD HARRISON, F.R.C.S.——Venereal Diseases. By ALFRED COOPER, F.R.C.S.——Diseases of Women. By D. BERRY HART, M. D., F. R. C. P. Ed.——Midwifery. By GEORGE ERNEST HERMAN, M. B., F. R. C. P.——Skin. By MALCOLM MORRIS, F. R. C. S. Ed.——Eye. By HENRY POWER, M. B., F. R. C. S.——Ear. By GEORGE P. FIELD, M. R. C. S.——Throat and Nose. By BARCLAY J. BARON, M. B., C. M. Edin.——Summary of the Therapeutics of the Year 1889–90. By WALTER G. SMITH, M. D.

THE YEAR-BOOKS of TREATMENT for '86, '87 and '90
Similar to above. 12mo., 320–341 pages. Limp cloth, $1.25 each.

SCHREIBER, JOSEPH, M. D.
A Manual of Treatment by Massage and Methodical Muscle Exercise. Translated by WALTER MENDELSON, M. D., of New York. In one handsome octavo volume of 274 pages, with 117 fine engravings. Cloth, $2.75.

STURGES' INTRODUCTION TO THE STUDY OF CLINICAL MEDICINE. Being a Guide to the Investigation of Disease. In one handsome 12mo. volume of 127 pages. Cloth, $1.25.

DAVIS' CLINICAL LECTURES ON VARIOUS IMPORTANT DISEASES. By N. S. DAVIS, M. D. Edited by FRANK H. DAVIS, M. D. Second edition. 12mo. 287 pages. Cloth, $1.75.

TODD'S CLINICAL LECTURES ON CERTAIN ACUTE DISEASES. In one octavo volume of 320 pages. Cloth. $2.50.

PAVY'S TREATISE ON THE FUNCTION OF DIGESTION; its Disorders and their Treatment. From the second London edition. In one octavo volume of 238 pages. Cloth, $2.00.

BARLOW'S MANUAL OF THE PRACTICE OF MEDICINE. With additions by D. F. CONDIE, M. D. 1 vol. 8vo., pp. 603. Cloth, $2.50.

CHAMBERS' MANUAL OF DIET AND REGIMEN IN HEALTH AND SICKNESS. In one handsome octavo volume of 302 pp. Cloth, $2.75.

HOLLAND'S MEDICAL NOTES AND REFLECTIONS. 1 vol. 8vo., pp. 493. Cloth, $3.50.

FULLER ON DISEASES OF THE LUNGS AND AIR-PASSAGES. Their Pathology, Physical Diagnosis, Symptoms and Treatment. From the second and revised English edition. In one octavo volume of 475 pages. Cloth, $3.50.

WALSHE ON THE DISEASES OF THE HEART AND GREAT VESSELS. Third American edition. In 1 vol. 8vo., 416 pp. Cloth, $3.00.

SLADE ON DIPHTHERIA; its Nature and Treatment, with an account of the History of its Prevalence in various Countries. Second and revised edition. In one 12mo. vol., 158 pp. Cloth, $1.25.

SMITH ON CONSUMPTION; its Early and Remediable Stages. 1 vol. 8vo., 253 pp. Cloth, $2.25.

LA ROCHE ON PNEUMONIA. 1 vol. 8vo. of 490 pages. Cloth, $3.00.

WILLIAMS ON PULMONARY CONSUMPTION; its Nature, Varieties and Treatment. With an analysis of one thousand cases to exemplify its duration. In one 8vo. vol. of 303 pp. Cloth, $2.50.

18 LEA BROTHERS & CO.'S PUBLICATIONS—Throat, Lungs, Heart, Nerves.

FLINT, AUSTIN, M. D., LL. D.,
Professor of the Principles and Practice of Medicine in Bellevue Hospital Medical College, N. Y.

A Manual of Auscultation and Percussion; Of the Physical Diagnosis of Diseases of the Lungs and Heart, and of Thoracic Aneurism. New (fifth) edition. Edited by James C. Wilson, M. D., Jefferson Medical College, Philadelphia. In one handsome royal 12mo. volume of 274 pages, with 12 illustrations. Cloth, $1.75. *Just ready.*

This little book through its various editions has probably done more to advance the science of physical exploration of the chest than any other dissertation upon the subject, and now in its fifth edition it is as near perfect as it can be. The rapidity with which previous editions were sold shows how the profession appreciated the thoroughness of Prof. Flint's investigations. For students it is excellent. Its value is shown both in the arrangement of the material and in the clear, concise style of expression. For the practitioner it is a ready manual for reference.—*North American Practitioner*, January, 1891.

BY THE SAME AUTHOR.

A Practical Treatise on the Physical Exploration of the Chest and the Diagnosis of Diseases Affecting the Respiratory Organs. Second and revised edition. In one handsome octavo volume of 591 pages. Cloth, $4.50.

Phthisis: Its Morbid Anatomy, Etiology, Symptomatic Events and Complications, Fatality and Prognosis, Treatment and Physical Diagnosis; In a series of Clinical Studies. In one octavo volume of 442 pages. Cloth, $3.50.

A Practical Treatise on the Diagnosis, Pathology and Treatment of Diseases of the Heart. Second revised and enlarged edition. In one octavo volume of 550 pages, with a plate. Cloth, $4.

Essays on Conservative Medicine and Kindred Topics. In one very handsome royal 12mo. volume of 210 pages. Cloth, $1.38.

BROWNE, LENNOX, F. R. C. S., E.,
Senior Physician to the Central London Throat and Ear Hospital.

A Practical Guide to Diseases of the Throat and Nose, including Associated Affections of the Ear. New (third) and enlarged edition. In one imperial octavo volume of 734 pages, with 120 illustrations in color, and 235 engravings on wood. Cloth, $6.50.

The third edition of Mr. Lennox Browne's instructive and artistic work on "The Throat and Its Diseases" appears under the title of "The Throat and Nose and Their Diseases." This change has been rendered desirable by the advances made during the last decade in rhinology. The nasal sections, which extend to upwards of 100 pages, give in a short space the best account of the present position of rhinology with which we are acquainted. The engravings in this handsome volume are of the same high order as heretofore, and more numerous than ever; they cannot fail to be of the greatest assistance to senior students and practitioners. The instruments, either figured or described, are those which, as the result of experience, Mr. Browne has found to be of the greatest utility in diagnosis and treatment; they are most simple, inexpensive and easily kept aseptic—points of much importance. We have on a former occasion eulogised the beautiful and typical colored plates drawn on stone by the author-artist himself, and forming in themselves a valuable and instructive atlas, the equal of which is not to be found in any modern work, treating of these subjects. Mr. Lennox Browne is to be congratulated on having produced the best practical text-book on diseases of the throat and nose extant. We are glad to learn that it is being translated into French and German.—*The Provincial Medical Journal*, August 1, 1890.

BY THE SAME AUTHOR.

Koch's Remedy in Relation to Throat Consumption. In one octavo volume of 121 pages, with 45 illustrations, 4 of which are colored, and 17 charts. Cloth, $1.50. *Just ready.*

SEILER, CARL, M. D.,
Lecturer on Laryngoscopy in the University of Pennsylvania.

A Handbook of Diagnosis and Treatment of Diseases of the Throat, Nose and Naso-Pharynx. New (third) edition. In one handsome royal 12mo. volume of 373 pages, with 101 illustrations and 2 colored plates. Cloth, $2.25.

Few medical writers surpass this author in ability to make his meaning perfectly clear in a few words, and in discrimination in selection, both of topics and methods. The book deserves a large sale, especially among general practitioners—*Chicago Medical Journal and Examiner*, April, 1889.

COHEN, J. SOLIS, M. D.,
Lecturer on Laryngoscopy and Diseases of the Throat and Chest in the Jefferson Medical College.

Diseases of the Throat and Nasal Passages. A Guide to the Diagnosis and Treatment of Affections of the Pharynx, Œsophagus, Trachea, Larynx and Nares. Third edition, thoroughly revised and rewritten, with a large number of new illustrations. In one very handsome octavo volume. *Preparing.*

GROSS, S. D., M.D., LL.D., D.C.L. Oxon., LL.D. Cantab.

A Practical Treatise on Foreign Bodies in the Air-passages. In one octavo volume of 452 pages, with 59 illustrations. Cloth, $2.75.

ROSS, JAMES, M. D., F. R. C. P., LL. D.,
Senior Assistant Physician to the Manchester Royal Infirmary.

A Handbook on Diseases of the Nervous System. In one octavo volume of 725 pages, with 184 illustrations. Cloth, $4.50; leather, $5.50.

The book before us is entitled to the highest consideration; it is painstaking, scientific and exceedingly comprehensive.—*New York Medical Journal*, July 10, 1886.

The author has rendered a great service to the profession by condensing into one volume the principal facts pertaining to neurology and nervous diseases as understood at the present time, and he has succeeded in producing a work at once brief and practical yet scientific, without entering into the discussion of theorists, or burdening the mind with mooted questions.—*Pacific Medical and Surgical Journal and Western Lancet*, May, 1886.

This admirable work is intended for students of medicine and for such medical men as have no time for lengthy treatises. In the present instance the duty of arranging the vast store of material at the disposal of the author, and of abridging the description of the different aspects of nervous diseases, has been performed with singular skill, and the result is a concise and philosophical guide to the department of medicine of which it treats. Dr. Ross holds such a high scientific position that any writings which bear his name are naturally expected to have the impress of a powerful intellect. In every part this handbook merits the highest praise, and will no doubt be found of the greatest value to the student as well as to the practitioner.—*Edinburgh Medical Journal*, Jan. 1887.

HAMILTON, ALLAN McLANE, M. D.,
Attending Physician at the Hospital for Epileptics and Paralytics, Blackwell's Island, N. Y.

Nervous Diseases; Their Description and Treatment. Second edition, thoroughly revised and rewritten. In one octavo volume of 598 pages, with 72 illustrations. Cloth, $4.

When the first edition of this good book appeared we gave it our emphatic endorsement, and the present edition enhances our appreciation of the book and its author as a safe guide to students of clinical neurology. One of the best and most critical of English neurological journals, *Brain*, has characterized this book as the best of its kind in any language, which is a handsome endorsement from an exalted source. The improvements in the new edition, and the additions to it, will justify its purchase even by those who possess the old.—*Alienist and Neurologist*, April, 1882.

TUKE, DANIEL HACK, M. D.,
Joint Author of The Manual of Psychological Medicine, etc.

Illustrations of the Influence of the Mind upon the Body in Health and Disease. Designed to elucidate the Action of the Imagination. New edition. Thoroughly revised and rewritten. In one 8vo. vol. of 467 pp., with 2 col. plates. Cloth, $3.

It is impossible to peruse these interesting chapters without being convinced of the author's perfect sincerity, impartiality, and thorough mental grasp. Dr. Tuke has exhibited the requisite amount of scientific address on all occasions, and the more intricate the phenomena the more firmly has he adhered to a physiological and rational method of interpretation. Guided by an enlightened deduction, the author has reclaimed for science a most interesting domain in psychology, previously abandoned to charlatans and empirics. This book, well conceived and well written, must commend itself to every thoughtful understanding.—*New York Medical Journal*, September 6, 1884.

GRAY, LANDON CARTER, M. D.,
Professor of Diseases of the Mind and Nervous System in the New York Polyclinic.

A Practical Treatise on Diseases of the Nervous System. *Preparing.*

CLOUSTON, THOMAS S., M. D., F. R. C. P., L. R. C. S.,
Lecturer on Mental Diseases in the University of Edinburgh.

Clinical Lectures on Mental Diseases. With an Appendix, containing an Abstract of the Statutes of the United States and of the Several States and Territories relating to the Custody of the Insane. By CHARLES F. FOLSOM, M. D., Assistant Professor of Mental Diseases, Med. Dep. of Harvard Univ. In one handsome octavo volume of 541 pages, with eight lithographic plates, four of which are beautifully colored. Cloth, $4.

The practitioner as well as the student will accept the plain, practical teaching of the author as a forward step in the literature of insanity. It is refreshing to find a physician of Dr. Clouston's experience and high reputation giving the bedside notes upon which his experience has been founded and his mature judgment established. Such clinical observations cannot but be useful to the general practitioner in guiding him to a diagnosis and indicating the treatment, especially in many obscure and doubtful cases of mental disease. To the American reader Dr. Folsom's *Appendix* adds greatly to the value of the work, and will make it a desirable addition to every library.—*American Psychological Journal*, July, 1884.

☞ Dr. Folsom's *Abstract* may also be obtained separately in one octavo volume of 108 pages. Cloth, $1.50.

SAVAGE, GEORGE H., M. D.,
Lecturer on Mental Diseases at Guy's Hospital, London.

Insanity and Allied Neuroses, Practical and Clinical. In one 12mo. vol. of 551 pages, with 18 illus. Cloth, $2.00. See *Series of Clinical Manuals*, page 31.

As a handbook, a guide to the practitioner and student, the book fulfils an admirable purpose. The many forms of insanity are described with characteristic clearness, the illustrative cases are carefully selected, and as regards treatment sound common sense is everywhere apparent. Dr. Savage has written an excellent manual for the practitioner and student.—*Amer. Jour. of Insan.*, Apr. '85.

PLAYFAIR, W. S., M. D., F. R. C. P.
The Systematic Treatment of Nerve Prostration and Hysteria. In one handsome small 12mo. volume of 97 pages. Cloth, $1.00.

BLANDFORD ON INSANITY AND ITS TREATMENT. Lectures on the Treatment, Medical and Legal, of Insane Patients. In one very handsome octavo volume.

JONES' CLINICAL OBSERVATIONS ON FUNCTIONAL NERVOUS DISORDERS. Second American Edition. In one handsome octavo volume of 340 pages. Cloth, $3.25.

ROBERTS, JOHN B., M. D.,
Professor of Anatomy and Surgery in the Philadelphia Polyclinic. Professor of the Principles and Practice of Surgery in the Woman's Medical College of Pennsylvania. Lecturer in Anatomy in the University of Pennsylvania.

The Principles and Practice of Modern Surgery. For the use of Students and Practitioners of Medicine and Surgery. In one very handsome octavo volume of 780 pages, with 501 illustrations. Cloth, $4.50; leather, $5.50. *Just ready.*

In this work the author has endeavored to give to the profession in a condensed form the doctrines and procedures of Modern Surgery. He has made it a work devoted more especially to the practice than to the theory of surgery. His own large experience has added many valuable features to the work. It contains many practical points in diagnosis, which render it the more valuable to the practitioner; and the systematization which pervades the whole work, together with its perspicuity, enhance its value as a student's manual. The fact that this work is eminently practical cannot be too strongly emphasized. It is modern, and as its teaching is that generally accepted and such that affords little opportunity for discussion, it will be lasting. It is clear and concise, yet full. The book is entitled to a place in modern surgical literature.—*Annals of Surgery*, Jan. 1891.

This work is a very comprehensive manual upon general surgery, and will doubtless meet with a favorable reception by the profession. It has a thoroughly practical character, the subjects are treated with rare judgment, its conclusions are in accord with those of the leading practitioners of the art, and its literature is fully up to all the advanced doctrines and methods of practice of the present day. Its general arrangement follows this rule, and the author in his desire to be concise and practical is at times almost dogmatic, but this is entirely excusable considering the admirable manner in which he has thus increased the usefulness of his work.—*Medical Record*, Jan. 17, 1891.

ASHHURST, JOHN, Jr., M. D.,
Barton Prof. of Surgery and Clin. Surgery in Univ. of Penna., Surgeon to the Penna. Hosp., etc.

The Principles and Practice of Surgery. New (fifth) edition, enlarged and thoroughly revised. In one large and handsome octavo volume of 1144 pages, with 642 illustrations. Cloth, $6; leather, $7.

A complete and most excellent work on surgery. It is only necessary to examine it to see at once its excellence and real merit either as text-book for the student or a guide for the general practitioner. It fully considers in detail every surgical injury and disease to which the body is liable, and every advance in surgery worth noting is to be found in its proper place. It is unquestionably the best and most complete single volume on surgery, in the English language, and cannot but receive that continued appreciation which its merits justly demand.—*Southern Practitioner*, Feb. 1890.

This is one of the most popular and useful of the many well-known treatises on general surgery. It furnishes in a concise manner a clear and comprehensive description of the modes of practice now generally employed in the treatment of surgical affections, with a plain exposition of the principles on which those modes of practice are based. The entire work has been carefully revised, and a number of new illustrations introduced that greatly enhance the value of the book.—*Cincinnati Lancet-Clinic*, Dec. 14, 1889.

DRUITT, ROBERT, M. R. C. S., etc.
Manual of Modern Surgery. Twelfth edition, thoroughly revised by STANLEY BOYD, M.B., B.S., F.R.C.S. In one 8vo. volume of 965 pages, with 373 illustrations. Cloth, $4; leather, $5.

It is essentially a new book, rewritten from beginning to end. The editor has brought his work up to the latest date, and nearly every subject on which the student and practitioner would desire to consult a surgical volume, has found its place here. The volume closes with about twenty pages of formulæ covering a broad range of practical therapeutics. The student will find that the new Druitt is to this generation what the old one was to the former, and no higher praise need be accorded to any volume.—*North Carolina Medical Journal*, October, 1887.

Druitt's Surgery has been an exceedingly popular work in the profession. It is stated that 50,000 copies have been sold in England, while in the United States, ever since its first issue, it has been used as a text-book to a very large extent. During the late war in this country it was so highly appreciated that a copy was issued by the Government to each surgeon. The present edition, while it has the same features peculiar to the work at first, embodies all recent discoveries in surgery, and is fully up to the times.—*Cincinnati Medical News*, September, 1887.

GANT, FREDERICK JAMES, F. R. C. S.,
Senior Surgeon to the Royal Free Hospital, London.

The Student's Surgery. A Multum in Parvo. In one square octavo volume of 848 pages, with 159 engravings. Cloth, $3.75.

GROSS, S. D., M. D., LL. D., D. C. L. Oxon., LL. D. Cantab.,
Emeritus Professor of Surgery in the Jefferson Medical College of Philadelphia.

A System of Surgery: Pathological, Diagnostic, Therapeutic and Operative. Sixth edition, thoroughly revised and greatly improved. In two large and beautifully printed imperial octavo volumes containing 2382 pages, illustrated by 1623 engravings. Strongly bound in leather, raised bands, $15.

BALL, CHARLES B., M. Ch., Dub., F. R. C. S., E.,
Surgeon and Teacher at Sir P. Dun's Hospital, Dublin.

Diseases of the Rectum and Anus. In one 12mo. volume of 417 pp., with 54 cuts, and 4 colored plates Cloth, $2.25. See *Series of Clinical Manuals* 31.

GIBNEY, V. P., M. D.,
Surgeon to the Orthopædic Hospital, New York, etc.

Orthopædic Surgery. For the use of Practitioners and Students. In one handsome octavo volume, profusely illustrated. *Preparing.*

ERICHSEN, JOHN E., F. R. S., F. R. C. S.,
Professor of Surgery in University College, London, etc.
The Science and Art of Surgery; Being a Treatise on Surgical Injuries, Diseases and Operations. From the eighth and enlarged English edition. In two large and beautiful octavo volumes of 2316 pages, illustrated with 984 engravings on wood. Cloth, $9; leather, raised bands, $11.

We have always regarded "The Science and Art of Surgery" as one of the best surgical text-books in the English language, and this eighth edition only confirms our previous opinion. We take great pleasure in cordially commending it to our readers.—*The Medical News*, April 11, 1885.
For many years this classic work has been made by preference of teachers the principal text-book on surgery for medical students, while through translations into the leading continental languages it may be said to guide the surgical teachings of the civilized world. No excellence of the former edition has been dropped and no discovery, device or improvement which has marked the progress of surgery during the last decade has been omitted. The illustrations are many and executed in the highest style of art. —*Louisville Medical News*, Feb. 14, 1885.
We cannot speak too highly of this excellent work. It represents the most advanced and settled views in regard to the science of surgery, and will ever be found a faithful guide and counsellor in practice.—*Canada Lancet*, May, 1885.
It appears simultaneously in England, America, Spain and Italy, and is too well known as a safe guide and familiar friend to need further comment.—*New York Medical Journal*, March 28, 1885.

BRYANT, THOMAS, F. R. C. S.,
Surgeon and Lecturer on Surgery at Guy's Hospital, London.
The Practice of Surgery. Fourth American from the fourth and revised English edition. In one large and very handsome imperial octavo volume of 1040 pages, with 727 illustrations. Cloth, $6.50; leather, $7.50.

The fourth edition of this work is fully abreast of the times. The author handles his subjects with that degree of judgment and skill which is attained by years of patient toil and varied experience. The present edition is a thorough revision of those which preceded it, with much new matter added. His diction is so graceful and logical, and his explanations are so lucid, as to place the work among the highest order of text-books for the medical student. Almost every topic in surgery is presented in such a form as to enable the busy practitioner to review any subject in every-day practice in a short time. No time is lost with useless theories or superfluous verbiage. In short, the work is eminently clear, logical and practical.—*Chicago Medical Journal and Examiner*, April, 1886.
This book is essentially what it purports to be, viz.: a manual for the *practice* of surgery. It is peculiarly well fitted for the student or busy general practitioner.—*The Medical News*, August 15, 1885.

TREVES, FREDERICK, F. R. C. S.,
Hunterian Professor at the Royal College of Surgeons of England.
A Manual of Surgery. In Treatises by Various Authors. In three 12mo. volumes, containing 1866 pages, with 213 engravings. Price per volume, cloth, $2. See *Students' Series of Manuals*, page 31.

We have here the opinions of thirty-three authors, in an encyclopædic form for easy and ready reference. The three volumes embrace every variety of surgical affections likely to be met with, the paragraphs are short and pithy, and the salient points and the beginnings of new subjects are always printed in extra-heavy type, so that a person may find whatever information he may be in need of at a moment's glance.—*Cincinnati Lancet-Clinic*, August 21, 1886.

HOLMES, TIMOTHY, M. A.,
Surgeon and Lecturer on Surgery at St. George's Hospital, London.
A System of Surgery; Theoretical and Practical. IN TREATISES BY VARIOUS AUTHORS. AMERICAN EDITION, THOROUGHLY REVISED AND RE-EDITED by JOHN H. PACKARD, M. D., Surgeon to the Episcopal and St. Joseph's Hospitals, Philadelphia, assisted by a corps of thirty-three of the most eminent American surgeons. In three large imperial octavo volumes containing 3137 double-columned pages, with 979 illustrations on wood and 13 lithographic plates, beautifully colored. Price per set, cloth, $18.00; leather, $21.00. *Sold only by subscription.*

MARSH, HOWARD, F. R. C. S.,
Senior Assistant Surgeon to and Lecturer on Anatomy at St. Bartholomew's Hospital, London.
Diseases of the Joints. In one 12mo. volume of 468 pages, with 64 woodcuts and a colored plate. Cloth, $2.00. See *Series of Clinical Manuals*, page 31.

BUTLIN, HENRY T., F. R. C. S.,
Assistant Surgeon to St. Bartholomew's Hospital, London.
Diseases of the Tongue. In one 12mo. volume of 456 pages, with 8 colored plates and 3 woodcuts. Cloth, $3.50. See *Series of Clinical Manuals*, page 31.

TREVES, FREDERICK, F. R. C. S.,
Surgeon to and Lecturer on Surgery at the London Hospital.
Intestinal Obstruction. In one pocket-size 12mo. volume of 522 pages, with 60 illustrations. Limp cloth, blue edges, $2.00. See *Series of Clinical Manuals*, page 31.

GOULD, A. PEARCE, M. S., M. B., F. R. C. S.,
Assistant Surgeon to Middlesex Hospital.
Elements of Surgical Diagnosis. In one pocket-size 12mo. volume of 589 pages. Cloth, $2.00. See *Students' Series of Manuals*, page 31.

PIRRIE'S PRINCIPLES AND PRACTICE OF SURGERY. Edited by JOHN NEILL, M. D. In one 8vo. vol. of 784 pp. with 316 illus. Cloth, $3.75.
MILLER'S PRINCIPLES OF SURGERY. Fourth American from the third Edinburgh edition. In one 8vo. vol. of 638 pages, with 340 illustrations. Cloth, $3.75.
MILLER'S PRACTICE OF SURGERY. Fourth and revised American edition. In one large 8vo. vol. of 682 pp., with 364 illustrations. Cloth, $3.75.

SMITH, STEPHEN, M. D.,
Professor of Clinical Surgery in the University of the City of New York.

The Principles and Practice of Operative Surgery. New (second) and thoroughly revised edition. In one very handsome octavo volume of 892 pages, with 1005 illustrations. Cloth, $4.00; leather, $5.00.

This excellent and very valuable book is one of the most satisfactory works on modern operative surgery yet published. Its author and publisher have spared no pains to make it as far as possible an ideal, and their efforts have given it a position prominent among the recent works in this department of surgery. The book is a compendium for the modern surgeon. The present, the only *revised* edition since 1879, presents many changes from the original manual. The volume is much enlarged, and the text has been thoroughly revised, so as to give the most improved methods in aseptic surgery, and the latest instruments known for operative work. It can be truly said that as a handbook for the student, a companion for the surgeon, and even as a book of reference for the physician not especially engaged in the practice of surgery, this volume will long hold a most conspicuous place, and seldom will its readers, no matter how unusual the subject, consult its pages in vain. Its compact form, excellent print, numerous illustrations, and especially its decidedly practical character, all combine to commend it.—*Boston Medical and Surgical Journal*, May 10, 1888.

HOLMES, TIMOTHY, M. A.,
Surgeon and Lecturer on Surgery at St. George's Hospital, London.

A Treatise on Surgery; Its Principles and Practice. New American from the fifth English edition, edited by T. PICKERING PICK, F. R. C. S., Surgeon and Lecturer on Surgery at St. George's Hospital, London. In one octavo volume of 997 pages, with 428 illustrations. Cloth, $6; leather, $7.

To the younger members of the profession and to others not acquainted with the book and its merits, we take pleasure in recommending it as a surgery complete, thorough, well-written, fully illustrated, modern, a work sufficiently voluminous for the surgeon specialist, adequately concise for the general practitioner, teaching those things that are necessary to be known for the successful prosecution of the physician's career, imparting nothing that in our present knowledge is considered unsafe, unscientific or inexpedient.—*Pacific Medical Journal*, July, 1889.

HAMILTON, FRANK H., M. D., LL. D.,
Surgeon to Bellevue Hospital, New York.

A Practical Treatise on Fractures and Dislocations. New (8th) edition, revised and edited by STEPHEN SMITH, A. M., M. D., Professor of Clinical Surgery in the University of the City of New York. In one very handsome octavo volume of 832 pages, with 507 illustrations. Cloth, $5.50; leather, $6.50. *Just ready.*

It has received the highest endorsement that a work upon a department of surgery can possibly receive. It is used as a text-book in every medical college of this country, and the publishers have been called upon to print eight editions of it. What more can be said in commendation of it? It has been said with truth that it is doubtful if any surgical work has appeared during the last half century which more completely filled the place for which it was designed. As Dr. Smith says, its great merits appear most conspicuously in its clear, concise, and yet comprehensive statement of principles, which renders it an admirable text-book for teacher and pupil, and in its wealth of clinical materials, which adapts it to the daily necessities of the practitioner. Fractures and dislocations are injuries which the general practitioner, in his character as a surgeon, is most called upon to treat. They form a part of surgery that he cannot avoid taking charge of. Under the circumstances, therefore, he needs all the aid he can secure. But what better assistance can he seek than a work that is devoted exclusively to treating fractures and dislocations, and consequently contains full information, in plain language, for the management of every emergency that is likely to be met with in such injuries? The country is filled with railroads and manufactories where accidents are constantly occurring, and to which general practitioners, and not distinguished surgeons, are constantly liable to be called. We consider that the work before us should be in the library of every practitioner.—*Cincinnati Medical News*, February, 1891.

STIMSON, LEWIS A., B. A., M. D.,
Surgeon to the Presbyterian and Bellevue Hospitals, Professor of Clinical Surgery in the Medical Faculty of Univ. of City of N. Y., Corresponding Member of the Societe de Chirurgie of Paris.

A Manual of Operative Surgery. New (second) edition. In one very handsome royal 12mo. volume of 503 pages, with 342 illustrations. Cloth, $2.50.

There is always room for a good book, so that while many works on operative surgery must be considered superfluous, that of Dr. Stimson has held its own. The author knows the difficult art of condensation. Thus the manual serves as a work of reference, and at the same time as a handy guide. It teaches what it professes, the steps of operations. In this edition Dr. Stimson has sought to indicate the changes that have been effected in operative methods and procedures by the antiseptic system, and has added an account of many new operations and variations in the steps of older operations. We do not desire to extol this manual above many excellent standard British publications of the same class, still we believe that it contains much that is worthy of imitation.—*British Medical Journal*, Jan. 22, 1887.

By the same Author.

A Treatise on Fractures and Dislocations. In two handsome octavo volumes. Vol. I., FRACTURES, 582 pages, 360 beautiful illustrations. Vol. II., DISLOCATIONS, 540 pages, with 163 illustrations. Complete work, cloth, $5.50; leather, $7.50. Either volume separately, cloth, $3.00; leather, $4.00.

The appearance of the second volume marks the completion of the author's original plan of preparing a work which should present in the fullest manner all that is known on the cognate subjects of Fractures and Dislocations. The volume on Fractures assumed at once the position of authority on the subject, and its companion on Dislocations will no doubt be similarly received. The closing volume of Dr. Stimson's work exhibits the surgery of Dislocations as it is taught and practised by the most eminent surgeons of the present time. Containing the results of such extended researches it must for a long time be regarded as an authority on all subjects pertaining to dislocations. Every practitioner of surgery will feel it incumbent on him to have it for constant reference.—*Cincinnati Medical News*, May, 1888.

PICK, T. PICKERING, F. R. C. S.,
Surgeon to and Lecturer on Surgery at St. George's Hospital, London.

Fractures and Dislocations. In one 12mo. volume of 530 pages, with 93 illustrations. Limp cloth, $2.00. See *Series of Clinical Manuals*, page 31.

BURNETT, CHARLES H., A. M., M. D.,
Professor of Otology in the Philadelphia Polyclinic; President of the American Otological Society.

The Ear, Its Anatomy, Physiology and Diseases. A Practical Treatise for the use of Medical Students and Practitioners. Second edition. In one handsome octavo volume of 580 pages, with 107 illustrations. Cloth, $4.00 ; leather, $5.00.

We note with pleasure the appearance of a second edition of this valuable work. When it first came out it was accepted by the profession as one of the standard works on modern aural surgery in the English language; and in his second edition Dr. Burnett has fully maintained his reputation, for the book is replete with valuable information and suggestions. The revision has been carefully carried out, and much new matter added. Dr. Burnett's work must be regarded as a very valuable contribution to aural surgery, not only on account of its comprehensiveness, but because it contains the results of the careful personal observation and experience of this eminent aural surgeon. —*London Lancet,* Feb. 21, 1885.

BERRY, GEORGE A., M. B., F. R. C. S., Ed.,
Ophthalmic Surgeon, Edinburgh Royal Infirmary.

Diseases of the Eye. A Practical Treatise for Students of Ophthalmology. In one octavo volume of 683 pages, with 144 illustrations, 62 of which are beautifully colored. Cloth, $7.50.

This newest candidate for favor among ophthalmological students is designed to be purely clinical in character and the plan is well adhered to. We have been forcibly struck by the rare good taste in the selection of what is essential which pervades the book. The author seems to have the uncommon faculty of viewing his subject as a whole and seizing the salient points and not confusing his reader—presumably a student and a novice—with a mass of details with no key to their unravelling. It is apparent that the literature of each subject has been gone over in a very thorough manner. The fact that he was writing a clinical treatise for beginners and not an encyclopædia has always been present with the author. The number and excellence of the colored illustrations in the text deserve more than a passing notice.—*Archives of Ophthalmology,* Sept. 1889.

NETTLESHIP, EDWARD, F. R. C. S.,
Ophthalmic Surgeon at St. Thomas' Hospital, London. Surgeon to the Royal London (Moorfields) Ophthalmic Hospital.

Diseases of the Eye. New (fourth) American from the fifth English edition, thoroughly revised. With a Supplement on the Detection of Color Blindness, by WILLIAM THOMSON, M. D., Professor of Ophthalmology in the Jefferson Medical College. In one 12mo. volume of 500 pages, with 164 illustrations, selections from Snellen's test-types and formulæ, and a colored plate. Cloth, $2.00. *Just ready.*

This is a well-known and a valuable work. It was primarily intended for the use of students, and supplies their needs admirably, but it is as useful for the practitioner, or indeed more so. It does not presuppose the large amount of recondite knowledge to be present which seems to be assumed in some of our larger works, is not tedious from over-conciseness, and yet covers the more important parts of clinical ophthalmology. A supplement is made to the present edition on the practical examination of railroad employés as to color-blindness and acuteness of vision and hearing. This is well written, and contains good suggestions for those who may be called on to make such examinations.—*New York Medical Journal,* December 13, 1890.

JULER, HENRY E., F. R. C. S.,
Senior Ass't Surgeon, Royal Westminster Ophthalmic Hosp.; late Clinical Ass't, Moorfields, London.

A Handbook of Ophthalmic Science and Practice. Handsome 8vo. volume of 460 pages, with 125 woodcuts, 27 colored plates, selections from Test-types of Jaeger and Snellen, and Holmgren's Color-blindness Test. Cloth, $4.50 ; leather, $5.50.

It presents to the student concise descriptions and typical illustrations of all important eye affections, placed in juxtaposition, so as to be grasped at a glance. Beyond a doubt it is the best illustrated handbook of ophthalmic science which has ever appeared. Then, what is still better, these illustrations are nearly all original. We have examined this entire work with great care, and it represents the commonly accepted views of advanced ophthalmologists. We can most heartily commend this book to all medical students, practitioners and specialists.—*Detroit Lancet,* Jan. '85.

NORRIS, WM. F., M. D., and OLIVER, CHAS. A., M. D.
Clin. Prof. of Ophthalmology in Univ. of Pa.

A Text-Book of Ophthalmology. In one octavo volume of about 500 pages, with illustrations. *Preparing.*

CARTER, R. BRUDENELL, & FROST, W. ADAMS,
F. R. C. S.,   F. R. C. S.,
Ophthalmic Surgeon to and Lect. on Ophthalmic Surgery at St. George's Hospital, London.   *Ass't Ophthalmic Surgeon and Joint Lect. on Oph. Sur., St. George's Hosp., London.*

Ophthalmic Surgery. In one 12mo. volume of 559 pages, with 91 woodcuts, color-blindness test, test-types and dots and appendix of formulæ. Cloth, $2.25. See *Series of Clinical Manuals,* page 31.

WELLS ON THE EYE. In one octavo volume. LAURENCE AND MOON'S HANDY BOOK OF OPHTHALMIC SURGERY, for the use of Practitioners. Second edition. In one octavo volume of 227 pages, with 65 illus. Cloth, $2.75.

LAWSON ON INJURIES TO THE EYE, ORBIT AND EYELIDS: Their Immediate and Remote Effects. In one octavo volume of 404 pages, with 92 illustrations. Cloth, $3.50.

ROBERTS, WILLIAM, M. D.,
Lecturer on Medicine in the Manchester School of Medicine, etc.

A Practical Treatise on Urinary and Renal Diseases, including Urinary Deposits. Fourth American from the fourth London edition. In one handsome octavo volume of 609 pages, with 81 illustrations. Cloth, $3.50.

It may be said to be the best book in print on the subject of which it treats.—*The American Journal of the Medical Sciences*, Jan. 1886.

The peculiar value and finish of the book are in a measure derived from its resolute maintenance of a clinical and practical character. It is an unrivalled exposition of everything which relates directly or indirectly to the diagnosis, prognosis and treatment of urinary diseases, and possesses a completeness not found elsewhere in our language in its account of the different affections.—*The Manchester Medical Chronicle*, July, 1885.

The value of this treatise as a guide book to the physician in daily practice can hardly be overestimated. That it is fully up to the level of our present knowledge is a fact reflecting great credit upon Dr. Roberts, who has a wide reputation as a busy practitioner.—*Medical Record*, July 31, 1886.

PURDY, CHARLES W., M. D., Chicago.

Bright's Disease and Allied Affections of the Kidneys. In one octavo volume of 288 pages, with illustrations. Cloth, $2.

The object of this work is to "furnish a systematic, practical and concise description of the pathology and treatment of the chief organic diseases of the kidney associated with albuminuria, which shall represent the most recent advances in our knowledge on these subjects;" and this definition of the object is a fair description of the book. The work is a useful one, giving in a short space the theories, facts and treatments, and going more fully into their later developments. On treatment the writer is particularly strong, steering clear of generalities, and seldom omitting, what text-books usually do, the unimportant items which are all important to the general practitioner.—*The Manchester Medical Chronicle*, Oct. 1886.

MORRIS, HENRY, M. B., F. R. C. S.,
Surgeon to and Lecturer on Surgery at Middlesex Hospital, London.

Surgical Diseases of the Kidney. In one 12mo. volume of 554 pages, with 40 woodcuts, and 6 colored plates. Limp cloth, $2.25. See *Series of Clinical Manuals*, page 31.

In this manual we have a distinct addition to surgical literature, which gives information not elsewhere to be met with in a single work. Such a book was distinctly required, and Mr. Morris has very diligently and ably performed the task he took in hand. It is a full and trustworthy book of reference, both for students and practitioners in search of guidance. The illustrations in the text and the chromo-lithographs are beautifully executed.—*The London Lancet*, Feb. 26, 1886.

LUCAS, CLEMENT, M. B., B. S., F. R. C. S.,
Senior Assistant Surgeon to Guy's Hospital, London.

Diseases of the Urethra. In one 12mo. volume. *Preparing.* See *Series of Clinical Manuals*, page 4.

THOMPSON, SIR HENRY,
Surgeon and Professor of Clinical Surgery to University College Hospital, London.

Lectures on Diseases of the Urinary Organs. Second American from the third English edition. In one 8vo. volume of 203 pp., with 25 illustrations. Cloth, $2.25.

By the Same Author.

On the Pathology and Treatment of Stricture of the Urethra and Urinary Fistulæ. From the third English edition. In one octavo volume of 359 pages, with 47 cuts and 3 plates. Cloth, $3.50.

THE AMERICAN SYSTEM OF DENTISTRY.

In Treatises by Various Authors. Edited by WILBUR F. LITCH, M. D., D. D. S., Professor of Prosthetic Dentistry, Materia Medica and Therapeutics in the Pennsylvania College of Dental Surgery. In three very handsome octavo volumes containing 3160 pages, with 1863 illustrations and 9 full-page plates. Per volume, cloth, $6; leather, $7; half Morocco, gilt top, $8. The complete work is *now ready*. *For sale by subscription only.*

As an encyclopædia of Dentistry it has no superior. It should form a part of every dentist's library, as the information it contains is of the greatest value to all engaged in the practice of dentistry.—*American Jour. Dent. Sci.*, Sept. 1886.

A grand system, big enough and good enough and handsome enough for a monument (which doubtless it is), to mark an epoch in the history of dentistry. Dentists will be satisfied with it and proud of it—they must. It is sure to be precisely what the student needs to put him and keep him in the right track, while the profession at large will receive incalculable benefit from it.—*Odontographic Journal*, Jan. 1887.

COLEMAN, A., L. R. C. P., F. R. C. S., Exam. L. D. S.,
Senior Dent. Surg. and Lect. on Dent. Surg. at St. Bartholomew's Hosp. and the Dent. Hosp., London.

A Manual of Dental Surgery and Pathology. Thoroughly revised and adapted to the use of American Students, by THOMAS C. STELLWAGEN, M. A., M. D., D. D. S., Prof. of Physiology in the Philadelphia Dental College. In one handsome octavo volume of 412 pages, with 331 illustrations. Cloth, $3.25.

It should be in the possession of every practitioner in this country. The part devoted to first and second dentition and irregularities in the permanent teeth is fully worth the price. In fact, price should not be considered in purchasing such a work. If the money put into some of our so-called standard text-books could be converted into such publications as this, much good would result.—*Southern Dental Journal*, May, 1882.

The author brings to his task a large experience acquired under the most favorable circumstances. There have been added to the volume a hundred pages by the American editor, embodying the views of the leading home teachers in dental surgery. The work, therefore, may be regarded as strictly abreast of the times, and as a very high authority on the subjects of which it treats.—*American Practitioner*, July, 1882.

BASHAM ON RENAL DISEASES: A Clinical Guide to their Diagnosis and Treatment. In one 12mo. vol. of 304 pages, with 21 illustrations. Cloth, $2.00.

LEA BROTHERS & CO.'S PUBLICATIONS—Venereal, Impotence. 25

GROSS, SAMUEL W., A. M., M. D., LL. D.,
Professor of the Principles of Surgery and of Clinical Surgery in the Jefferson Medical College of Phila.

A Practical Treatise on Impotence, Sterility, and Allied Disorders of the Male Sexual Organs. New (4th) edition, thoroughly revised by F. R. STURGIS, M. D., Prof. of Diseases of the Genito-Urinary Organs and of Venereal Diseases, N. Y. Post Grad. Med. School. In one very handsome octavo volume of 165 pages, with 18 illustrations. Cloth, $1.50. *Just ready.*

Without question this is the most valuable book ever written on this subject. Dr. Gross has devoted time and talent to the study of these disorders, with the result of immensely increasing our intimate knowledge as to the nature of the complaints and teaching how best to remedy them. The book is reliable and authoritative.—*Pacific Medical Journal*, Dec. 1890.

It has been the aim of the author to supply in a compact form, practical and strictly scientific information especially adapted to the wants of the general practitioner in regard to a class of common and grave disorders. The work contains very many facts in regard to the sexual disorders of men, of the most interesting character. We commend the study of it to every professional man,

and especially to those engaged in the general practice of medicine.—*Cincinnati Medical News*, Jan. 1891.

The work before us has become a standard text-book on the subjects of which it treats. In the present edition the author's work has been considerably augmented by Dr. Sturgis, whose contributions and views are to be seen everywhere. They contain many valuable suggestions and are the fruit of a ripe experience which cannot but enhance the original text. The profession is quick to appreciate succinct treatises which are full and complete, more especially when the authors are known to be worthy of respect and confidence.—*St. Louis Medical and Surgical Journal*, Feb. 1891.

TAYLOR, R. W., A. M., M. D.,
Clinical Professor of Genito-Urinary Diseases in the College of Physicians and Surgeons, New York, Prof. of Venereal and Skin Diseases in the University of Vermont,

The Pathology and Treatment of Venereal Diseases. Including the results of recent investigations upon the subject. Being the sixth edition of Bumstead and Taylor. Entirely rewritten by Dr. Taylor. Large 8vo. volume, about 900 pages, with about 150 engravings, as well as numerous chromo-lithographs. *In active preparation.*

A few notices of the previous edition are appended.

It is a splendid record of honest labor, wide research, just comparison, careful scrutiny and original experience, which will always be held as a high credit to American medical literature. This is not only the best work in the English language upon the subjects of which it treats, but also one which has no equal in other tongues for its clear, comprehensive and practical handling of its themes.—*Am. Jour. of the Med. Sciences*, Jan. 1884.

The character of this standard work is so well known that it would be superfluous here to pass in review its general or special points of excellence.

The verdict of the profession has been passed; it has been accepted as the most thorough and complete exposition of the pathology and treatment of venereal diseases in the language. Admirable as a model of clear description, an exponent of sound pathological doctrine, and a guide for rational and successful treatment, it is an ornament to the medical literature of this country. The additions made to the present edition are eminently judicious, from the standpoint of practical utility.—*Journal of Cutaneous and Venereal Diseases*, Jan. 1884.

CORNIL, V.,
Professor to the Faculty of Medicine of Paris, and Physician to the Lourcine Hospital.

Syphilis, its Morbid Anatomy, Diagnosis and Treatment. Specially revised by the Author, and translated with notes and additions by J. HENRY C. SIMES, M. D., Demonstrator of Pathological Histology in the Univ. of Pa., and J. WILLIAM WHITE, M. D., Lecturer on Venereal Diseases, Univ. of Pa. In one handsome octavo volume of 461 pages, with 84 very beautiful illustrations. Cloth, $3.75.

The anatomy, the histology, the pathology and the clinical features of syphilis are represented in this work in their best, most practical and most instructive form, and no one will rise from its

perusal without the feeling that his grasp of the wide and important subject on which it treats is a stronger and surer one.—*The London Practitioner*, Jan. 1882.

HUTCHINSON, JONATHAN, F. R. S., F. R. C. S.,
Consulting Surgeon to the London Hospital.

Syphilis. In one 12mo. volume of 542 pages, with 8 chromo-lithographs. Cloth, $2.25. See *Series of Clinical Manuals*, page 31.

Those who have seen most of the disease and those who have felt the real difficulties of diagnosis and treatment will most highly appreciate the facts and suggestions which abound in these pages. It is a worthy and valuable record, not only of Mr. Hutchinson's very large experience

and power of observation, but of his patience and assiduity in taking notes of his cases and keeping them in a form available for such excellent use as he has put them to in this volume.—*London Medical Record*, Nov. 12, 1887.

GROSS, S. D., M. D., LL. D., D. C. L., etc.

A Practical Treatise on the Diseases, Injuries and Malformations of the Urinary Bladder, the Prostate Gland and the Urethra. Third edition, thoroughly revised by SAMUEL W. GROSS, M. D. In one octavo volume of 574 pages, with 170 illustrations. Cloth, $4.50.

CULLERIER, A., & BUMSTEAD, F. J., M.D., LL.D.,
Surgeon to the Hôpital du Midi.   Late Professor of Venereal Diseases in the College of Physicians and Surgeons, New York.

An Atlas of Venereal Diseases. Translated and edited by FREEMAN J. BUMSTEAD, M. D. In one imperial 4to. volume of 328 pages, double-columns, with 26 plates, containing about 150 figures, beautifully colored, many of them the size of life. Strongly bound in cloth, $17.00. A specimen of the plates and text sent by mail, on receipt of 25 cts.

HILL ON SYPHILIS AND LOCAL CONTAGIOUS DISORDERS. In one 8vo vol. of 479 p. Cloth, $3.25.
LEE'S LECTURES ON SYPHILIS AND SOME

FORMS OF LOCAL DISEASE AFFECTING PRINCIPALLY THE ORGANS OF GENERATION. In one 8vo. vol. of 246 pages. Cloth, $2.25.

TAYLOR, ROBERT W., A. M., M. D.,
Clinical Professor of Genito-Urinary Diseases in the College of Physicians and Surgeons, New York; Surgeon to the Department of Venereal and Skin Diseases of the New York Hospital; President of the American Dermatological Association.

A Clinical Atlas of Venereal and Skin Diseases: Including Diagnosis, Prognosis and Treatment. In eight large folio parts, measuring 14 x 18 inches, and comprising 58 beautifully-colored plates with 213 figures, and 431 pages of text with 85 engravings. Complete work *just ready*. Price per part, $2.50. Bound in one volume, half Russia, $27; half Turkey Morocco, $28. *For sale by subscription only.* Specimen plates sent on receipt of 10 cents. A full prospectus sent to any address on application.

The completion of this monumental work is a subject of congratulation, not only to the author and publishers, but to the profession at large; indeed it is to the latter that it directly appeals as a wonderfully clear exposition of a confessedly difficult branch of medicine. Good literature has joined hands with good art with highly satisfactory results for both. There are altogether 213 figures, many of which are life size, and represent the highest perfection of the chromo-lithographic art, and scattered throughout the text are innumerable engravings. Quite a proportion of these illustrations are from the author's own collection, while on the other hand the best atlases of the world have been drawn upon for the most typical and successful pictures of the many different types of venereal and skin disease. We think we may say without undue exaggeration that the reproductions, both in color and in black and white, are almost invariably successful. The text is practical, full of therapeutical suggestions, and the clinical accounts of disease are clear and incisive. Dr. Taylor is, happily, an eminent authority in both departments, and we find as a consequence that the two divisions of this work possess an equal scientific and literary merit. We have already passed the limits allotted to a notice of this kind, and while we have nothing but praise for this admirable atlas, it must be said in justification that it is more than warranted by the merits of the work itself.—*The Medical News*, Dec. 14, 1889.

It would be hard to use words which would perspicuously enough convey to the reader the great value of this *Clinical Atlas*. This Atlas is more complete even than an ordinary course of clinical lectures, for in no one college or hospital course is it at all probable that all of the diseases herein represented would be seen. It is also more serviceable to the majority of students than attendance upon clinical lectures, for most of the students who sit on remote seats in the lecture hall cannot see the subject as well as the office student can examine these true-to-life chromo-lithographs. Comparing the text to a lecturer, it is more satisfactory in exactness and fullness than he would be likely to be in lecturing over a single case. Indeed, this *Atlas* is invaluable to the general practitioner, for it enables the eye of the physician to make diagnosis of a given case of skin manifestation by comparing the case with the picture in the *Atlas*, where will be found also the text of diagnosis, pathology, and full sections on treatment.—*Virginia Medical Monthly*, Dec. 1889.

HYDE, J. NEVINS, A. M., M. D.,
Professor of Dermatology and Venereal Diseases in Rush Medical College, Chicago.

A Practical Treatise on Diseases of the Skin. For the use of Students and Practitioners. New (second) edition. In one handsome octavo volume of 676 pages, with 2 colored plates and 85 beautiful and elaborate illustrations. Cloth, $4.50; leather, $5.50.

We can heartily commend it, not only as an admirable text-book for teacher and student, but in its clear and comprehensive rules for diagnosis, its sound and independent doctrines in pathology, and its minute and judicious directions for the treatment of disease, as a most satisfactory and complete practical guide for the physician.—*American Journal of the Medical Sciences*, July, 1888.

A useful glossary descriptive of terms is given. The descriptive portions of this work are plain and easily understood, and above all are very accurate. The therapeutical part is abundantly supplied with excellent recommendations. The picture part is well done. The value of the work to practitioners is great because of the excellence of the descriptions, the suggestiveness of the advice, and the correctness of the details and the principles of therapeutics impressed upon the reader.—*Virginia Med. Monthly*, May, 1888.

The second edition of his treatise is like his clinical instruction, admirably arranged, attractive in diction, and strikingly practical throughout. The chapter on general symptomatology is a model in its way; no clearer description of the various primary and consecutive lesions of the skin is to be met with anywhere. Those on general diagnosis and therapeutics are also worthy of careful study. Dr. Hyde has shown himself a comprehensive reader of the latest literature, and has incorporated into his book all the best of that which the past years have brought forth. The prescriptions and formulæ are given in both common and metric systems. Text and illustrations are good, and colored plates of rare cases lend additional attractions. Altogether it is a work exactly fitted to the needs of a general practitioner, and no one will make a mistake in purchasing it.—*Medical Press of Western New York*, June, 1888.

FOX, T., M. D., F.R.C.P., and FOX, T.C., B.A., M.R.C.S.,
Physician to the Department for Skin Diseases, University College Hospital, London. *Physician for Diseases of the Skin to the Westminster Hospital, London.*

An Epitome of Skin Diseases. With Formulæ. For Students and Practitioners. Third edition, revised and enlarged. In one 12mo. vol. of 238 pp. Cloth, $1.25.

The third edition of this convenient handbook calls for notice owing to the revision and expansion which it has undergone. The arrangement of skin diseases in alphabetical order, which is the method of classification adopted in this work, becomes a positive advantage to the student. The book is one which we can strongly recommend, not only to students but also to practitioners who require a compendious summary of the present state of dermatology.—*British Medical Journal*, July 2, 1883.

We cordially recommend Fox's *Epitome* to those whose time is limited and who wish a handy manual to lie upon the table for instant reference. Its alphabetical arrangement is suited to this use, for all one has to know is the name of the disease, and here are its description and the appropriate treatment at hand and ready for instant application. The present edition has been very carefully revised and a number of new diseases are described, while most of the recent additions to dermal therapeutics find mention, and the formulary at the end of the book has been considerably augmented.—*The Medical News*, December, 1883.

WILSON, ERASMUS, F. R. S.

The Student's Book of Cutaneous Medicine and Diseases of the Skin. In one handsome small octavo volume of 535 pages. Cloth, $3.50.

HILLIER'S HANDBOOK OF SKIN DISEASES; for Students and Practitioners. Second American edition. In one 12mo. volume of 353 pages, with plates. Cloth, $2.25.

The American Systems of Gynecology and Obstetrics.

Systems of Gynecology and Obstetrics, in Treatises by American Authors. Gynecology edited by MATTHEW D. MANN, A. M., M. D., Professor of Obstetrics and Gynecology in the Medical Department of the University of Buffalo; and Obstetrics edited by BARTON COOKE HIRST, M. D., Associate Professor of Obstetrics in the University of Pennsylvania, Philadelphia. In four very handsome octavo volumes, containing 3612 pages, 1092 engravings and 8 plates. Complete work *just ready*. Per volume: Cloth, $5.00; leather, $6.00; half Russia, $7.00. *For sale by subscription only. Address the Publishers.* Full descriptive circular free on application.

LIST OF CONTRIBUTORS.

WILLIAM H. BAKER, M. D.,
ROBERT BATTEY, M. D.,
SAMUEL C. BUSEY, M. D.,
JAMES C. CAMERON, M. D.,
HENRY C. COE, A. M., M. D.,
EDWARD P. DAVIS, M. D.,
G. E. DE SCHWEINITZ, M. D.,
E. C. DUDLEY, A. B., M. D.,
B. McE. EMMET, M. D.,
GEORGE J. ENGELMANN, M. D.,
HENRY J. GARRIGUES, A. M., M. D.,
WILLIAM GOODELL, A. M., M. D.,
EGBERT H. GRANDIN, A. M., M. D.,
SAMUEL W. GROSS, M. D.,
ROBERT P. HARRIS, M. D.,
GEORGE T. HARRISON, M. D.,
BARTON C. HIRST, M. D.,
STEPHEN Y. HOWELL, M. D.,
A. REEVES JACKSON, A. M., M. D.,
W. W. JAGGARD, M. D.,
EDWARD W. JENKS, M. D., LL. D.,
HOWARD A. KELLY, M. D.,
CHARLES CARROLL LEE, M. D.,
WILLIAM T. LUSK, M. D., LL. D.,
J. HENDRIE LLOYD, M. D.,
MATTHEW D. MANN, A. M., M. D.,
H. NEWELL MARTIN, F. R. S., M. D., D. Sc., M. A.,
RICHARD B. MAURY, M. D.,
C. D. PALMER, M. D.,
ROSWELL PARK, M. D.,
THEOPHILUS PARVIN, M. D., LL. D.,
R. A. F. PENROSE, M. D., LL. D.,
THADDEUS A. REAMY, A. M., M. D.,
J. C. REEVE, M. D.,
A. D. ROCKWELL, A. M., M. D.,
ALEXANDER J. C. SKENE, M. D.,
J. LEWIS SMITH, M. D.,
STEPHEN SMITH, M. D.,
R. STANSBURY SUTTON, A. M., M. D., LL. D.,
T. GAILLARD THOMAS, M. D., LL. D.,
ELY VAN DE WARKER, M. D.,
W. GILL WYLIE, M. D.

This is volume two of The American System of Obstetrics, completing the wonderfully full series issued from the house of Lea Brothers & Co. during the past two years. Two magnificent volumes devoted to gynecology, and now two like volumes embracing everything pertaining to obstetrics. These volumes are the indubitable product of the most eminent gentlemen of this country in these departments of the profession. Each contributor presents a monograph upon his special topic, apparently without restriction in space, so that everything in the way of history, theory, methods, and results is presented to our fullest need. The work will long remain as a monument of great industry and good judgment. As a work of general reference, it will be found remarkably full and instructive in every direction of inquiry.—*The Obstetric Gazette*, September, 1889.

There can be but little doubt that this work will find the same favor with the profession that has been accorded to the "System of Medicine by American Authors," and the "System of Gynecology by American Authors." One is at a loss to know what to say of this volume, for fear that just and merited praise may be mistaken for flattery. The subjects of some of the papers are discussed in various works on obstetrics, though not to the full extent that is found in this volume. The papers of Drs. Engelmann, Martin, Hirst, Jaggard and Reeve are incomparably beyond anything that can be found in obstetrical works. Certainly the Editor may be congratulated for having made such a wise selection of his contributors.—*Journal of the American Medical Association*, Sept. 8, 1888.

In our notice of the "System of Practical Medicine by American Authors," we made the following statement:—"It is a work of which the profession in this country can feel proud. Written exclusively by American physicians who are acquainted with all the varieties of climate in the United States, the character of the soil, the manners and customs of the people, etc., it is peculiarly adapted to the wants of American practitioners of medicine, and it seems to us that every one of them would desire to have it." Every word thus expressed in regard to the "American System of Practical Medicine" is applicable to the "System of Gynecology by American Authors," which we desire now to bring to the attention of our readers. It, like the other, has been written exclusively by American physicians who are acquainted with all the characteristics of American people, who are well informed in regard to the peculiarities of American women, their manners, customs, modes of living, etc. As every practising physician is called upon to treat diseases of females, and as they constitute a class to which the family physician must give attention, and cannot pass over to a specialist, we do not know of a work in any department of medicine that we should so strongly recommend medical men generally purchasing.—*Cincinnati Med. News*, July, 1887.

THOMAS, T. GAILLARD, M. D., LL. D.,
Professor of Diseases of Women in the College of Physicians and Surgeons, N. Y.

A Practical Treatise on the Diseases of Women. Fifth edition, thoroughly revised and rewritten. In one large and handsome octavo volume of 810 pages, with 266 illustrations. Cloth, $5.00; leather, $6.00.

That the previous editions of the treatise of Dr. Thomas were thought worthy of translation into German, French, Italian and Spanish, is enough to give it the stamp of genuine merit. At home it has made its way into the library of every obstetrician and gynæcologist as a safe guide to practice. No small number of additions have been made to the present edition to make it correspond to recent improvements in treatment.—*Pacific Medical and Surgical Journal*, Jan. 1881.

EDIS, ARTHUR W., M. D., Lond., F. R. C. P., M. R. C. S.,
Assist. Obstetric Physician to Middlesex Hospital, late Physician to British Lying-in Hospital.

The Diseases of Women. Including their Pathology, Causation, Symptoms, Diagnosis and Treatment. A Manual for Students and Practitioners. In one handsome octavo volume of 576 pages, with 148 illustrations. Cloth, $3.00; leather, $4.00.

It is a pleasure to read a book so thoroughly good as this one. The special qualities which are conspicuous are thoroughness in covering the whole ground, clearness of description and conciseness of statement. Another marked feature of the book is the attention paid to the details of many minor surgical operations and procedures, as, for instance, the use of tents, application of leeches, and use of hot water injections. These are among the more common methods of treatment, and yet very little is said about them in many of the text-books. The book is one to be warmly recommended especially to students and general practitioners, who need a concise but complete *résumé* of the whole subject. Specialists, too, will find many useful hints in its pages.—*Boston Med. and Surg. Journ.*, March 2, 1882.

EMMET, THOMAS ADDIS, M. D., LL. D.,
Surgeon to the Woman's Hospital, New York, etc.

The Principles and Practice of Gynæcology; For the use of Students and Practitioners of Medicine. New (third) edition, thoroughly revised. In one large and very handsome octavo volume of 880 pages, with 150 illustrations. Cloth, $5; leather, $6; very handsome half Russia, raised bands, $6.50.

We are in doubt whether to congratulate the author more than the profession upon the appearance of the third edition of this well-known work. Embodying, as it does, the life-long experience of one who has conspicuously distinguished himself as a bold and successful operator, and who has devoted so much attention to the specialty, we feel sure the profession will not fail to appreciate the privilege thus offered them of perusing the views and practice of the author. His earnestness of purpose and conscientiousness are manifest. He gives not only his individual experience but endeavors to represent the actual state of gynæcological science and art.—*British Medical Journal,* May 16, 1885.

TAIT, LAWSON, F. R. C. S.,
Professor of Gynæcology in Queen's College, Birmingham; late President of the British Gynecological Society; Fellow American Gynecological Society.

Diseases of Women and Abdominal Surgery. In two very handsome octavo volumes. Volume I., 554 pages, 62 engravings and 3 plates. Cloth, $3. *Now ready.* Volume II., *preparing.*

The plan of the work does not indicate the regular system of a text book, and yet nearly everything of disease pertaining to the various organs receives a fair consideration. The description of diseased conditions is exceedingly clear, and the treatment, medical or surgical, is very satisfactory. Much of the text is abundantly illustrated with cases, which add value in showing the results of the suggested plans of treatment. We feel confident that few gynecologists of the country will fail to place the work in their libraries.—*The Obstetric Gazette,* March, 1890.

DAVENPORT, F. H., M. D.,
Assistant in Gynæcology in the Medical Department of Harvard University, Boston.

Diseases of Women, a Manual of Non-Surgical Gynæcology. Designed especially for the Use of Students and General Practitioners. In one handsome 12mo. volume of 317 pages, with 105 illustrations. Cloth, $1.50. *Just ready.*

We agree with the many reviewers whose notices we have read in other journals congratulating Dr. Davenport on the success which he has attained. He has tried to write a book for the student and general practitioner which would tell them just what they ought to know without distracting their attention with a lot of compilations for which they could have no possible use. In this he has been eminently successful. There is not even a paragraph of useless matter. Everything is of the newest, freshest and most practical, so much so that we have recommended it to our class of gynecology students. What the author advises in the way of treatment has all been practically tested by himself, and each method receives only so much commendation as he has found that it deserves. We are sure that these good qualities will command for it a large sale.—*Canada Medical Record,* Dec. 1889.

MAY, CHARLES H., M. D.,
Late House Surgeon to Mount Sinai Hospital, New York.

A Manual of the Diseases of Women. Being a concise and systematic exposition of the theory and practice of gynecology. New (2d) edition, edited by L. S. Rau, M. D., Attending Gynecologist at the Harlem Hospital, N. Y. In one 12mo. volume of 360 pages, with 31 illustrations. Cloth, $1.75. *Just ready.*

This is a manual of gynecology in a very condensed form, and the fact that a second edition has been called for indicates that it has met with a favorable reception. It is intended, the author tells us, to aid the student who after having carefully perused larger works desires to review the subject, and he adds that it may be useful to the practitioner who wishes to refresh his memory rapidly but has not the time to consult larger works. We are much struck with the readiness and convenience with which one can refer to any subject contained in this volume. Carefully compiled indexes and ample illustrations also enrich the work. This manual will be found to fulfil its purposes very satisfactorily.—*The Physician and Surgeon,* June, 1890.

DUNCAN, J. MATTHEWS, M.D., LL. D., F. R. S. E., etc.

Clinical Lectures on the Diseases of Women; Delivered in Saint Bartholomew's Hospital. In one handsome octavo volume of 175 pages. Cloth, $1.50.

They are in every way worthy of their author; indeed, we look upon them as among the most valuable of his contributions. They are all upon matters of great interest to the general practitioner. Some of them deal with subjects that are not, as a rule, adequately handled in the text-books; others of them, while bearing upon topics that are usually treated of at length in such works, yet bear such a stamp of individuality that they deserve to be widely read.—*N. Y. Medical Journal,* March, 1880.

HODGE ON DISEASES PECULIAR TO WOMEN. Including Displacements of the Uterus. Second edition, revised and enlarged. In one beautifully printed octavo volume of 519 pages, with original illustrations. Cloth, $4.50.

RAMSBOTHAM'S PRINCIPLES AND PRACTICE OF OBSTETRIC MEDICINE AND SURGERY. In reference to the Process of Parturition. A new and enlarged edition, thoroughly revised by the Author. With additions by W. V. KEATING, M. D., Professor of Obstetrics, etc., in the Jefferson Medical College of Philadelphia. In one large and handsome imperial octavo volume of 640 pages, with 64 full page plates and 43 woodcuts in the text, containing in all nearly 200 beautiful figures. Strongly bound in leather, with raised bands, $7.

WEST'S LECTURES ON THE DISEASES OF WOMEN. Third American from the third London edition. In one octavo volume of 543 pages. Cloth, $3.75; leather, $4.75.

PARVIN, THEOPHILUS, M. D., LL. D.,
Prof. of Obstetrics and the Diseases of Women and Children in Jefferson Med. Coll., Phila.

The Science and Art of Obstetrics. New (2d) edition. In one handsome 8vo. volume of 701 pages, with 239 engravings and a colored plate. Cloth, $4.25; leather, $5.25. *Just ready.*

The second edition of this work is fully up to the present state of advancement of the obstetric art. The author has succeeded exceedingly well in incorporating new matter without apparently increasing the size of his work or interfering with the smoothness and grace of its literary construction. He is very felicitous in his descriptions of conditions, and proves himself in this respect a scholar and a master. Rarely in the range of obstetric literature can be found a work which is so comprehensive and yet compact and practical. In such respect it is essentially a text book of the first merit. The treatment of the subjects gives a real value to the work—the individualities of a practical teacher, a skilful obstetrician, a close thinker and a ripe scholar.—*Medical Record*, Jan. 17, 1891.

PLAYFAIR, W. S., M. D., F. R. C. P.,
Professor of Obstetric Medicine in King's College, London, etc.

A Treatise on the Science and Practice of Midwifery. New (fifth) American, from the seventh English edition. Edited, with additions, by ROBERT P. HARRIS, M. D. In one handsome octavo volume of 664 pages, with 207 engravings and 5 plates. Cloth, $4.00; leather, $5.00. *Just ready.*

Truly a wonderful book; an epitome of all obstetrical knowledge, full, clear and concise. In thirteen years it has reached seven editions. It is perhaps the most popular work of its kind ever presented to the profession. Beginning with the anatomy and physiology of the organs concerned, nothing is left unwritten that the practical accoucheur should know. It seems that every conceivable physiological or pathological condition from the moment of conception to the time of complete involution has had the author's patient attention. The plates and illustrations, carefully studied, will teach the science of midwifery. The reader of this book will have before him the very latest and best of obstetric practice, and also of all the coincident troubles connected therewith.—*Southern Practitioner*, Dec., 1889.

KING, A. F. A., M. D.,
Professor of Obstetrics and Diseases of Women in the Medical Department of the Columbian University, Washington, D. C., and in the University of Vermont, etc.

A Manual of Obstetrics. New (fourth) edition. In one very handsome 12mo. volume of 432 pages, with 140 illustrations. Cloth, $2.50.

Dr. King, in the preface to the first edition of this manual, modestly states that "its purpose is to furnish a good groundwork to the student in the beginning of his obstetric studies." Its purpose is attained; it *will* furnish a good groundwork to the student who carefully reads it; and further, the busy practitioner should not scorn the volume because written for students, as it contains much valuable obstetric knowledge, some of which is not found in more elaborate textbooks. The chapters on the anatomy of the female generative organs, menstruation, fecundation, the signs of pregnancy, and the diseases of pregnancy, are all excellent and clear; but it is in the description of labor, both normal and abnormal, that Dr. King is at his best. Here his style is so concise, and the illustrations are so good, that the veriest tyro could not fail to receive a clear conception of labor, its complications and treatment. Of the 141 illustrations it may be safely said that they all *illustrate*, and that the engraver's work is excellent. The name of the publishers is a sufficient guarantee that the work is presented in an attractive form, and from every standpoint we can most heartily recommend the book both to practitioner and student.—*The Medical News*, Dec. 7, 1889.

BARNES, ROBERT, M. D., and FANCOURT, M. D.,
Phys. to the General Lying-in Hosp., Lond.   *Obstetric Phys. to St. Thomas' Hosp., Lond.*

A System of Obstetric Medicine and Surgery, Theoretical and Clinical. For the Student and the Practitioner. The Section on Embryology by Prof. Milnes Marshall. In one 8vo. volume of 872 pp., with 231 illustrations. Cloth, $5; leather, $6.

The immediate purpose of the work is to furnish a handbook of obstetric medicine and surgery for the use of the student and practitioner. It is not an exaggeration to say of the book that it is the best treatise in the English language yet published, and this will not be a surprise to those who are acquainted with the work of the elder Barnes. Every practitioner who desires to have the best obstetrical opinions of the time in a readily accessible and condensed form, ought to own a copy of the book.—*Journal of the American Medical Association*, June 12, 1886. The Authors have made a text-book which is in every way quite worthy to take a place beside the best treatises of the period.—*New York Medical Journal*, July 2, 1887.

BARKER, FORDYCE, A. M., M. D., LL. D., Edin.,
Clinical Professor of Midwifery and the Diseases of Women in the Bellevue Hospital Medical College, New York, Honorary Fellow of the Obstetrical Societies of London and Edinburgh, etc., etc.

Obstetrical and Clinical Essays. 12mo., about 300 pages. *Preparing.*

WINCKEL, F.
A Complete Treatise on the Pathology and Treatment of Childbed, For Students and Practitioners. Translated, with the consent of the Author, from the second German edition, by J. R. CHADWICK, M. D. Octavo 484 pages. Cloth, $4.00.

ASHWELL'S PRACTICAL TREATISE ON THE DISEASES PECULIAR TO WOMEN. Third American from the last and revised London edition. In one 8vo. vol., pp. 520. Cloth, $3.50.
PARRY ON EXTRA-UTERINE PREGNANCY; Its Clinical History, Diagnosis, Prognosis and Treatment. Octavo, 272 pages. Cloth, $2.50.

TANNER ON PREGNANCY. Octavo, 490 pages, colored plates, 16 cuts. Cloth, $4.25
CHURCHILL ON THE PUERPERAL FEVER AND OTHER DISEASES PECULIAR TO WOMEN. In one 8vo. vol. of 464 pages. Cloth, $2.50.
MEIGS ON THE NATURE, SIGNS AND TREATMENT OF CHILDBED FEVER. In one 8vo. volume of 346 pages. Cloth, $2.00.

SMITH, J. LEWIS, M. D.,
Clinical Professor of Diseases of Children in the Bellevue Hospital Medical College, N. Y.

A Treatise on the Diseases of Infancy and Childhood. New (seventh) edition, thoroughly revised and rewritten. In one handsome octavo volume of 881 pages, with 51 illustrations. Cloth, $4.50; leather, $5.50. *Just ready.*

Notwithstanding the many excellent volumes that have been issued recently on diseases of children, the work of Dr. J. Lewis Smith easily holds a front place. Its several editions have all been thoroughly revised. In the present one we notice that many of the chapters have been entirely rewritten. Full notice is taken of all the recent advances that have been made. As its author states in the preface, the necessary revision has virtually produced a new work. In the amount of information presented the work may properly be considered to have doubled in size, but by condensation and the exclusion of all obsolete material the volume has not been rendered inconveniently large. Many diseases not previously treated of have received special chapters. The work is a very practical one. Especial care has been taken that the directions for treatment shall be particular and full. In no other work are such careful instructions given in the details of infant hygiene and the artificial feeding of infants.—*Montreal Medical Journal*, Feb. 1891.

Every department shows that it has been thoroughly revised, and that every advantage has been taken of recent advance in knowledge to bring it completely up to the times. What makes the work of Dr. Smith of especial value is the attention paid to diagnosis and the careful detail of treatment. It is undoubtedly one of the best treatises on children's diseases, and as a text book for students and practitioners it is unsurpassed.—*Buffalo Medical and Surgical Journal*, Jan. 1891.

All the important pertinent facts that modern research has brought to light are embodied in the present volume, thus bringing it up to date and giving it the dignity of ultimate authority upon the subjects of which it treats.—*New York Medical Journal*, Dec. 6, 1890.

LEISHMAN, WILLIAM, M. D.,
Regius Professor of Midwifery in the University of Glasgow, etc.

A System of Midwifery, Including the Diseases of Pregnancy and the Puerperal State. Fourth edition. Octavo, 220 engravings. *Shortly.*

LANDIS, HENRY G., A. M., M. D.,
Professor of Obstetrics and the Diseases of Women in Starling Medical College, Columbus, O.

The Management of Labor, and of the Lying-in Period. In one handsome 12mo. volume of 334 pages, with 28 illustrations. Cloth, $1.75.

The author has designed to place in the hands of the young practitioner a book in which he can find necessary information in an instant. As far as we can see, nothing is omitted. The advice is sound, and the proceedures are safe and practical. *Centralblatt für Gynakologie*, December 4, 1886.

This is a book we can heartily recommend. the author goes much more practically into the details of the management of labor than most text-books, and is so readable throughout as to tempt any one who should happen to commence the book to read it through. The author presupposes a theoretical knowledge of obstetrics, and has consistently excluded from this little work everything that is not of practical use in the lying-in room. We think that if it is as widely read as it deserves, it will do much to improve obstetric practice in general.—*New Orleans Medical and Surgical Journal*, Mar. 1886.

OWEN, EDMUND, M. B., F. R. C. S.,
Surgeon to the Children's Hospital, Great Ormond St., London.

Surgical Diseases of Children. In one 12mo. volume of 525 pages, with 4 chromo-lithographic plates and 85 woodcuts. Cloth, $2. See *Series of Clinical Manuals*, page 31.

One is immediately struck on reading this book with its agreeable style and the evidence it everywhere presents of the practical familiarity of its author with his subject. The book may be honestly recommended to both students and practitioners. It is full of sound information, pleasantly given.—*Annals of Surgery*, May, 1886.

STUDENTS' SERIES OF MANUALS.

A Series of Fifteen Manuals, for the use of Students and Practitioners of Medicine and Surgery, written by eminent Teachers or Examiners, and issued in pocket-size 12mo volumes of 300-540 pages, richly illustrated and at a low price. The following volumes are now ready: TREVES' *Manual of Surgery*, by various writers, in three volumes, each, $2; BELL's *Comparative Physiology and Anatomy*, $2; GOULD's *Surgical Diagnosis*, $2; ROBERTSON's *Physiological Physics*, $2; BRUCE's *Materia Medica and Therapeutics* (4th edition), $1.50; POWER's *Human Physiology* (2d edition), $1.50; CLARKE and LOCKWOOD's *Dissectors' Manual*, $1.50; RALFE's *Clinical Chemistry*, $1.50; TREVES' *Surgical Applied Anatomy*, $2; PEPPER's *Surgical Pathology*, $2; and KLEIN's *Elements of Histology* (4th edition), $1.75. The following is in press: PEPPER's *Forensic Medicine*. For separate notices see index on last page.

SERIES OF CLINICAL MANUALS.

In arranging for this Series it has been the design of the publishers to provide the profession with a collection of authoritative monographs on important clinical subjects in a cheap and portable form. The volumes will contain about 550 pages and will be freely illustrated by chromo-lithographs and woodcuts. The following volumes are now ready: YEO on *Food in Health and Disease*, $2; BROADBENT on the *Pulse*, $1.75; CARTER & FROST's *Ophthalmic Surgery*, $2.25; HUTCHINSON on *Syphilis*, $2.25; BALL on the *Rectum and Anus*, $2.25; MARSH on the *Joints*, $2; OWEN on *Surgical Diseases of Children*, $2; MORRIS on *Surgical Diseases of the Kidney*, $2.25; PICK on *Fractures and Dislocations*, $2; SAVAGE on *Insanity and Allied Neuroses*, $2. The *Tongue*, $3.50; TREVES on *Intestinal Obstruction*, $2; and SAVAGE on *Insanity and Allied Neuroses*, $2. The following is in active preparation: LUCAS on *Diseases of the Urethra*. For separate notices see index on last page.

CONDIE'S PRACTICAL TREATISE ON THE DISEASES OF CHILDREN. Sixth edition, revised and augmented. In one octavo volume of 779 pages. Cloth, $5.25; leather, $6.25.

WEST ON SOME DISORDERS OF THE NERVOUS SYSTEM IN CHILDHOOD. In one small 12mo. volume of 127 pages. Cloth, $1.00.

TIDY, CHARLES MEYMOTT, M. B., F. C. S.,
Professor of Chemistry and of Forensic Medicine and Public Health at the London Hospital, etc.

Legal Medicine. VOLUME II. Legitimacy and Paternity, Pregnancy, Abortion, Rape, Indecent Exposure, Sodomy, Bestiality, Live Birth, Infanticide, Asphyxia, Drowning, Hanging, Strangulation, Suffocation. Making a very handsome imperial octavo volume of 529 pages. Cloth, $6.00; leather, $7.00.

VOLUME I. Containing 664 imperial octavo pages, with two beautiful colored plates. Cloth, $6.00; leather, $7.00.

The satisfaction expressed with the first portion of this work is in no wise lessened by a perusal of the second volume. We find it characterized by the same fulness of detail and clearness of expression which we had occasion so highly to commend in our former notice, and which render it so valuable to the medical jurist. The copious tables of cases appended to each division of the subject must have cost the author a prodigious amount of labor and research, but they constitute one of the most valuable features of the book, especially for reference in medico-legal trials.—*American Journal of the Medical Sciences*, April, 1884.

TAYLOR, ALFRED S., M. D.,
Lecturer on Medical Jurisprudence and Chemistry in Guy's Hospital, London.

Poisons in Relation to Medical Jurisprudence and Medicine. Third American, from the third and revised English edition. In one large octavo volume of 788 pages. Cloth, $5.50; leather, $6.50.

By the Same Author.
A Manual of Medical Jurisprudence. Eighth American from the tenth London edition, thoroughly revised and rewritten. Edited by JOHN J. REESE, M. D. In one large octavo volume.

PEPPER, AUGUSTUS J., M. S., M. B., F. R. C. S.,
Examiner in Forensic Medicine at the University of London.

Forensic Medicine. In one pocket-size 12mo. volume. *Preparing.* See *Students' Series of Manuals,* below.

LEA, HENRY C., LL. D.
Chapters from the Religious History of Spain.—Censorship of the Press.—Mystics and Illuminati.—The Endemoniadas.—El Santo Nino de la Guardia.—Brianda de Bardaxi. In one 12mo. volume of 522 pages. Cloth, $2.50. *Just ready.*

The width, depth and thoroughness of research which have earned Dr. Lea a high European place as the ablest historian the Inquisition has yet found are here applied to some side-issues of that great subject. We have only to say of this volume that it worthily complements the author's earlier studies in ecclesiastical history. His extensive and minute learning, much of it from inedited manuscripts in Mexico, appears on every page.—*London Antiquary*, Jan. 1891.

After attentively reading the work one does not know whether the author is a Catholic, a Protestant or a free-thinker. This moderation deprives the indictment of none of its force. The facts and the documents, of which the number and novelty attest a patient erudition, are grouped in luminous order and produce on the reader an effect all the more powerful in that it seems the less designed. When we add that the style is in every way excellent, that it is clear, sober and precise, we do full justice to a work which reflects the highest honor on the talents of the writer and on the method of the modern school of history.—*Revue Critique d'Histoire et de Littérature*, Paris, Jan. 1891.

By the same Author.
Superstition and Force: Essays on The Wager of Law, The Wager of Battle, The Ordeal and Torture. Third revised and enlarged edition. In one handsome royal 12mo. volume of 552 pages. Cloth, $2.50.

Mr. Lea's curious historical monographs, of which one of the most important is here produced in an enlarged form, have given him a unique position among English and American scholars. He is distinguished for his recondite and affluent learning, his power of exhaustive historical analysis, the breadth and accuracy of his researches among the rarer sources of knowledge, the gravity and temperance of his statements, combined with singular earnestness of conviction, and his warm attachment to the cause of freedom and intellectual progress.—*N. Y. Tribune*, August 9, 1878.

By the Same Author.
Studies in Church History. The Rise of the Temporal Power—Benefit of Clergy—Excommunication—The Early Church and Slavery. Second and revised edition. In one royal octavo volume of 605 pages. Cloth, $2.50.

The author is preëminently a scholar; he takes up every topic allied with the leading theme and traces it out to the minutest detail with a wealth of knowledge and impartiality of treatment that compel admiration. The amount of information compressed into the book is extraordinary, and the profuse citation of authorities and references makes the work particularly valuable to the student who desires an exhaustive review from original sources. In no other single volume is the development of the primitive church traced with so much clearness and with so definite a perception of complex or conflicting forces.—*Boston Traveller.*

Mr. Lea is *facile princeps* among American scholars in the history of the Middle Ages, and, indeed, we know of no European writer who has shown such research, accuracy and grasp in investigating important and out-of-the-way topics connected with the history of Europe in the Middle Ages.—*N. Y. Times.*

It is some years since we read the first edition of this work by Mr. Lea, and the impression made by it on us at the time is confirmed by reperusal of it in this enlarged and improved form; namely, that it is a book of great research and accuracy, full of varied information on very interesting phases of church life and history. It discusses each subject with a rare fulness of dates and facts, and a curious conscientiousness of verification and citation of authorities.—*Edinburgh Scotsman.*

Allen's Anatomy	6	Hyde on the Diseases of the Skin	26
American Journal of the Medical Sciences	2	Jones (C. Handfield) on Nervous Disorders	19
American System of Gynæcology and Obstetrics	27	Juler's Ophthalmic Science and Practice	23
American System of Practical Medicine	15	King's Manual of Obstetrics	29
American System of Dentistry	24	Klein's Histology	13, 30
Ashhurst's Surgery	20	Landis on Labor	30
Ashwell on Diseases of Women	29	La Roche on Pneumonia, Malaria, etc.	17
Attfield's Chemistry	9	La Roche on Yellow Fever	16
Ball on the Rectum and Anus	20, 30	Laurence and Moon's Ophthalmic Surgery	23
Barker's Obstetrical and Clinical Essays,	29	Lawson on the Eye, Orbit and Eyelid	23
Barlow's Practice of Medicine	15	Lea's Chapters from Religious History of Spain	31
Barnes' System of Obstetric Medicine	29	Lea's Studies in Church History	31
Bartholow on Electricity	17	Lea's Superstition and Force	31
Basham on Renal Diseases	24	Lee on Syphilis	25
Bell's Comparative Physiology and Anatomy	7, 30	Lehmann's Chemical Physiology	8
Bellamy's Surgical Anatomy	6	Leishman's Midwifery	30
Berry on the Eye	23	Lucas on Diseases of the Urethra	24, 30
Billings' National Medical Dictionary	4	Ludlow's Manual of Examinations	3
Blandford on Insanity	19	Lyons on Fever	16
Bloxam's Chemistry	9	Maisch's Organic Materia Medica	11
Bristowe's Practice of Medicine	14	Marsh on the Joints	21, 30
Broadbent on the Pulse	16, 30	May on Diseases of Women	28
Browne on Koch's Remedy	18	Medical News	1
Browne on the Throat, Nose and Ear	18	Medical News Visiting List	3
Bruce's Materia Medica and Therapeutics	12, 30	Medical News Physicians' Ledger	3
Brunton's Materia Medica and Therapeutics	11	Meigs on Childbed Fever	21
Bryant's Practice of Surgery	21	Miller's Practice of Surgery	21
Bumstead and Taylor on Venereal. See *Taylor*.	25	Miller's Principles of Surgery	21
Burnett on the Ear	23	Morris on Diseases of the Kidney	24, 30
Butlin on the Tongue	21, 30	National Dispensatory	12
Carpenter on the Use and Abuse of Alcohol	8	National Medical Dictionary	4
Carpenter's Human Physiology	8	Neill and Smith's Compendium of Med. Sci.	15
Carter & Frost's Ophthalmic Surgery	23, 30	Nettleship on Diseases of the Eye	23
Chambers on Diet and Regimen	17	Norris and Oliver on the Eye	23
Chapman's Human Physiology	8	Owen on Diseases of Children	30
Charles' Physiological and Pathological Chem.	10	Parrish's Practical Pharmacy	11
Churchill on Puerperal Fever	29	Parry on Extra-Uterine Pregnancy	29
Clarke and Lockwood's Dissectors' Manual	6, 30	Parvin's Midwifery	29
Classen's Quantitative Analysis	10	Pavy on Digestion and its Disorders	17
Cleland's Dissector	6	Payne's General Pathology	13
Clouston on Insanity	19	Pepper's System of Medicine	15
Clowes' Practical Chemistry	10	Pepper's Forensic Medicine	30, 31
Coats' Pathology	13	Pepper's Surgical Pathology	13, 30
Cohen on the Throat	18	Pick on Fractures and Dislocations	22, 30
Coleman's Dental Surgery	24	Pirrie's System of Surgery	21
Condie on Diseases of Children	30	Playfair on Nerve Prostration and Hysteria	19
Cornil on Syphilis	25	Playfair's Midwifery	29
Dalton on the Circulation	7	Power's Human Physiology	8, 30
Dalton's Human Physiology	8	Purdy on Bright's Disease and Allied Affections	24
Davenport on Diseases of Women	28	Ralfe's Clinical Chemistry	10, 30
Davis' Clinical Lectures	17	Ramsbotham on Parturition	28
Draper's Medical Physics	7	Remsen's Theoretical Chemistry	10
Druitt's Modern Surgery	20	Reynolds' System of Medicine	14
Duncan on Diseases of Women	28	Richardson's Preventive Medicine	17
Dunglison's Medical Dictionary	4	Roberts on Urinary Diseases	24
Edes' Materia Medica and Therapeutics	12	Roberts' Compend of Anatomy	7
Edis on Diseases of Women	27	Roberts' Surgery	20
Ellis' Demonstrations of Anatomy	6	Robertson's Physiological Physics	7, 30
Emmet's Gynæcology	28	Ross on Nervous Diseases	19
Erichsen's System of Surgery	21	Savage on Insanity, including Hysteria	19, 30
Farquharson's Therapeutics and Mat. Med.	12	Schäfer's Essentials of Histology,	13
Finlayson's Clinical Diagnosis	15	Schreiber on Massage	7
Flint on Auscultation and Percussion	18	Seiler on the Throat, Nose and Naso-Pharynx	18
Flint on Phthisis	18	Senn's Surgical Bacteriology	13
Flint on Respiratory Organs	18	Series of Clinical Manuals	4
Flint on the Heart	18	Simon's Manual of Chemistry	8
Flint's Essays	18	Slade on Diphtheria	17
Flint's Practice of Medicine	14	Smith (Edward) on Consumption	17
Folsom's Laws of U. S. on Custody of Insane	19	Smith (J. Lewis) on Children	30
Foster's Physiology	8	Smith's Operative Surgery	22
Fothergill's Handbook of Treatment	16	Stillé on Cholera	14
Fownes' Elementary Chemistry	9	Stillé & Maisch's National Dispensatory	12
Fox on Diseases of the Skin	26	Stillé's Therapeutics and Materia Medica	11
Frankland and Japp's Inorganic Chemistry	9	Stimson on Fractures and Dislocations	22
Fuller on the Lungs and Air Passages	17	Stimson's Operative Surgery	22
Gant's Student's Surgery	20	Students' Series of Manuals	4
Gibbes' Practical Pathology	13	Sturges' Clinical Medicine	15
Gibney's Orthopædic Surgery	20	Tait's Diseases of Women and Abdom. Surgery	28
Gould's Surgical Diagnosis	21, 30	Tanner on Signs and Diseases of Pregnancy	29
Gray's Anatomy	5	Tanner's Manual of Clinical Medicine	16
Gray on Nervous Diseases	19	Taylor's Atlas of Venereal and Skin Diseases	26
Greene's Medical Chemistry	9	Taylor on Venereal Diseases	25
Green's Pathology and Morbid Anatomy	13	Taylor on Poisons	31
Griffith's Universal Formulary	12	Taylor's Medical Jurisprudence	31
Gross on Foreign Bodies in Air-Passages	16	Thomas on Diseases of Women	27
Gross on Impotence and Sterility	25	Thompson on Stricture	24
Gross on Urinary Organs	24	Thompson on Urinary Organs	24
Gross System of Surgery	20	Tidy's Legal Medicine	17
Habershon on the Abdomen	16	Todd on Acute Diseases	16
Hamilton on Fractures and Dislocations	22	Treves' Manual of Surgery	21, 30
Hamilton on Nervous Diseases	19	Treves' Surgical Applied Anatomy	6, 30
Hare's Practical Therapeutics	11	Treves on Intestinal Obstruction	21, 30
Hartshorne's Anatomy and Physiology	6	Tuke on the Influence of Mind on the Body	19
Hartshorne's Conspectus of the Med. Sciences	3	Vaughan & Novy's Ptomaines and Leucomaines	16
Hartshorne's Essentials of Medicine	14	Visiting List, The Medical News	3
Hermann's Experimental Pharmacology	11	Walshe on the Heart	17
Hill on Syphilis	25	Watson's Practice of Physic	14
Hillier's Handbook of Skin Diseases	26	Wells on the Eye	23
Hoblyn's Medical Dictionary	3	West on Diseases of Women	28
Hodge on Women	28	West on Nervous Disorders in Childhood	30
Hofmann and Power's Chemical Analysis	10	Williams on Consumption	17
Holden's Landmarks	6	Wilson's Handbook of Cutaneous Medicine	26
Holland's Medical Notes and Reflections	17	Wilson's Human Anatomy	5
Holmes' Principles and Practice of Surgery	22	Winckel on Pathol. and Treatment of Childbed	29
Holmes' System of Surgery	21	Wöhler's Organic Chemistry	9
Horner's Anatomy and Histology	6	Woodhead's Practical Pathology	13
Hudson on Fever	16	Year-Books of Treatment for 1886, '87, '89 and '90	17
Hutchinson on Syphilis	25, 30	Yeo on Food in Health and Disease	17, 30

LEA BROTHERS & CO., Philadelphia.

www.ingramcontent.com/pod-product-compliance
Lightning Source LLC
Chambersburg PA
CBHW030755230426
43667CB00007B/971